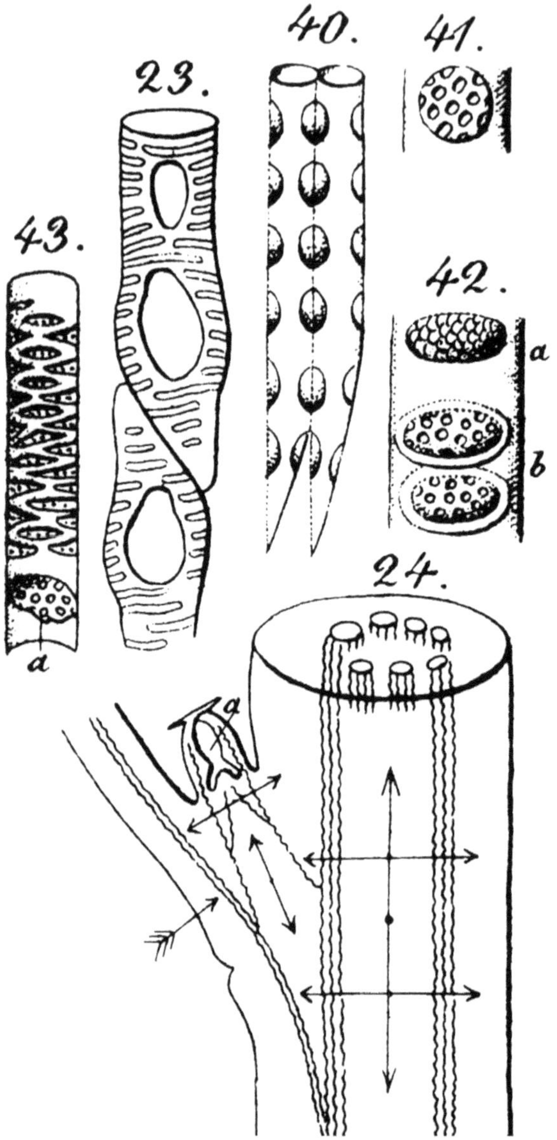

A selection of Hartig's (1837) drawings illustrating his *Vergleichende Untersuchungen über die Organisation des Stammes der einheimischen Waldbäume: Fig. 23.* Xylem element (species not mentioned); *Fig. 24.* course of vascular bundles in a nodal area (species not mentioned); *Fig. 40.* sieve fibers ("Siebfasern") in a peripheral section of *Pinus sylvestris; Fig. 41.* sieve pores viewed from the ray side *(Pinus sylvestris); Fig. 42.* sieve pores in sieve fibers of *Acer, Aesculus, Carpinus; a.* sieve area of young sieve fiber of *Carpinus; Fig. 43.* part of a sieve tube of *Acer; a.* sieve-like perforated cross walls in sieve tubes of *Acer, Robinia* (from pp 157–158; Hartig did not provide legends)

H.-D. Behnke R. D. Sjolund (Eds.)

Sieve Elements

Comparative Structure, Induction and Development

With 266 Figures

Springer-Verlag Berlin Heidelberg New York
London Paris Tokyo Hong Kong

Professor Dr. H.-D. Behnke

Zellenlehre, Universität Heidelberg
Im Neuenheimer Feld 230
D-6900 Heidelberg, FRG

Professor Dr. R. D. Sjolund

Department of Botany
University of Iowa
Iowa City, Iowa 52242, USA

Cover: Longitudinal section through the phloem of *Cheiranthus cheiri* with mature sieve element in center. (From Behnke 1981, Nordic J Bot 1: 381-400)

ISBN-13: 978-3-642-74447-1 e-ISBN-13: 978-3-642-74445-7
DOI: 10.1007/ 978-3-642-74445-7

Softcover reprint of the hardcover 1st edition 1990

Typesetting : Appl, Wemding
2131/3145-543210 Printed on acid-free paper

Preface

As part of his *Comparative Investigations of the Organization of the Trunk of the Native Forest Trees* (Theodor Hartig 1837, Vergleichende Untersuchungen über die Organisation des Stammes der einheimischen Waldbäume. Jahresberichte über die Fortschritte der Forstwissenschaften und forstlichen Naturkunde 1: 125-168) Hartig gives an anatomical description of the "composition and nature" of the then "completely uninvestigated elementary organs" of what he called the "sap skin" (Safthaut) of trees, a tissue for which Nägeli later (1858) coined the term phloem. Within the "Safthaut" Hartig describes three cell types in detail, (1) "Siebfasern", (2) "Siebröhren", and (3) "keulenförmige Saftröhren" (club-shaped sap-tubes). While the description of the latter refers to laticifers in *Euphorbia* and resin ducts in *Acer* and *Robinia,* "Siebfasern" and "Siebröhren" comprise the sieve elements.

A literal translation of the more significant parts of the description of these cell types demonstrates that his "Siebröhren" entirely correspond to what has later been defined as "sieve tubes" but that his "Siebfasern" are less well-defined and in addition to what we call "sieve cells" also include small sieve tubes as well as spindle-shaped cells of cambium, phloem parenchyma and sclerenchyma. Both in his "Siebfasern" and "Siebröhren" Hartig describes sieve areas (his expression is "lense-shaped cavities") and sieve pores (Siebporen).
from page 157:
"1) Sieve fibers
... We are able to shortly characterize them, if we say that they correspond most to xylem fibers of the conifers both in form and in position. Comparable to these, they show on the sides facing the rays lense-shaped cavities in the intercellular space ... which are penetrated by a great number of smaller pores. Fig. 40 shows sieve fibers in a peripheral section from *Pinus sylvestris.* Fig. 41 shows sieve pores as viewed from the ray. ... Their formation within conifers and deciduous trees basically is the same. In *Alnus, Aesculus, Carpinus* they are distinguished by size of both the sieve fibers and the sieve pores, Fig. 42. In *Carpinus* it can be seen that in those sieve fibers which are closest to the wood the concavity of the cell wall, it is true, is already present, and even more depressed than at a later time, but no porosity is displayed, instead, however, the surface of the concavity appears to be occupied with small warts, Fig. 42 a. Are they perhaps small, in the pores lying vesicles? ..."
from page 158:
"2) Sieve tubes
The sieve tube is to be paralleled to the xylem vessel as is the sieve fiber to the

xylem fiber. They are delicate tubes of considerable diameter which are densely covered by sieve pores at sides not adjacent to rays, while at sides facing the rays only roundish cavities, contoured by faint shadows, are shown; a detail by which they are distinguished from sieve fibers, the pores of which are always on the ray side. Their nature and their relationship to xylem vessels may best be seen in the chestnut and in the maple from which I have depicted a part of a tube, Fig. 44." [This obviously is a misprint and should refer to Fig. 43, since according to Hartig's earlier description Fig. 44 refers to sieve fibers of *Acer* showing a "transition to pachydermal cells".] "At places where the sieve pores are arranged very densely, the tube looks like a netlike spiral vessel with pitted meshes; however, it can be distinctly recognized that the network is formed by the close proximity of the sieve pores only. Without any visible segmentation of the sieve tube into single members, sieve-like perforated cross walls can be seen within the inner lumen at rather wide distances, Fig. 43 a *(Acer, Robinia).* The pores are almost regularly hexagonal and sit so close together that the remaining cell wall looks like a network. ..." (See frontispiece for reproduction of figures)

In the 150 years since the publication of Hartig's description of the phloem, our knowledge about the structure and the function of sieve elements has increased greatly. The improved resolution made possible initially by refinements in the light microscope, and more recently through the development of the electron microscope, has provided modern investigators a tool with which to extend Hartig's early and limited observations down to the current level of organelles, membranes and macromolecules. This book reviews the structure of the conducting elements of the major plant groups, from the algae to the flowering plants, from extant and extinct groups (Chaps. 1 to 7 and 14), revealing both common features and divergent solutions to the problem of nutrient transport. Additional information about phloem structure and function is provided in the coverage of special cases of phloem development (Chaps. 8 and 12), experimental systems that provide new insight in sieve element induction and formation (Chaps. 9 to 11), and of specific phloem contents (Chap. 13).

From these pages it will become clear, however, that, despite 150 years of progress, many fundamental questions concerning sieve-element structure, development and function remain to be resolved.

Portions of this book were presented during the XIVth International Botanical Congress, Berlin 1987, in a symposium commemorating Theodor Hartig's first description of sieve elements, 150 years ago.

October, 1989

H.-Dietmar Behnke, Heidelberg
and Richard D. Sjolund, Iowa City, Iowa

Contents

Contributors

H.-DIETMAR BEHNKE
Zellenlehre
Universität Heidelberg
Im Neuenheimer Feld 230
D-6900 Heidelberg, FRG

JAMES CRONSHAW
Department of Biological Sciences
University of California
Santa Barbara, CA 93106 USA

INGE DÖRR
Botanisches Institut der Universität
Olshausenstr. 40
D-2300 Kiel, FRG

ELEFTHERIOS P. ELEFTHERIOU
Department of Botany
University of Thessaloniki
GR-540 06 Thessaloniki

RAY F. EVERT
Department of Botany
University of Wisconsin
430 Lincoln Drive
Madison, WI 53706, USA

CHRISTL GLOCKMANN
Botanisches Institut der Universität
Olshausenstr. 40
D-2300 Kiel, FRG

RAINER KOLLMANN
Botanisches Institut der Universität
Olshausenstr. 40
D-2300 Kiel, FRG

DINKAR D. SABNIS
Department of Plant Science
University of Aberdeen
St. Machar Drive
Aberdeen AB9 2UD, UK

DANIEL C. SCHEIRER
Department of Biology
Northeastern University
360 Huntington Avenue
Boston, MA 02115, USA

KLAUS SCHMITZ
Botanisches Institut der Universität
Gyrhofstr. 15
D-5000 Köln 41, FRG

ALEXANDER SCHULZ
Zellenlehre
Universität Heidelberg
Im Neuenheimer Feld 230
D-6900 Heidelberg, FRG

RICHARD D. SJOLUND
Department of Botany
University of Iowa
Iowa City, Iowa 52242, USA

EDITH L. TAYLOR
Department of Botany
Ohio State University
Columbus, Ohio 43210, USA

1 Algae

KLAUS SCHMITZ

1.1 Requirement for Medium-Distance and Long-Distance Transport in Algae

Algae are a diverse group of eukaryotic plant organisms which show much variation in morphology, reproduction, and cellular fine structure. Although the thalli of most multicellular algae are rather simply constructed, some of the sporophytes of advanced brown algae reveal a structurally and functionally specialized plant body, composed of rhizoids, cauloids and phylloids and various categories of cells and tissues which resemble cormophytic organization.

Although most algae are able to exchange nutrients directly with their environment, the structural and functional differentiation in some algae causes spatial separation of sites of photoassimilate formation, mineral nutrient acquisition, storage and consumption. This organization imposes requirements for medium-distance and long-distance transport of minerals and organic material.

1.2 Medium-Distance Transport

The thallus of multicellular algae is composed of two compartments, the apoplast and the symplast. Circumstantial evidence for transport in the apoplast compartment is provided by the demonstration that lanthanum nitrate, when introduced through the base of excised branches of *Cystoseira nodicaulis,* will slowly move up the axis and can be localized in the apoplast of the meristoderm and the first cortical cell layers, but not in the medulla (Penot 1986).

Regulated transport in algae seems to be confined to the symplast. Although symplastic transport over distances of tens to hundreds of micrometers may be achieved by diffusion, symplastic cell-to-cell transport is facilitated by plasmodesmata, which provide a conductive pathway of least resistance (Tyree 1970). Movement through plasmodesmata seems to be limited, however, to compounds with a molecular weight of 800 or less (Tucker 1982; Goodwin 1983). The distribution and frequency of plasmodesmata is indicative of symplastic, intercellular solute fluxes.

1.3 Long-Distance Transport

Transport over a distance of millimeters to meters, along the symplast of a file of elongated cells, connected end-to-end via plasmodesmata, requires other means of solute movement, operating in addition to diffusion. One of these mechanisms for symplastic transport is cytoplasmic streaming, as revealed by Charophyceae. The most effective way to translocate a high quantity of solute with high velocity, however, is by means of mass flow of a solution along a sieve tube-like cell system, as is found in certain brown algae.

Conducting cells for symplastic, long-distance transport are in general characterized by several cytological attributes: they are usually thick-walled and very elongated, and their cytoplasmic, intercellular connections (plasmodesmata or pores) are concentrated and restricted to defined cell wall areas (terminal end walls, lateral cross-connecting wall areas).

In addition to the tendency in these cells toward larger pores and reduced resistance for solute flow, there is also a modified fine structure and a rearrangement of the protoplast. The large central vacuole is either lost entirely, or is replaced by numerous small vacuoles. There is also a decrease in the number and complexity of chloroplasts, a decrease in ribosomes and storage compounds, and a tendency toward nuclear disintegration.

Specialized conducting cells resembling sieve elements are not known among green algae, but they have been reported to occur in red algae and have been described for brown algae.

1.4 Conducting Cells of Red Algae

With the exception of host-parasite nutritional relationships (Evans et al. 1973; Callow et al. 1979; Goff 1979; Kremer 1983), intercellular transport in red algae remains a poorly investigated process and is still a matter of debate. Ultrastructural evidence indicates that rhodophycean pit connections (Fig. 1) do not provide symplastic continuity between cells since the septal pore is occluded by a pit plug. Typically, these plugs consist of a core of protein and a pair of plug caps which are at least partly composed of polysaccharides, and bracket the plug core (Pueschel 1980a). The fine structure and morphology of pit connections varies widely. The presence or absence of plug caps and the number of cap layers may be useful for elucidating taxonomic relationships (Pueschel and Cole 1982).

The septal pore (Fig. 1.1) is always lined by a plasmalemma but a pit (cap) membrane cannot always be demonstrated (Pueschel 1987). If present, pit membranes are claimed to be continuous with the plasmalemma (Pueschel 1977). This junction of the pit membrane and plasmalemma implies membrane bifurcation and compartmentation of the apoplast. Even though the fine structure of

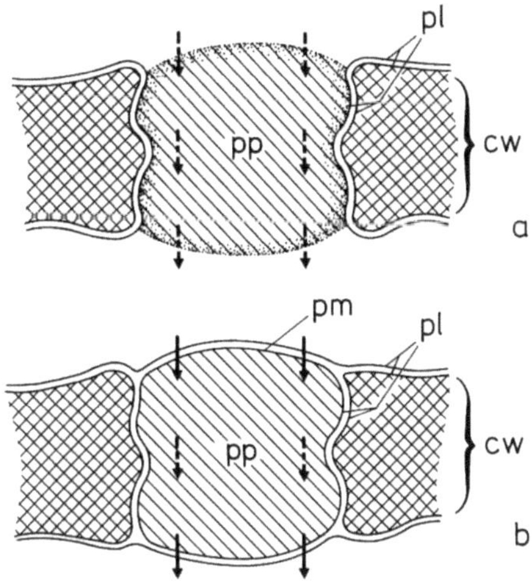

Fig. 1.1 a, b. Diagrams of red algal pit connections. **a** Pit plug (pp) not delimited by a pit membrane. Intercellular transport by diffusion (--→) through the permeable pit plug; *pl* plasmalemma; *cw* cell wall; **b** Pit plug separated from cytoplasm by a plug membrane *(pm)*. Intercellular transport must pass two membranes (→) and either the cell wall or the pit plug

pit plugs varies between taxonomic groups and within the same species (with regard to size and form of the pit plug, number of cap layers and presence or absence of pit membranes), the general implications of pit plug fine structure on intercellular transport for these red algae are evident (Fig. 1.1).

If a pit membrane is lacking and a gradient of concentration, or a chemical potential gradient, exists between adjacent cells, transport by diffusion is conceivable as long as the plug core is permeable (Fig. 1.1 a). If the pit plug is separated from the cytoplasm by a pit membrane, intercellular transport must necessarily pass two membranes and either the cell wall or the pit plug. Passage through the pit plug seems advantageous, since it provides a more regulated pathway than the passage through the cell wall. In addition, pit membranes could have different permeability properties compared to the plasmalemma, but this is most speculative (Fig. 1.1 b).

Modified pit connections have been studied in two species of *Cryptonemia*. There is an increase of width, volume and electron transparency of the pit connections with an increase in the distance from the thallus surface and there is a large increase of the convoluted pit membrane. These modifications suggest a role in enhancing intercellular transport (Wetherbee and Kraft 1981; Scott et al. 1982).

Whether pit connections are structurally specialized sites for intercellular transport or not is a matter of debate (Wetherbee 1979; Pueschel 1980b). An electrophysiological investigation by Bauman and Jones (1986) has shown, however, that adjacent cells in a filament of *Griffithsia pacifica,* with cross wall pit connections, are electrically coupled, which is only appreciable if there is a connection between cells allowing ions to flow. Pit connections of *Griffithsia pacifica* were not permeable for the fluorochrome lucifer yellow CH and seem to have a lower molecular size limit for cell-to-cell transport than that reported for the plasmodesmata of higher plant cells (Bauman and Jones 1986).

While medium-distance transport from cell to cell seems probable, long-distance transport of organic compounds, with velocities up to 63 cm^{-1}, as reported by Hartmann and Eschrich (1969), is highly unusual. The driving forces and the conducting pathway for such translocation still remain obscure and require further investigation.

Based on characteristic structural features which can be observed by light microscopy, axial cells and pericentral cells of uniaxial types of Florideophycidae have been described as conducting cells (Celan 1939, 1941; Feldmann 1939; Feldmann and Feldmann 1946). A diagram of axial cells which show different degrees of structural differentiation, indicating a conductive function, is presented in Fig. 1.2. The common cytological features are: a tubular, elongated form of cells; large or numerous pit connections in the cross walls; polar

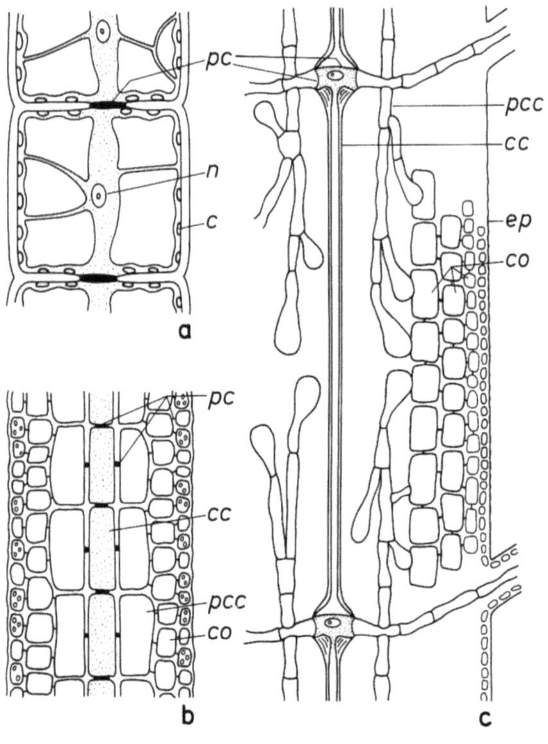

Fig. 1.2a–c. Diagrams illustrating developmental steps of transport specialized axial and pericentral cells. **a** *Crouania attenuata*. A central thread of cytoplasm attached to the pit connections *(pc)* of cross walls may provide a route for enhanced transport; *n* nucleus; *c* chloroplast; **b** *Halopitys incurvus*. Structure of central cells *(cc)* indicates transport function. Cross walls of the axial cells reveal large pit connections *(pc)*. There are pit connections between central cells *(cc)*, pericentral cells *(pcc)* and cortex cells *(co)*; **c** *Asparagopsis armata*. The transport specialized file of central cells is connected to a branched pericentral cell system which collects assimilate from the cortex and epidermis *(ep)*. Pit connections in axial cells and bordering pericentral cells are large and occupy nearly the whole cross walls. (Redrawn and modified after Feldmann and Feldmann 1946)

arrangement of the protoplast; very large and often hypertrophic nuclei; absence or scarcity of floridean starch; and modified chloroplasts with poorly developed pigmentation. Unfortunately, the ideas illustrated in Fig. 1.2 have never been adequately followed up with modern techniques for fine structure investigation.

The overall morphology of the red algae, the spatial and temporal separation of sites of photoassimilation, growth and storage, as well as the above mentioned observations, are highly indicative of intercellular transport of information and nutrients. Cell-to-cell transport, however, urgently needs to be proven in a more direct way. The existence of a specialized conducting cell system for long-distance transport in the red algae also remains unproven and requires further cytological examination.

1.5 Conducting Cells in Brown Algae

1.5.1 General Remarks

Aspects of the biochemistry and physiology of translocation in brown algae were reviewed by Schmitz (1981, 1982), Buggeln (1983) and Raven (1984). A survey of the older literature on the anatomy and early electron microscopic work of conducting cells may be found in Fritsch (1945) and Esau (1969). More recent publications on the fine structure of sieve elements of Laminariales were summarized by Schmitz (1981).

1.5.2 Conducting Cells in Dictyotales, Scytosiphonales, Desmarestiales and Fucales

Dictyotales. Dictyopteris membranacea was investigated by Katsaros and Galatis (1988), Medullary cells in the midribs of *D. membranacea* resemble, in some aspects, sieve elements of Laminariales. They are very elongated cells, with a diameter of about 30 μm and a length of up to 300 μm, and are arranged end-to-end. Transverse cell walls are perforated by numerous pores (2000-3000 μm^{-2}) with a diameter of 60-100 nm, which is slightly larger than the width of the plasmodesmata in the cortex. Longitudinal walls are thick and exhibit primary pit fields. Callose was not detected. Medullary cells retain their nuclei and organelles but signs of degeneration have been observed (Katsaros and Galatis 1988). The fine structure of these cells suggests that they may be involved in translocation, although this is not proven.

Scytosiphonales. Although the Scytosiphonales are not known to contain conducting cells, a newly described member of this order, *Jolyna laminarioides* reported by Guimarães et al. (1986), is characterized by medullary filamentous cells, resembling sieve elements of Laminariales (see Sect. 1.5.3). The formalin-preserved material, however, did not allow the study of the fine structure of these cells. The terminology and description of the "structural elements" in the medulla of *Jolyna* is confusing. Guimarães et al. (1986) apparently distinguished "fiber-like cells" measuring up to 1.4 cm in length and 14 μm in width, with thick cell walls, comparable to the "solenocysts" of *Saccorhiza* (Sauvageau 1918; Emerson et al. 1982) from "differentiated medullary filaments" resembling filaments of sieve elements in other Laminariales. "Differentiated medullary filaments" are composed of cell elements, 50-150 μm long and 10-20 μm wide, which have thickened longitudinal walls and are arranged end-to-end. Pores have not been detected in these cross walls by light microscopy, however. The authors speculate that the "fiber-like cells" provide mechanical strength to the thallus and that they may also function in long-distance transport.

Desmarestiales. Members of the Desmarestiales have attracted much attention since it was recognized that they are the most abundant, dominating seaweeds of the Antarctic. The most conspicuous species is *Himantothallus grandifolius,* composed of a holdfast, a short flat stipe and one to several thick undivided blades which are up to 10 m in length and 1 m in width. The following description is based upon the work of Moe and Silva (1981): the medulla is composed of a central layer of mainly longitudinally-arranged, filamentous cells. Among those, files of "trumpet hypha" run the length of the blade and form a sieve tube-like symplast. They are sheathed by intertwined, filamentous cells which are not connected with the sieve elements.

The end walls of the sieve elements are perforated by pores, approximately 400 nm wide, at a density of 2.5-4 μm^{-2}. Callose was detected at the perforated end walls, but may also extend for some distance along the longitudinal walls. The fine structure of the protoplast was not studied.

Sheathed sieve elements in the medulla are claimed to be a common anatomical feature in the genera *Desmarestia, Himantothallus* and *Phaeurus* and they seem to be especially well developed in perennial Antarctic species of *Desmarestia* (Moe and Silva 1981). A longitudinal section of sieve elements in the veins of a ligulate lateral of *Desmarestia ligulata* is shown in Fig. 1.3. In addition to the well-known trumpet shape and cell wall structure of brown-algal sieve elements, i. e., thick longitudinal walls and thin perforated end walls, a protoplast mainly composed of numerous small vesicles is seen.

The morphology of the cell, as observed in the electron microscope, changes during development. The nucleus shows signs of disintegration and the ribosomes, dictyosomes, mitochondria and plastids are mainly found at the cell periphery. There is no published experimental evidence, however, for the function of sieve elements of Desmarestiales as a conducting system for translocation.

Fucales. Long-distance transport of organic and inorganic compounds in various members of the Fucales, with a velocity faster than can be accounted for

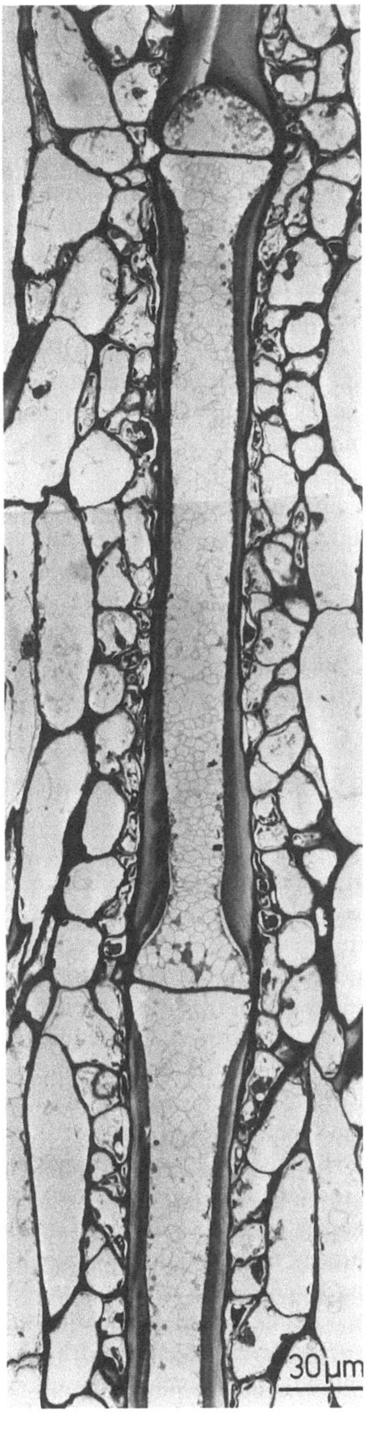

Fig. 1.3. *Desmarestia ligulata*. Longitudinal section through a file of sieve elements. Note the thick longitudinal walls and thin perforated cross walls. Cells are trumpet-shaped and the protoplast is composed of numerous small vacuoles and the usual organelles of eukaryotic cells

by diffusion (2–4 cm h^{-1}), is well established (Bil' et al. 1982; Floc'h 1982; Diouris and Floc'h 1984; Penot et al. 1985) but the organization, anatomy and fine structure of the conducting cell system for translocation is poorly understood. Moos (1983) investigated six genera of the Fucales and recognized that all of them have a system of interconnected conducting cells, very similar to those of Laminariales. The central medulla is made up of sieve elements with well-defined "sieve plates" on their thin (1.0–1.5 μm) end walls. Pore diameters are approximately 100 nm or less. Although some variation of pore size and distribution was recorded, the "sieve plate" structure was very similar in all the species examined. The results published by Moss (1983) have been disputed by Fielding et al. (1987).

In addition to the genera *Pelvetia, Fucus, Ascophyllum, Himanthalia, Halidrys* and *Bifurcaria,* all investigated by Moss (1983), the pore diameter and frequency in end walls of conducting cells of *Cystoseira nodicaulis* were determined by Penot et al. (1985). The average diameter was 70 nm and the pore frequency was 50–60 μm^{-2}. The protoplasts of sieve elements revealed a polar arrangement with cellular contents banked up on one side of a "sieve plate". Sieve elements of the Fucales contain numerous mitochondria and modified chloroplasts. The very irregularly lobed nucleus seems to be retained throughout the life of the cell (Moss 1983).

It was confirmed by histoautoradiography that the long-distance transport of ^{14}C-assimilates occurs through the medulla, most likely through the longitudinal files of symplastically interconnected sieve elements (Diouris and Floc'h 1984).

1.5.3 Sieve Elements in Laminariales

Cell Types and Interconnections as Related to Transport. A general feature of the large sporophytes of the Laminariales is a well-developed conducting system of sieve elements for long-distance transport of organic and inorganic nutrients. These cells are arranged end-to-end, are symplastically connected through perforated end walls and form files of sieve elements which resemble sieve tubes. Longitudinal files of sieve elements are cross-connected by interposed sieve elements, establishing a three-dimensional conducting network in the medulla. The diameter and the length of the sieve elements, as well as the cell shape and the width of sieve pores, are very variable, depending on the plant species in question (Table 1.1) and the state of differentiation.

The medulla of the Laminariales is not only composed of sieve elements but also of hyphae, the latter being more numerous in stipes and midribs than in the blades. Hyphal cells form slender, frequently branched and irregularly bent threads, extending mainly radially from the cortex where they originate through the intercellular wall matrix of the medulla (Steinbiss and Schmitz 1974). According to Parker and Fu (1965), Parker and Huber (1965) and Nicholson (1976), sieve elements and hyphae do not communicate and may be separate cell systems. Hyphal cells are symplastically connected with each other and

Table 1.1. Pore diameter and pore frequency of perforated cross walls of sieve elements and sieve element-like transport cells of brown algae

Species	Pore diameter (nm)	Pore frequency (number · μm^{-2})	References
Dictyotales			
Dictyopteris membranacea	60–100	2000–3000	Katsaros and Galatis (1988)
Desmarestiales			
Himantothallus grandifolius	400	2.5–4.0	Moe and Silva (1981)
Fucales			
Fucus vesiculosus,	100 or less	–	Moss (1983)
F. spiralis, *F. serratus*		–	
and other genera,			
F. vesiculosus, *F. spiralis*,	37.5–50		Fielding et al. (1987)
F. serratus,			
Cystoseira nodicaulis	70	50–60	Penot et al. (1985)
Laminariales			
Saccorhiza dermatodea	50	100	Emerson et al. (1982)
Laminaria saccharina	50–90	37[a]	Sideman and Scheirer (1977)
L. saccharina, *L. digitata*, *L. hyperborea*	40–90	50–60	Schmitz (1981)
L. groenlandica	40–80	30–60	Schmitz and Srivastava (1974)
Alaria marginata	100–300	18	Schmitz and Srivastava (1975)
Nereocystis lütkeana			
blade	100–400	0.5	Schmitz and Srivastava (1976)
stipe	400–900	0.05	Schmitz and Srivastava (1976)
Pelagophycus porra	350–650	1.4–1.8[b]	Parker and Fu (1965)
Pelagophycus giganteus	500–1300	0.6–1.3[b]	Parker and Fu (1965)
Macrocystis laevis	2000	0.09–0.11	Hay (1986)
Macrocystis integrifolia	700–4500	0.07–0.15	Schmitz (1981)
Macrocystis pyrifera	2400–2600	0.03–0.08[c]	Parker and Huber (1965)

[a] Calculated from Sideman and Scheirer (1977), Fig. 9.
[b] Calculated from Parker and Fu (1965).
[c] Calculated from Parker and Huber (1965).

with the cortex parenchyma cells by frequent plasmodesmata (van Went et al. 1973), however, their function is not clear.

Cytoplasmic, intercellular connections between sieve elements and cortex parenchyma cells are scarce. They may be absent in stipes (Nicholson 1976), but are present in *Saccorhiza dermatodea* (Emerson et al. 1982) and in blades of several *Laminariales* (Schmitz 1984 and unpublished results). It is a crucial point for a more profound understanding of sieve element loading and unloading, to learn more about cytoplasmic interconnections between sieve elements and cortex cells. Since the main capacity for photosynthetic carbon fixation is confined to the meristoderm and the outermost cortex cells (Hellebust and

Haug 1972; Kremer 1980) and since histoautoradiography has clearly shown that sieve elements in the medulla, separated by several cell layers from the source tissue, are the conduits for long-distance transport (Steinbiss and Schmitz 1973; Schmitz and Srivastava 1974; Emerson et al. 1982; Shih et al. 1983), a medium-distance transversal assimilate transfer is mandatory. In a detailed study of the fine structure, distribution and frequency of pits and plasmodesmata in the cortex of *Laminaria hyperborea* and *L. saccharina,* a continuous symplastic pathway between the meristoderm and the innermost cortex cells was traced (Schmitz and Kühn 1982). Circumstantial evidence, based on the anatomy and physiological transport data, supports the assumption that transport between the meristoderm and sieve elements is entirely symplastic (for comprehensive discussion see Buggeln 1983).

The terminology used to describe conducting cells in Laminariales is still non uniform. Suggestions have been made by Esau (1969) and Schmitz and Srivastava (1974). A special nomenclature for the genus *Saccorhiza* was originally proposed by Sauvageau (1918), and his system was adopted by Emerson et al. (1982).

Aspects of Development. The early development of sporophytes and the origin of tissue systems as well as the formation of sieve elements, connecting sieve elements and hyphal cells in the growing region of brown algae were studied by several anatomists at the beginning of this century (for literature review see Fritsch 1945; Esau 1969; Schmitz 1981). Detailed observations by Killian (1911) on *Laminaria digitata* were basically confirmed for the genus *Laminaria* by Steinbiss and Schmitz (1974), Schmitz and Srivastava (1974), Sideman and Scheirer (1977). Although developmental aspects were briefly covered by other authors as well (Schmitz and Srivastava 1975 for *Alaria;* Schmitz and Srivastava 1976 for *Nereocystis;* Emerson et al. 1982 for *Saccorhiza;* Shih et al. 1983 for *Macrocystis*), several other aspects of development have not satisfactorily been studied.

Sieve elements which are formed early in sporophyte development are generally smaller and not as highly differentiated as those formed later. These "proto" sieve elements stop growing and can therefore not cope with thallus growth. They are passively stretched and acquire the typical trumpet shape ("trumpet filaments", "trumpet hyphae") but may later be pulled out to a length of >1 mm and become nonfunctional. Van Went and Tammes (1973) claimed that the trumpet shape of such sieve elements in *Laminaria* is a fixation artifact. New cells are contributed to the outer part of the medulla from inner cortex cell layers which differentiate into files of sieve elements and connecting sieve elements; cortex and medulla are therefore not sharply delimited (Schmitz and Srivastava 1974, 1975; Shih et al. 1983). Sieve element formation from cortex cells occurs mainly at the growing regions of a thallus but differentiation into mature sieve elements continues for a considerable distance above or below zones of growth (Shih et al. 1983). Differentiation of sieve elements in perennials has not been studied carefully. In cross-sections of stipes, earlier formed sieve elements, which may be collapsed and nonfunctional, occur in the central part of the medulla, while ontogenetically younger but more highly developed sieve elements are found at the periphery. In *Macrocystis,* such sieve elements are typically arranged in radial rows (Parker 1971; Shih et al. 1983). In the thin medulla of phylloids, young and old sieve elements are interwoven and their spacial arrangement can hardly be traced in sectioned material (Schmitz 1984).

Cell Walls and Wall Perforations. Most descriptions of sieve elements primarily focus on cell wall structure. Although these cell walls can easily be depicted, no detailed analysis of chemical composition or structural arrangement and development of the sieve element cell walls has been done. It is agreed, however, that sieve elements have special cell walls, usually much thickened, longitudinal walls and relatively thin, perforated end walls (Fig. 1.4). Apart from the intercellular space, probably mainly composed of sulfated fucans (Mabeau and Kloareg 1987), a fibrillar wall with cellulose and alginic acid is recognized. At least two fibrillar wall layers can be distinguished: an outer wall layer with longitudinal orientation of fibrillar wall material and an inner wall layer with fibrils oriented transversely to the long axis of the cell. As cells mature, cell walls thicken by secondary wall deposition which is more extensive at the longitudinal than at the end walls. Sieve elements can be recognized by their thick secondary walls, those of "solenocysts" are typically multilayered (Emerson et al. 1982).

The most characteristic structural feature of the sieve element is its perforated end walls (Figs. 1.4, 1.6, 1.8). Depending on the plant species, perforations may be as small as plasmodesmata or as wide as the sieve pores of vascular plants (Table 1.1). Perforated end walls of *Laminaria* were interpreted as large single primary pit fields by Ziegler and Ruck (1967) but electron microscopy has revealed that they are *specialized* primary pit fields and should therefore be considered sieve areas (Schmitz and Srivastava 1974; Sideman and Scheirer 1977).

Cross-connecting sieve elements are also linked to longitudinal files of sieve elements by sieve areas. According to old light microscopic observations, the pores in "lateral sieve areas" of *Macrocystis* are usually smaller than those in the sieve-element end walls. Although this observation has not been confirmed, it strengthens the case for considering such files of sieve elements as true sieve tubes and the perforated end walls as true sieve plates.

Callose was detected in older sieve elements of most of the *Laminariales* but seems to be absent in "solenocysts" of *Saccorhiza dermatodea* (Emerson et al. 1982). It was confirmed by Ziegler (1963) and Parker (1964) that callose of *Macrocystis* is chemically very similar to the callose of sieve tubes of vascular plants. Callose could not be detected in young sieve elements, even though their pores were well developed, as demonstrated for *Macrocystis* (Ziegler 1963) and *Nereocystis* (Schmitz and Srivastava 1976). It is not clear from the work of Shih et al. (1983) on *Macrocystis integrifolia* whether the mature sieve elements with wide open pores which have been shown to be involved in translocation, have callose-lined sieve pores. In *Macrocystis pyrifera,* callose is first deposited as cal-

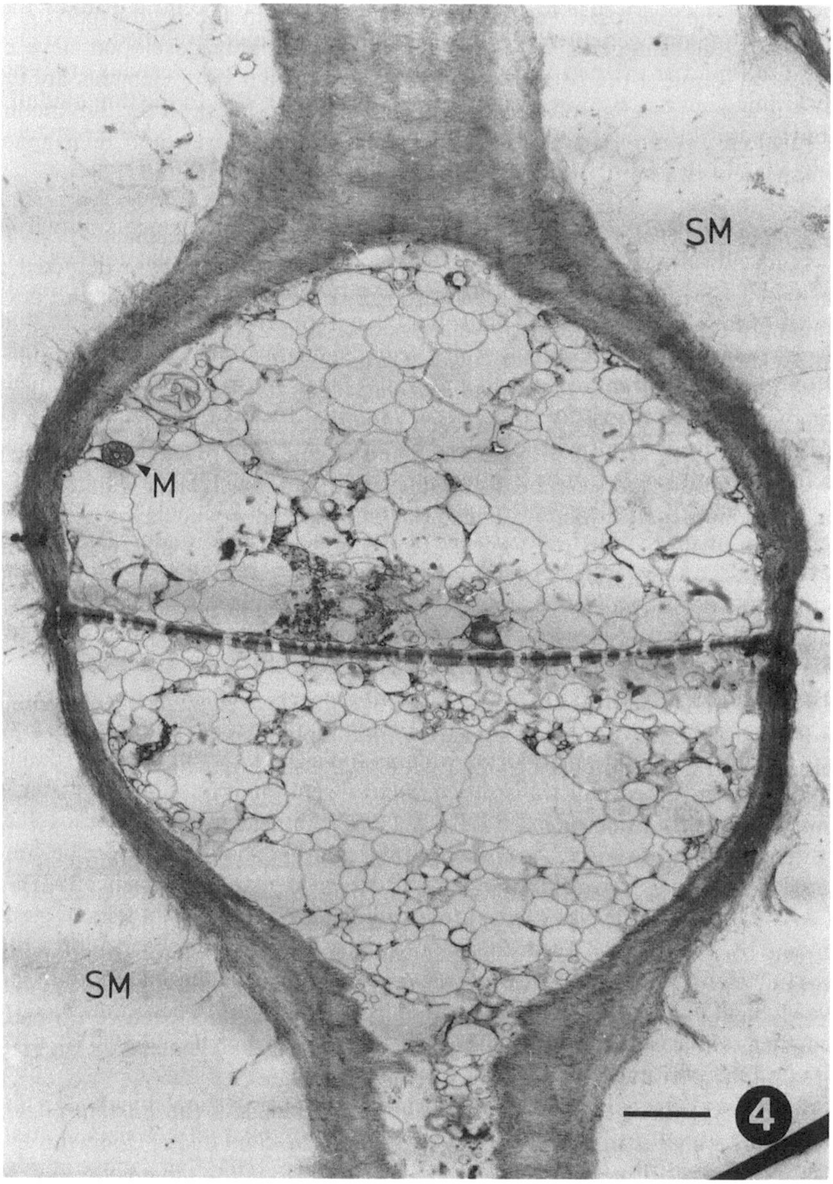

Fig. 1.4. *Nereocystis lütkeana.* Longitudinal section through part of adjacent sieve elements in the pneumatocyst. Files of sieve elements are laterally separated by a voluminous space matrix *(SM).* Note the thick longitudinal walls and the thin perforated cross wall. The dominating cytoplasmic structures are numerous vesicles and small vacuoles. *M* mitochondrium; *Bar line* = 2 μm

lose cylinders, lining each sieve pore (Ziegler 1963; Parker 1971). Later on, massive deposits of callose may cover the sieve areas and interrupt translocation. Callose may also extend along the longitudinal walls and may finally completely fill the lumen of a sieve element (Parker 1971; Schmitz and Srivastava 1974, 1975, 1976; Shih et al. 1983). Sideman and Scheirer (1977) could not detect callose in *Nereocystis* and *Laminaria saccharina,* whereas Nicholson (1976) reported presence of callose at sieve element end walls of *Nereocystis.* This is also documented in Fig. 1.6.

Sieve Element Protoplasts. The fine structure of the sieve element protoplast as well as its differentiation and its structural change during development are still imperfectly known. This is mainly due to inadequate fixation procedures. Sieve elements of Laminariales are never associated with companion cells and they contain all the organelles normally found in eukaryotic cells. There is no information about microtubules and microfilaments; P-protein was not observed.

A nucleus is always present in young sieve elements, and such nuclei are often lobed and contain a distinct nucleolus (Fig. 1.5). No nucleus could be detected in mature sieve elements of *Macrocystis* (Ziegler 1963; Parker and Huber 1965; Parker 1971; Shih et al. 1983) nor in the sieve elements in the stipe of *Nereocystis* (Schmitz and Srivastava 1976). Nuclei, along with other organelles, are decomposed and degenerate during the obliteration of sieve elements of Laminariales.

Mitochondria and plastids are always present in abundance. Their shape and structure are most variable and change significantly during development (Figs. 1.5, 1.7). Their appearance in young sieve elements corresponds to the organization of these organelles in other tissues of brown algae. Plastids change their outer form and internal arrangement of structural components during sieve element differentiation. Plastids progressively degenerate as cells grow older. There is no information on the function of sieve element plastids. Mitochondria, on the other hand, may also change their outer form but seem to retain their internal structure and function as long as the cell is actively translocating. When sieve elements are obliterated, the mitochondria progessively degenerate (Schmitz and Srivastava 1974, 1975, 1976; Sideman and Scheirer 1977).

The most characteristic structural component of sieve element protoplasts are numerous small vacuoles and vesicles of varying sizes that fill the cell lumen of younger sieve elements near the end walls. This was clearly shown in vivo by hand sections as well as in sectioned chemically fixed material (Schmitz and Srivastava 1974, 1975, 1976; Sideman and Scheirer 1977; Shih et al. 1983). Vesicles form during sieve element differentiation and seem to be present in mature translocating cells of *Laminaria* species (Steinbiss and Schmitz 1973; Schmitz and Srivastava 1974; Sideman and Scheirer 1977). Mature sieve elements of *Nereocystis* reveal fewer and larger vacuoles (Schmitz and Srivastava 1976) and actively translocating mature sieve elements of *Macrocystis integrifolia* show

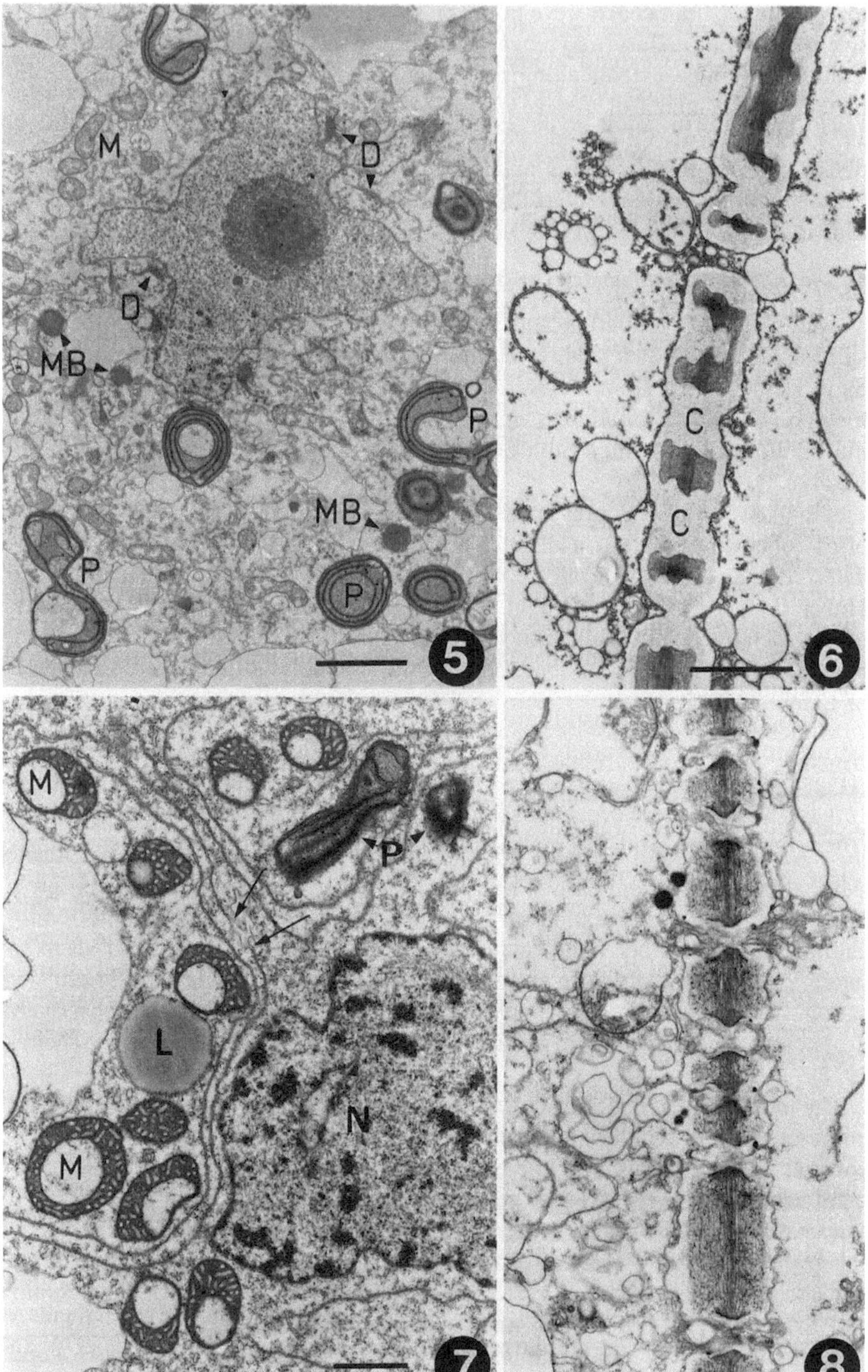
M
D
D
MB
MB
P
P
P
5
C
C
6
M
P
L
M
N
7
8

only a thin parietal layer of cytoplasm with the remaining cell lumen essentially free of vacuoles (Shih et al. 1983).

The ribosomes, dictyosomes and endoplasmic reticulum are also well developed during the early phase of sieve element differentiation when the growth of cytoplasm and the cell wall occurs (Figs. 1.5, 1.7). Microbodies have also been observed in such cells (Schmitz and Srivastava 1974). Cytoplasmic structures and vesicles break down during progressive differentiation and finally disappear or remain in a clumped, necrotic state. Shih et al. (1983) state for *Macrocystis integrifolia* that this process and the breakdown of the nucleus occur at the same time as the sieve pores enlarge.

Brown algae of the order Laminariales have evolved a very efficient long-distance transport system comparable to the sieve-element system of vascular plants. Sieve elements of Laminariales are subject to ontogenetic development with pronounced variations of fine structure. Three stages of differentiation can be recognized: (1) differentiation of parenchymatic precursor cells into sieve elements (increase of cytoplasm, multiplication of cell organelles, formation of small vacuoles and vesicles, cell growth in length and width, deposition of cell wall material); (2) formation of the mature sieve element, optimized for transport (progressive formation of a structure-free cell lumen by impoverishment of the cytoplasm, fusion of small vesicles and formation of fewer but larger vacuoles, disintegration of the nucleus, structural modification of plastids, enlargement of sieve pores); (3) degeneration of structure and loss of function (closure of sieve pores with callose, massive callose deposits at the end walls that may also extend along longitudinal walls, lysis and/or necrosis of protoplast structures).

Correlation of the structure and function of these cells in the algae needs further investigation. Based on sieve-element fine structure and transport physiology data, long-distance transport in *Macrocystis* can best be explained by a Münch-type pressure flow. The same mechanism, however, is hard to accept as the driving force in species of *Laminaria* and in Fucales, because the structure and organization of sieve elements in them are different (for discussion see Schmitz 1981, 1982; Buggeln 1983; Raven 1984, 1988).

Fig. 1.5. *Laminaria groenlandica.* Protoplast of a younger sieve element. A lobed nucleus with prominent nucleolus is surrounded by cytoplasm with dictyosomes *(D)*, mitochondria *(M)*, microbodies *(MB)* and structurally modified plastids *(P)*; *Bar line* = 2 μm. **Fig. 1.6.** *Nereocystis lütkeana.* Sieve area of adjacent older sieve elements. Sieve pores are partly plugged by callose *(C)*. Protoplasts of the sieve elements appear necrotic, the cell membrane and vesicles of different size persist longest; *Bar line* = 2 μm. **Fig. 1.7.** *Alaria marginata.* Protoplast of a younger sieve element with part of a nucleus *(N)*, ring-shaped mitochondria *(M)* and extended endoplasmatic reticulum. Vesiculation of ER *(unlabeled arrows)*; *L* lipid body; *P* plastid; *Bar line* = 1 μm. **Fig. 1.8.** *Nereocystis lütkeana.* Sieve area in a cross wall between sieve elements in the blade. Cytoplasm and plasmalemma of adjacent cells are continuous, tubular structures can be followed through the pores; *Bar line* = 1 μm

References

Bauman RW, Jones BR (1986) Electrophysiological investigations of the red alga *Griffithsia pacifica* Kyl. J Phycol 22: 49-56

Bil' KY, Medyannikov VM, Karpov EA (1982) Incorporation of ^{14}C into photosynthates and their redistribution in tissues of the thallus of *Sargassum pallidum*. Sov Plant Phys 28: 565-574

Buggeln RG (1983) Photoassimilate translocation in brown algae. Prog Phycol Res 2: 283-332

Callow JA, Callow ME, Evans LV (1979) Nutritional studies on the parasitic red alga *Choreocolax polysiphoniae*. New Phytol 83: 451-462

Celan M (1939) Sur les cellules conductrices des algues rouges. Bull Soc Bot 86: 297-300

Celan M (1941) Recherches cytologiques sur les algues rouges. Rev Cytol Cytophysiol Veg 5: 1-162

Diouris M, Floc'h JY (1984) Long-distance transport of ^{14}C-labelled assimilates in the *Fucales:* directionality, pathway and velocity. Mar Biol 78: 199-204

Emerson CJ, Buggeln RG, Bal AK (1982) Translocation in *Saccorhiza dermatodea* (Laminariales, Phaeophyceae): anatomy and physiology. Can J Bot 60: 2164-2184

Esau K (1969) The phloem. In: Zimmerman W, Ozenda P, Wulff HD (eds) Encyclopedia of plant anatomy, vol 5, pt 2. Borntraeger, Berlin Stuttgart, 505 pp

Evans LV, Callow JA, Callow ME (1973) Structural and physiological studies on the parasitic red alga *Holmsella*. New Phytol 72: 393-402

Feldmann J (1939) Sur la structure des cellules axiales de l'*Asparagopsis armata* Harv. C R Acad Sci 208: 1743-1745

Feldmann J, Feldmann G (1946) Recherches sur l'appareil conducteur des Floridées. Rev Cytol Cytophysiol Veg 8: 161-209

Fielding AH, Carter PL, Smith CA (1987) Sieve plates in *Fucus:* a reappraisal of size and pore distribution. Phycologia 26: 501-504

Floc'h JY (1982) Uptake of inorganic ions and their long-distance transport in *Fucales* and *Laminariales*. In: Srivastava LM (ed) Synthetic and degradative processes in marine macrophytes. De Gruyter, Berlin New York, pp 139-166

Fritsch FE (1945) The structure and reproduction of the algae, vol II. Cambridge Univ Press, London

Goff LJ (1979) The biology of *Harveyella mirabilis* (Cryptonemiales, Rhodophyceae). VI. Translocation of photoassimilated ^{14}C. J Phycol 15: 82-87

Goodwin PB (1983) Molecular size limit for movement in the symplast of the *Elodea* leaf. Planta 157: 124-130

Guimarães SMPB, Braga MRA, Cordeiro-Marino M, Pedrini AG (1986) Morphology and taxonomy of *Jolyna laminarioides,* a new member of the Scytosiphonales (Phaeophyceae) from Brazil. Phycologia 25: 99-108

Hartmann T, Eschrich W (1969) Stofftransport in Rotalgen. Planta 85: 303-312

Hay CH (1986) A new species of *Macrocystis* C. Ag. (Phaeophyta) from Marion Island, southern Indian Ocean. Phycologia 25: 241-252

Hellebust JA, Haug A (1972) Photosynthesis, translocation, and alginic acid synthesis in *Laminaria digitata* and *Laminaria hyperborea*. Can J Bot 50: 169-176

Katsaros C, Galatis B (1988) Thallus development in *Dictyopteris membranacea* (Phaeophyta, Dictyotales). Br Phycol J 23: 71-88

Killian K (1911) Beiträge zur Kenntnis der Laminarien. Z Bot 3: 433-494

Kremer B (1980) Transversal profiles of carbon assimilation in the fronds of three *Laminaria* species. Mar Biol 59: 95-103

Kremer B (1983) Carbon economy and nutrition of the alloparasitic red alga *Harveyella mirabilis*. Mar Biol 76: 231-239

Mabeau S, Kloareg B (1987) Isolation and analysis of the cell walls of brown algae: *Fucus spiralis, F. ceranoides, F. vesiculosus, F. serratus, Bifurcaria bifurcata* and *Laminaria digitata*. J Exp Bot 38: 1573-1580

Moe RL, Silva PC (1981) Morphology and taxonomy of *Himantothallus* (including *Phaeoglossum* and *Phyllogigas*), an antarctic member of the *Desmarestiales* (Phaeophyceae). J Phycol 17: 15-29
Moss BL (1983) Sieve elements in the *Fucales*. New Phytol 93: 433-437
Nicholson NL (1976) Anatomy of the medulla of *Nereocystis*. Bot Mar 19: 23-31
Parker BC (1964) Chemical nature of sieve tube callus in *Macrocystis*. Phycologia 4: 27-42
Parker BC (1971) The internal structure of *Macrocystis*. In: North WJ (ed) The biology of giant kelp beds *(Macrocystis)* in California. Nova Hedwigia 32: 99-121
Parker BC, Fu M (1965) The internal structure of the elk kelp (*Pelagophycus* species). Can J Bot 43: 1293-1305
Parker BC, Huber J (1965) Translocation in *Macrocystis*. II. Fine structure of sieve tubes. J Phycol 1: 172-179
Penot M (1986) Lanthanum as a tracer for apoplastic pathway in the axis of *Cystoseira nodicaulis* (Fucales, Cystoseiraceae). Cryptogam Algol 7: 189-192
Penot M, Dumay J, Pellegrini M (1985) Contribution à l'étude de la fixation et du transport du ^{14}C chez *Cystoseira nodicaulis* (Fucales, Cystoseiraceae). Phycologia 24: 93-102
Pueschel CM (1977) A freeze-etch study of the ultrastructure of red algal pit plugs. Protoplasma 91: 15-30
Pueschel CM (1980a) A reappraisal of the cytochemical properties of rhodophycean pit plugs. Phycologia 19: 210-217
Pueschel CM (1980b) Pit connections and translocation in red algae. Science 209: 422-423
Pueschel CM (1987) Absence of cap membranes as a characteristic of pit plugs of some red algal orders. J Phycol 23: 150-156
Pueschel CM, Cole KM (1982) Rhodophycean pit plugs: an ultrastructural survey with taxonomic implications. Am J Bot 69: 703-720
Raven JA (1984) Energetics and transport in aquatic plants. Liss, New York
Raven JA (1988) Algae. In: Baker DA, Hall JL (eds) Solute transport in plant cells and tissues. Longman, Harlow, pp 166-219
Sauvageau C (1918) Recherches sur les laminaires des côtes de France. Mem Acad Sci Paris 56: 1-240
Schmitz K (1981) Translocation. In: Lobban CS, Wynne MJ (eds) The biology of seaweeds. Botanical monographs, vol 17. Blackwell, Oxford, pp 534-558
Schmitz K (1982) Translocation of organic compounds in Laminariales. In: Srivastava LM (ed) Synthetic and degradative processes in marine macrophytes. De Gruyter, Berlin New York, pp 167-182
Schmitz K (1984) A cell system for symplastic transport of photoassimilate between meristoderm and sieve elements in *Alaria tenuifolia*. Mar Biol 78: 209-214
Schmitz K, Kühn R (1982) Fine structure, distribution and frequency of plasmodesmata and pits in the cortex of *Laminaria hyperborea* and *L. saccharina*. Planta 154: 383-392
Schmitz K, Srivastava LM (1974) Fine structure and development of sieve tubes in *Laminaria groenlandica* Rosenv. Cytobiol 10: 66-87
Schmitz K, Srivastava LM (1975) On the fine structure of sieve tubes and the physiology of assimilate transport in *Alaria marginata*. Can J Bot 53: 861-876
Schmitz K, Srivastava LM (1976) The fine structure of sieve elements of *Nereocystis lütkeana*. Am J Bot 63: 679-693
Scott FJ, Wetherbee R, Kraft GT (1982) The morphology and development of some prominently stalked Southern Australian *Halymeniaceae* (Cryptonemiales, Rhodophyta). I. *Cryptonemia kallymenioides* (Harvey) Karft Comb. Nov. and *C. undulata* Sonder. J Phycol 18: 245-257
Shih ML, Floc'h JY, Srivastava LM (1983) Localization of ^{14}C-labeled assimilates in sieve elements of *Macrocystis integrifolia* by histoautoradiography. Can J Bot 61: 157-163
Sideman EJ, Scheirer DC (1977) Some fine structural observations on developing and mature sieve elements in the brown alga *Laminaria saccharina*. Am J Bot 64: 649-657
Steinbiss HH, Schmitz K (1973) CO_2-Fixierung und Stofftransport in benthischen marinen

Algen. V. Zur autoradiographischen Lokalisation der Assimilattransportbahnen im Thallus von *Laminaria hyperborea*. Planta 112: 253-263

Steinbiss HH, Schmitz K (1974) Zur Entwicklung und funktionellen Anatomie des Phylloids von *Laminaria hyperborea*. Helgoländer Wiss Meeresunters 26: 134-152

Tucker EB (1982) Translocation in the staminal hairs of *Setcreasea purpurea*. I. A study of cell ultrastructure and cell-to-cell passage of molecular probes. Protoplasma 113: 193-201

Tyree MT (1970) The symplast concept. A general theory of symplastic transport according to the thermodynamics of irreversible processes. J Theor Biol 26: 181-214

Wetherbee R (1979) "Transfer connections": specialized pathways for nutrient translocation in a red alga? Science 204: 858-859

Wetherbee R, Kraft GT (1981) Morphological and fine structural features of pit connections in *Cryptonemia* sp., a highly differentiated marine red alga from Australia. Protoplasma 106: 167-172

Van Went JL, Tammes PML (1973) Trumpet filaments in *Laminaria digitata* as an artifact. Acta Bot Neerl 22: 112-119

Van Went JL, van Aelst AC, Tammes PML (1973) Transverse connections between cortex and translocating medulla in *Laminaria digitata*. Acta Bot Neerl 22: 77-78

Ziegler H (1963) Untersuchungen über die Feinstruktur des Phloems, II. Mitteilung: Die Siebplatten bei der Braunalge *Macrocystis pyrifera* (L.) AG. Protoplasma 57: 786-799

Ziegler H, Ruck I (1967) Untersuchungen über die Feinstruktur des Phloems, III. Mitteilung: Die "Trompetenzellen" von *Laminaria*-Arten. Planta 73: 62-73

2 Mosses

Daniel C. Scheirer

2.1 Introduction

2.1.1 Overview

Although the study of sieve element structure and function has its origins in Hartig's discovery of the sieve tube in 1837, the modern era of inquiry into the nature of the moss sieve element has a much more recent history. Starting in the mid-1960's with the work of Hébant (1964 a, b, 1966) and Eschrich and Steiner (1967, 1968 a, b), the structure and development of the phloem-like tissue of mosses has been shown to be quite similar to the sieve elements of vascular cryptogams.

Based on light microscopic studies, Esau (1969) reported that leptoids show a remarkable similarity with sieve elements of vascular plants. Since that time, the ultrastructural features of moss sieve elements have been detailed in a number of studies and reviewed periodically (Behnke 1975; Hébant 1977; Scheirer 1980; Schofield and Hébant 1984). The specific goals of this chapter are: (1) to review our knowledge of the structure of moss sieve elements beginning from the mid-1960's to the present, and (2) to interpret bryophyte sieve element structure in light of the vast body of data available for sieve elements of vascular cryptogams. This comparative approach will greatly aid our interpretation of the features of primitive sieve elements.

2.1.2 Terminology

Since mosses have been traditionally treated as nonvascular plants, a distinctive set of terms appears in the literature in reference to the moss sieve element (Hébant and Schäfer 1974). One example is the term *leptoid* (derived from leptome), first used by Tansley and Chick (1901) to describe the sieve tube-like cells in the rhizome of *Polytrichum commune*. This is the most commonly used term to describe the phloem-like cells found in mosses. However, in the present discussion, I will follow the recent proposal by Behnke (1986) that the general term *sieve element* be used with the addition of the characteristic higher taxon as specific epithet. For example, sieve elements of Bryophyta, or various modifications thereof, i. e., moss sieve elements, etc., will be used instead of the older term leptoid. Physiological data (Eschrich and Steiner 1967; Eschrich 1975; Thomas, et al. 1988) as well as structural evidence support the use of this term.

2.2 General Organization of Conducting Tissues in Mosses

2.2.1 The Gametophyte

Sieve elements are most highly developed in the Polytrichaceae and Dawsoniaceae, two families that contain anatomically complex gametophores (Scheirer 1972; Schofield and Hébant 1984). In the gametophytic stem of these mosses, the protostelic arrangement of the conducting strand is similar to the stelar organization in vascular cryptogams. In the Polytrichales, sieve elements and associated parenchyma are found as a sheath of cells which surround the central strand of hydroids (Scheirer 1973). In the rhizome, the sieve elements are located in three distinct regions or furrows, which alternate with the trilobed hydrome (Scheirer 1976). Sieve elements of the leaf have been referred to as leptoids, deuters, or socii, and there is disagreement as to whether these cells exhibit sufficient ultrastructural characteristics to warrant the term sieve element (Hébant 1972). A transection of the leaf shows a crescent-shaped row of large-diametered parenchymatic elements which extend the length of the blade and connect with the stem sieve elements via leaf traces (Scheirer 1983). Hébant (1972, 1977) preferred to use the term *deuter,* a name which has neither a physiological nor a cytological meaning, to describe these cells of the leaf bundle and trace because their cytoplasm is not as highly differentiated as the stem sieve elements. On the other hand, Wiencke and Schulz (1983) used the term *leptoid* in describing the assimilate-conducting cells in the leaf nerve of *Funaria.*

2.2.2 The Sporophyte

In the setae of Polytrichales, sieve elements are recognized as a layer of cells surrounding the central strand of hydroids (Eschrich 1975). Sieve elements in setae are not intermixed with parenchyma as is the case within the gametophyte. Hébant (1975 a, 1977) considered the setae sieve elements (leptoids, Hébant's term) not as highly developed as those that are found in the gametophyte and compared them with a pericycle or an endodermis which serve as a protective tissue for the hydroids. Few studies have been conducted on the sieve elements in the sporophyte and those that have been are ambiguous in their interpretations (Favali and Gianni 1973; Favali and Bassi 1974; see Hébant 1977 for a discussion of these papers). According to Hébant (1970 a), in certain taxa *(Funaria, Splachnum),* sieve elements are present in the seta but absent in the gametophyte. Hébant (1979) called these cells in the sporophyte "residual" leptoids. Further studies are certainly warranted in this area.

2.3 Structure of Sieve Elements

2.3.1 General Features of Moss Sieve Elements

Sieve elements, like most cells of the vascular tissue, are longitudinally elongated, their length usually being several times their width. The length of the sieve elements in the gametophytic stem of *Dendroligotrichum dendroides* can range from 300-500 μm (Scheirer 1974), while the sieve elements in *Funaria* leaflets are 120-160 μm in length (Wiencke and Schulz 1983). A general similarity exists between the sieve elements of mosses and those of vascular cryptogams (Hébant 1975 b). These similarities include: (1) their general size and appearance; (2) modifications of the protoplast as the cell differentiates, such as stacks of ER, small undifferentiated plastids, and the disappearance of dictyosomes; (3) the presence of a degenerated nucleus at maturity; (4) refractive spherules which develop in association with the ER; (5) oblique end walls with numerous small pores; and (6) frequent thickening of the lateral wall suggestive of the nacreous thickenings of fern sieve elements.

2.3.2 Differentiating and Mature Sieve Elements

The early development of the sieve elements is characterized by an active elongation of the cell. During this precocious elongation (Hébant 1977), the end walls become steeply inclined and are perforated by numerous plasmodesmata. In *Polytrichum commune,* the nucleus enlarges considerably and the nucleolus becomes quite prominent early in development (Fig. 2.1). Young sieve elements contain lipid droplets in the cytoplasm, but these disappear as the cells mature.

Nucleus. The retention of a degenerated nucleus in moss sieve elements is a feature that is consistent with sieve elements of certain vascular cryptogams (see Chap. 3). At maturity, the sieve element contains a spherical nucleus which may exhibit two types of degenerative phenomena that occur late in development. In the first type, the nucleus becomes pycnotic or condensed and greatly reduced in size (Fig. 2.2), while in the second type a dechromatinization or chromatolysis may be observed (Fig. 2.3) (Hébant 1969, 1970 b, 1973 b). In the latter case, the nucleus remains large or may even swell slightly and the nucleoplasm becomes progressively lighter in appearance (Fig. 2.3). In either case, the nuclear envelope remains intact. Dechromatinization seems to be the more frequent type of nuclear modification. Clumps of dense chromatin may persist which are preferentially appressed to the nuclear envelope (Hébant 1976 b). Both forms may occur in the sieve elements of the same leafy stem. Hallet (1972) demonstrated that the nucleus in sieve elements of *Polytrichum formosum* becomes endopolyploid, which may account for the precocious enlargement of the nucleus during early differentiation. Hébant (1973 b) demonstrated with *Polytrichum commune*

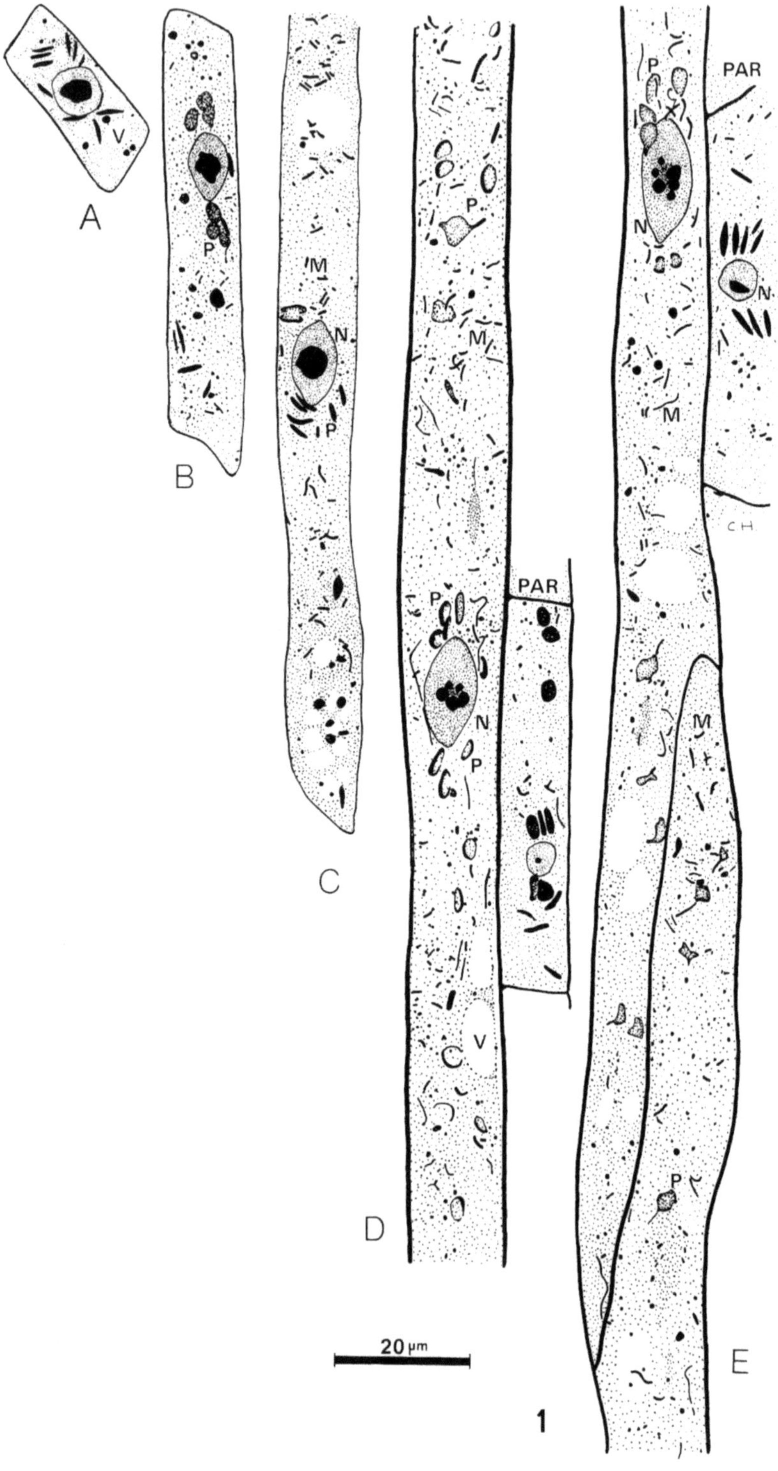

1

that the nucleus is truly inactive because of the failure to incorporate tritiated uridine, indicating the cessation of RNA synthesis.

Cell Wall and Plasmalemma. The cell wall of sieve elements begins to thicken early in development and may become quite pronounced in *Atrichum undulatum* and *Polytrichum juniperinum* (Fig. 2.4). The wall thickenings in these two species are suggestive of the nacreous walls of pteridophyte sieve elements (see Chap. 3). In another species, *Dendroligotrichum dendroides,* sieve elements exhibit unevenly thickened walls, with the thickenings appearing predominantly on the radial walls (Scheirer 1977). A histochemical study of these sieve elements revealed that the thickened walls are rich in cellulose with small amounts of polyuronides and pectins and are not lignified. Silver proteinate staining of the walls demonstrated a polylamellate structure. The nacreous walls in *Atrichum* are much more different than the wall described above for *D. dendroides* in that the *Atrichum* sieve element wall exhibits a diffuse microfibrillar inner layer (Fig. 2.4) (Stevenson 1974) much like sieve elements of polypodiaceous ferns. However, the sieve elements in the leaf of *Atrichum* do not have nacreous walls (Scheirer, unpublished observation). Dictyosomes increase in number during wall thickening and dictyosome derived vesicles fuse with the plasmalemma releasing their contents into the wall.

Another modification of the sieve element wall seen in the leaf of *Polytrichum commune* is the development of small wall ingrowths reminiscent of transfer cells (Fig. 2.5) (Scheirer 1983). The sieve elements (deuters or leptoids of some authors) develop walls with a dense outer layer and an inner layer consisting of a loosely arranged, fibrillar network (Fig. 2.5). The wall ingrowths are extensions of this inner wall layer and are lined with the plasmalemma (Fig. 2.5). No electron-lucent space was observed between the wall ingrowths and the plasmalemma, as has been frequently encountered in transfer cells (Gunning and Pate 1969). It appears that these wall ingrowths are a temporary feature of differentiating sieve elements because older sieve elements have a thickened inner wall layer without ingrowths. The differentiating leaf sieve elements also have distinctive cytoplasmic features such as numerous mitochondria, and a large population of polysomes and dictyosomes in addition to the convoluted wall-membrane apparatus. These features are consistent with the view that the sieve elements are involved in active uptake of photoassimilates from the apoplast of the leaf (Scheirer 1983; Reinhart and Thomas 1981). However, as the sieve elements reach maturity, their cytoplasm undergoes considerable modification (see below).

Fig. 2.1 A–E. Differentiating sieve elements in the leafy stem of *Polytrichum commune;* **2.1 A** Immature sieve element prior to elongation. The nucleus is large and a prominent nucleolus is evident; **2.1 B–C** Sieve elements become elongated while the nucleus and nucleolus show a considerable enlargement in comparison to neighboring parenchyma; **2.1 E** Nearly mature sieve elements with steeply inclined end walls. Nucleus remains prominent. *N* nucleus; *P* plastid; *PAR* parenchyma cell; *V* vacuole (All Figures from Hébant 1967)

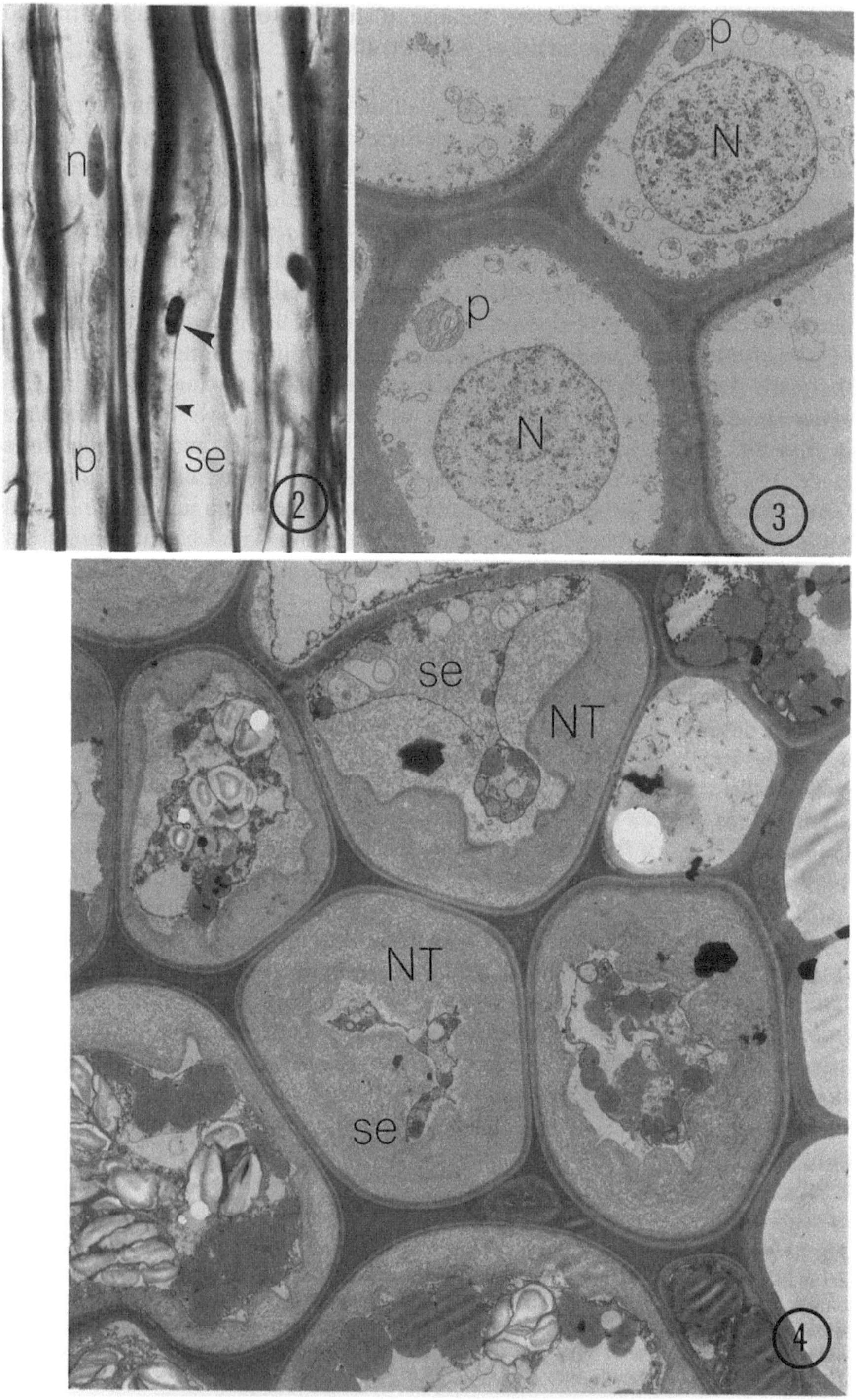
n
p
se
2
p
N
p
N
3
se
NT
NT
se
4

The plasmalemma in sieve elements is a normal-appearing unit membrane which is typically wavy or undulating. In rhizome sieve elements, looped invaginations of the plasmalemma are present. Whether this is a fixation artifact or whether it is related to translocation phenomena is not clear.

Microtubules and Microfilaments. Microtubules are reported to be a striking feature of the mature sieve elements of the *Funaria* leaflet (Wiencke and Schulz 1983). In addition, microtubules have been observed in differentiating sieve elements of the *Polytrichum* rhizome (Scheirer, unpublished data), but none have been observed in mature ones. Microtubules are known to disappear from sieve elements of angiosperms during differentiation. Microfilaments have not been reported in moss sieve elements, although they appear to be common in sieve elements of many plants (Parthasarathy and Pesacreta 1980).

Endoplasmic Reticulum. The endoplasmic reticulum of young sieve elements is in the common form of individual cisternae scattered in the cytoplasm. As differentiation proceeds, ribosomes are lost from the ER and smooth ER remains (Fig. 2.6). ER in the form of single tubules is frequently seen associated with plasmodesmata of the end wall (Fig. 2.7). The ER becomes extensively developed as differentiation continues and various forms of aggregated ER become progressively established. The various forms of ER may be stacked sheets (Figs. 2.6 and 2.8), convoluted tubules, or reticulate networks (Hébant 1969, 1973 a, 1976 a, b). These forms of ER are generally found in a parietal position as the sieve element matures.

Hébant (1974) reported that polarized accumulations of ER in the form of stacked sheets or tubular ER accumulate in the vicinity of sieve element end walls in *Polytrichadelphus magellanicus*. These accumulations of ER against the end wall and in relationship to the pores were not considered artifactual by Hébant (1974) because they were found even when whole leafy stems were fixed in the field. Hébant speculated that the polarized accumulations may have a role in translocation and that their presence would make the sieve elements of mosses distinctive from many other vascular plants in which "open" pores have been described. Perry and Evert (1975) reported similar occluded pores in *Psilotum* which were filled with tubular elements of ER. In lycopods and polypodiaceous ferns, the mature sieve pores are normally unoccluded by any cytoplasmic substances.

Another common feature of the developing ER is the presence of dense material which develops in the ER cisternae. Referred to as *refractive spherules,*

◁

Figs. 2.2–2.4. Mature sieve elements in mosses. **Fig. 2.2.** *Dendroligotrichum dendroides*, longitudinal section of stem. Sieve elements *(se)* with densely staining pynotic nucleus *(large arrowhead)* against oblique end wall *(small arrowhead)*. Feulgen reaction. *n* nucleus; *p* parenchyma. × 800. **Fig. 2.3.** Nuclei of mature sieve elements in rhizome of *Polytrichum commune*. *N* nucleus; *P* plastid. × 5900. **Fig. 2.4.** *Atrichum undulatum* stem. Sieve elements *(se)* with nacreous thickenings *(NT)*. × 2700 (Fig. 2.2 from Hébant 1973 a)

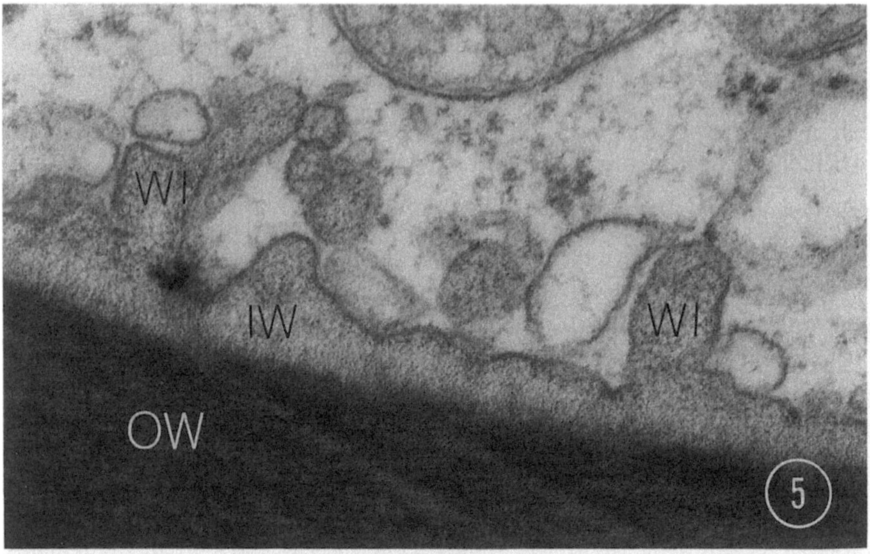
WI
WI
IW
OW
5

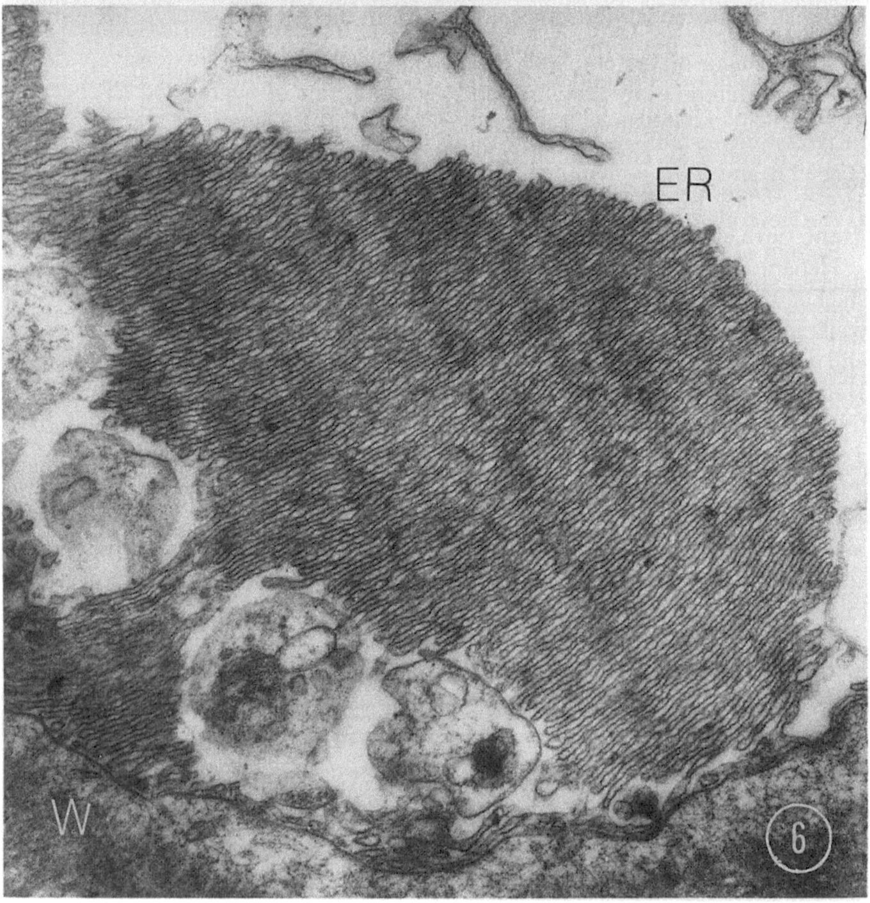
ER
W
6

they are known to arise in direct association with cisternae of smooth ER (Fig. 2.9) (Hébant 1976 a, b). Refractive spherules are delimited by a single membrane (Fig. 2.9) and are very similar in appearance to those found in pteridophyte sieve elements. Frequently dictyosomes can be seen in the vicinity of the developing spherules, but no evidence of direct participation of this organelle in their formation was reported (Hébant 1976 a). Fischer and Evert (1979) obtained evidence that implicated both the endoplasmic reticulum and the Golgi apparatus in the formation of refractive spherules in the fern *Davallia fijiensis*, while in other seedless vascular plants, the spherules arise in cisternae of smooth ER.

In a study of *Atrichum undulatum* sieve elements, Stevenson (1974) reported that the sieve elements are characterized by numerous membrane-bound protein bodies formed in close association with polysomes and ER. As the protein body ages, the membrane becomes disrupted, and the protein becomes dispersed into the cytoplasm and tends to plug the sieve area pores. These protein bodies may be equivalent to the refractive spherules found in other moss species, i. e., *Polytrichum commune*. In these cases *(Polytrichum commune)* the eventual fate or role of refractive spherules is not known, although in *Atrichum* it is known that they do disperse with the maturation of the sieve element (Stevenson 1974). P-protein is reported to be absent in moss sieve elements as well as in all vascular cryptogams.

Plastids, Dictyosomes, Mitochondria. Plastids in sieve elements are characterized by their small size and poorly developed grana (Fig. 2.3). The inner membrane may invaginate to form vesicle-like areas within the stroma. Occasionally, small starch grains persist in mature plastids, but their size never approaches those found in neighboring parenchyma. Because of the poor differentiation of the small plastids, they may resemble mitochondria in the sieve elements.

Dictyosomes are numerous as the sieve element differentiates. As the lateral wall thickens, dictyosomes and dictyosome-derived vesicles are seen in the vicinity of the wall. Although no detailed studies are available, it is assumed that the dictyosome vesicles fuse with the plasmalemma and discharge their contents into the developing wall. Dictyosomes rapidly disappear as the sieve element reaches maturity.

Mitochondria are present in both developing and mature sieve elements. Sieve elements of the leaf have numerous mitochondria during early stages of differentiation. This may be related to the presumed loading of sugars into the sieve elements (Reinhart and Thomas 1981; Scheirer 1983). Wall ingrowths are

Figs. 2.5 and 2.6. Mature sieve elements in *Polytrichum*. **Fig. 2.5.** *Polytrichum commune* leaf. Inner wall layer *(IW)* of leaf sieve element is invaginated forming wall ingrowths *(WI)*. Outer wall layer *(OW)* is seen as a distinct, dense layer. × 116900. **Fig. 2.6.** *Polytrichum commune* stem. Highly developed stacks of ER in fully mature sieve element. The stacks also show connections with the parietal system of sheets and tubules appressed to the plasmalemma. × 34000 (Fig. 2.6 from Hébant 1976 a)

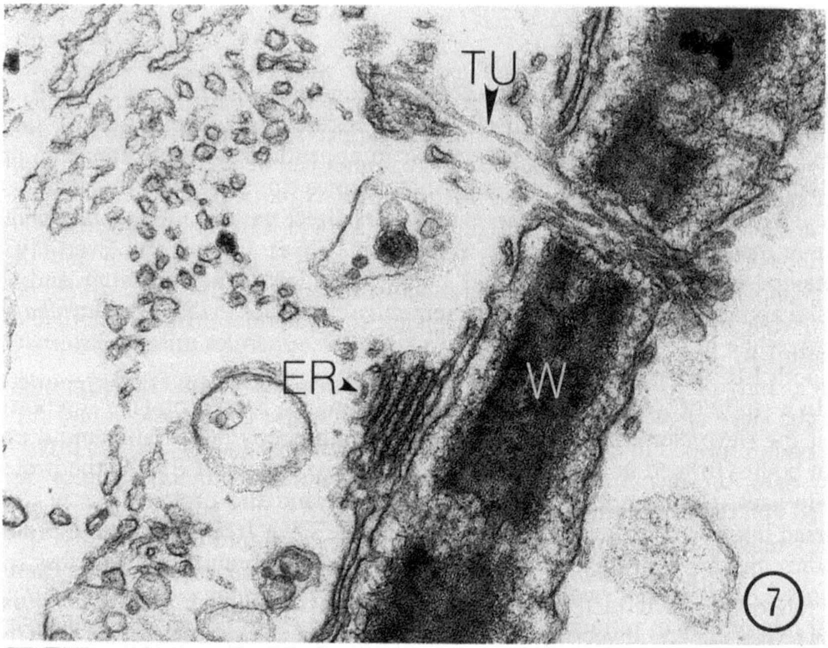
TU
ER
W
7

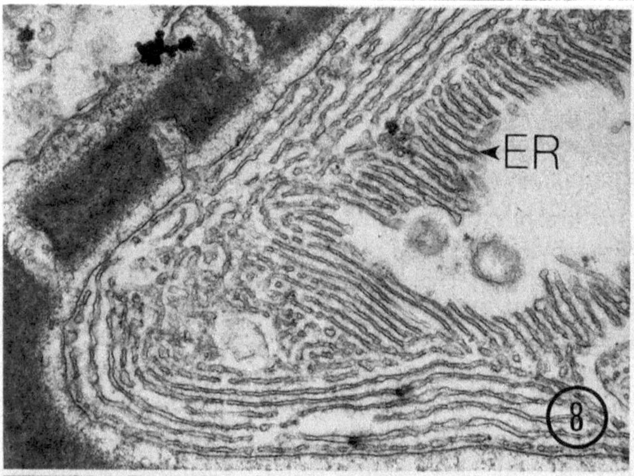
ER
8

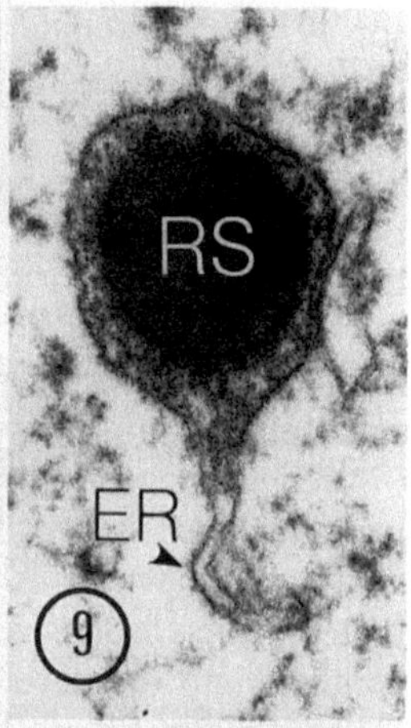
RS
ER
9

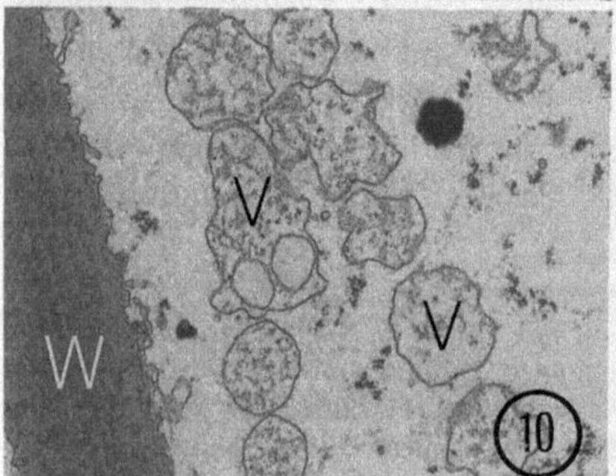
V
V
W
10

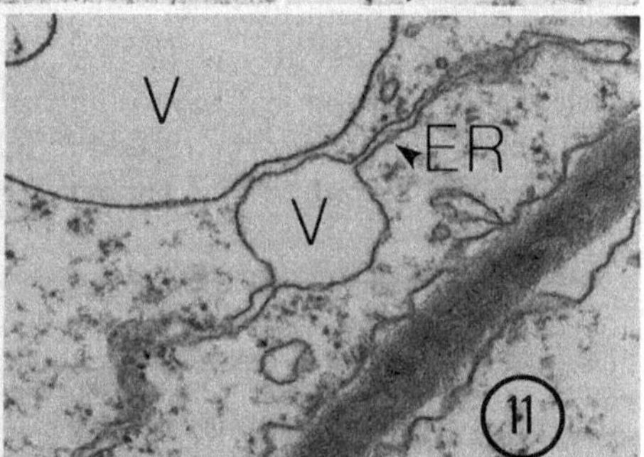
V
ER
V
11

also seen on the lateral walls of sieve elements during this stage of differentiation, especially in the leaf (Scheirer 1983).

Vacuoles. One feature that appears unique to moss sieve elements is the presence of small vacuoles which contain a fibrillar substance (Fig. 2.10). The vacuoles never fuse to form larger vacuoles although a mixture of small and medium-sized vacuoles is found in the sieve elements of the rhizome. The origin of vacuoles is from the dilation of smooth ER cisternae (Fig. 2.11). In sieve elements of *Funaria* leaflet, vacuoles and other small organelles (i.e., mitochondria) were only observed in the acropetal part of the cell giving it a distinct polarity (Wiencke and Schulz 1983). Although the tonoplast membranes persist in mature sieve elements, this fact does not halt the translocation of sugars. In a recent study of *Polytrichum commune,* Thomas et al. (1988) pulse-labeled isolated stems in clonal groupings with $^{14}CO_2$ and found radioactivity in the rhizome sieve elements 30 min after treatment. Increased amounts of translocated radioactivity appeared in starch and cell wall polysaccharide pools 1 week and 6 weeks after treatment. The authors concluded that long-distance transport of photoassimilate into new sister shoots occurs through sieve elements of perenniating rhizomes.

Development of Sieve Areas on End Walls. The development of sieve pores is confined to the highly inclined end wall of sieve elements. The appearance of end walls in longitudinal section is tibia-like (aspect "en-tibia", Hébant 1964 a) (Figs. 2.1, 2.2), making their appearance similar to sieve elements of *Psilotum* (Perry and Evert 1975) and *Equisetum* (Dute and Evert 1977). End walls have a uniform thickness, but are always thinner than the lateral walls. The end walls are perforated by a great number (15-20/μm^2) of small sieve pores which range in diameter from 0.1-0.2 μm (Figs. 2.12, 2.13) (Scheirer 1978). The pore sizes (0.1-0.2 μm) are similar to end wall sieve pores of *Psilotum*. Each pore begins as a single plasmodesma which soon becomes enlarged in the region of the middle lamella, forming a median nodule or cavity (Figs. 2.12, 2.13). ER tubules are commonly associated with the pores (Figs. 2.7, 2.12, 2.13). Both Hébant (1968, 1973 a) and Stevenson (1974), using fluorescence microscopy, reported the presence of callose associated with end walls, but callose has never been detected unequivocally at the ultrastructural level. In this regard moss sieve elements are

Figs. 2.7-2.11. Mature sieve elements in mosses. **Fig. 2.7.** *Dendroligotrichum dendroides* stem. Small sieve pore showing relationship with ER tubules *(TU)* in end wall of young sieve element. *ER* endoplasmic reticulum; *W* wall. × 64500. **Fig. 2.8.** *Dendroligotrichum dendroides* stem. Highly organized system of ER in an "accumulation" against end wall in a little-disturbed sieve element. × 38000. **Fig. 2.9.** *Polytrichum commune* stem. Refractive spherule *(RS)* apparently originating from ER in young sieve element. × 57500. **Fig. 2.10.** *Polytrichum commune* rhizome. Small vacuoles *(V)* in rhizome sieve element contain a fibrillar substance. *W* wall. × 13600. **Fig. 2.11.** *Dawsonia papuana* stem. Origin of vacuole *(V)* from the dilation of an ER cisterna. × 34900 (Figs. 2.7-2.9 from Hébant 1976 a)

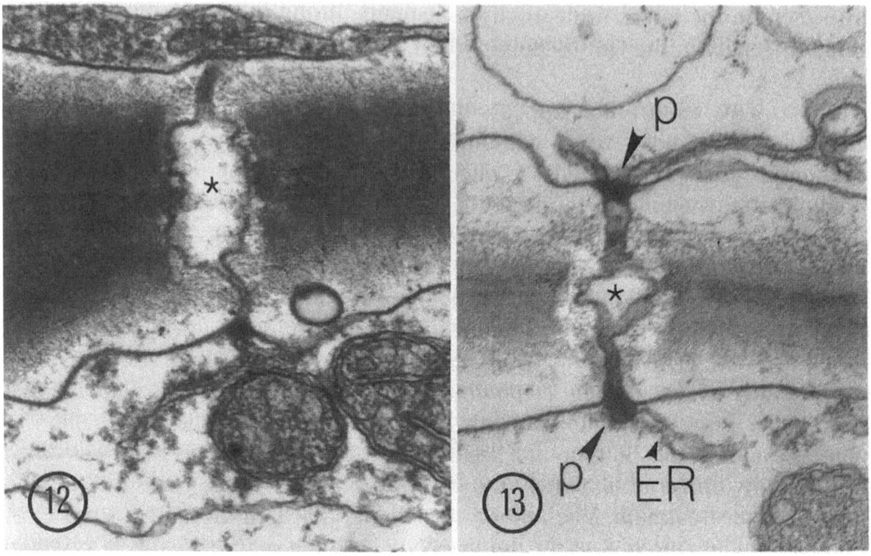

Figs. 2.12 and 2.13. Sieve pores in moss sieve elements. **Fig. 2.12.** *Dawsonia papuana* rhizome. Large median nodule (*) in sieve pore of end wall. × 55000. **Fig. 2.13.** *Dawsonia papuana* rhizome. ER and an unidentified dense plug *(p)* are seen at the oriface of a sieve pore. × 46700

similar to those found in *Psilotum* which has been shown to lack callose at all stages of sieve element development (Perry and Evert 1975).

Several workers (Stevenson 1974; Wiencke and Schulz 1983) reported dense material of a proteinaceous nature which appears to occlude the sieve pores (Fig. 2.13). This may be artifactual or poorly preserved membranous material. Stevenson (1974) argued that the protein plugs are probably an artifact of fixation and manipulation, but he concluded that their presence indicated that the sieve elements were translocating assimilates at the time.

It could be argued that the sieve pores are not specialized beyond the level of plasmodesmata. Hébant (1977) is equivocal about the nature of the pores, stating in one place that they represent enlarged plasmodesmata (Hébant 1974), while at other times he calls them small "pores" (Hébant 1976 a). Stevenson (1974) argued that in *Atrichum* they are sieve pores and not simply plasmodesmata. His argument was based on the distinctive characteristics of the sieve pores when compared with the structure of plasmodesmata that occur between contiguous parenchyma.

2.4 Associated Parenchyma

Parenchyma cells are associated with the sieve elements in the gametophytic stem, but not in the sporophyte. Parenchyma and sieve elements share a spatial relationship, but not an ontogenetic one, and hence, are not referred to as companion cells. These cells occur in vertical strands with their end walls transverse. Cytologically, the parenchyma cells contain numerous ribosomes, cisternoid ER, mitochondria, lipid globules and plastids. Plastids are larger than in sieve elements and have well-differentiated grana which generally contain starch. Lateral walls between parenchyma and sieve elements are connected with normal appearing plasmodesmata, but these are not numerous.

Acknowledgments. I would like to extend my gratitude and sincere thanks to Rose Hébant-Mauri, Université des Sciences et Techniques du Languedoc, Montpellier, France, for generously supplying the micrographs for Figs. 2.1, 2.2, 2.6, 2.7-2.9.

References

Behnke H-D (1975) Phloem tissue and sieve elements in algae, mosses and ferns. In: Arnoff S, Dainty J, Gorham PR, Srivastava LM, Swanson CA (eds) Phloem transport. Plenum, New York London, pp 187-210

Behnke H-D (1986) Sieve element characters and the systematic position of *Austrobaileya*, Austrobaileyaceae - with comments to the distinction and definition of sieve cells and sieve-tube members. Plant Syst Evol 152: 101-121

Dute JR, Evert RF (1977) Primitive-like metaphloem sieve elements in the aerial stem of *Equisetum hyemale*. Protoplasma 91: 257-266

Esau K (1969) The phloem. In: Zimmerman W, Ozenda P, Wulff HD (eds) Encyclopedia of plant anatomy, vol 5, pt 2. Borntraeger, Berlin Stuttgart, 505 pp

Eschrich W (1975) Bidirectional transport. In: Arnoff S, Dainty J, Gorham PR, Srivastava LM, Swanson CA (eds) Phloem transport. Plenum, New York London, pp 401-416

Eschrich W, Steiner M (1967) Autoradiographische Untersuchungen zum Stofftransport bei *Polytrichum commune*. Planta 74: 330-349

Eschrich W, Steiner M (1968 a) Die Struktur des Leitgewebesystems von *Polytrichum commune*. Planta 82: 33-49

Eschrich W, Steiner M (1968 b) Die submikroskopische Struktur der Assimilatleitbahnen von *Polytrichum commune*. Planta 82: 321-336

Evert RF (1976) Some aspects of sieve-element structure and development in *Botrychium virginianum*. Isr J Bot 25: 101-126

Favali M, Bassi M (1974) Seta ultrastructure in *Polytrichum commune* L. Nova Hedwigia 25: 451-463

Favali M, Gianni F (1973) Sporophyte ultrastructure in *Tortula muralis*. Österr Bot Z 122: 323-331

Fischer DG, Evert RF (1979) Endoplasmic reticulum-dictyosome involvement in the origin of refractive spherules in sieve elements of *Davallia fijiensis*. Ann Bot (London) 43: 255-258

Gunning BSE, Pate JS (1969) "Transfer cells". Plant cells with wall ingrowths, specialized in relation to short distance transport of solutes - their occurrence, structure and development. Protoplasma 68: 107-133

Hallet J-N (1972) Morphogenèse du gamétophyte feuillé du *Polytrichum formosum* Hedw. I. Etude histochimique, histoautoradiographique et cytophotométrique du point végétatif. Ann Sci Nat Bot Ser 12, 13: 19-118
Hébant C (1964 a) Sur les leptoïdes des Bryophytes. CR Acad Sci Paris Ser D 258: 3339-3341
Hébant C (1964 b) Signification et évolution des tissues conducteurs chez les Bryophytes. Nat Monspel Ser Bot 16: 79-86
Hébant C (1966) Différenciation des tissus conducteurs dans la tige feuillée de *Polytrichum juniperinum* Willd. (Mousses Polytrichales). CR Acad Sci Paris Ser D 262: 2585-2588
Hébant C (1967) Structure et différenciation des tissus conducteurs dans le gamétophyte des *Polytrichum*. Nat Monspel Ser Bot 18: 293-298
Hébant C (1968) Observations en microscopie à fluorescence sur la répartition de la callose. Nat Monspel Ser Bot 19: 75-78
Hébant C (1969) Nouvelles observations sur le leptome de la tige feuillée des *Polytrichum*. CR Acad Sci Paris Ser D 269: 2530-2533
Hébant C (1970 a) A new look at the conducting tissues of mosses (Bryopsida): their structure, distribution, and significance. Phytomorphology 20: 390-410
Hébant C (1970 b) Aspects infrastructuraux observés au cours de la différentiation du phloème (leptome) dans la tige feuillée de quelques mousses Polytrichales. CR Acad Sci Paris Ser D 271: 1361-1363
Hébant C (1972) Observations sur les traces foliaires des mousses s. str. II. Etude, chez quelques polytrichales, des éléments à caractères "phloémiens". Nova Hedwigia 23: 735-766
Hébant C (1973 a) Observations sur le leptome de *Dawsonia* et *Dendroligotrichum*. CR Acad Sci Paris Ser D 276: 3131-3134
Hébant C (1973 b) Le noyau des cellules "phloémiennes" (=leptoïdes) des Bryophytes est-il dégénérescent? CR Acad Sci Paris Ser D 277: 1445-1447
Hébant C (1974) Polarized accumulations of endoplasmic reticulum and other ultrastructural features of leptoids in *Polytrichadelphus magellanicus* gametophytes. Protoplasma 81: 373-382
Hébant C (1975 a) Organization of the conducting tissue-system in the sporophytes of *Dawsonia* and *Dendroligotrichum*. J Hattori Bot Lab 39: 235-254
Hébant C (1975 b) The phloem (leptom) of bryophytes. In: Arnoff S, Dainty J, Gorham PR, Srivastava LM, Swanson CA (eds) Phloem transport. Plenum, New York, pp 211-215
Hébant C (1976 a) Studies on the development of the conducting tissue-system in the gametophytes of some Polytrichales. III. Further observations on leptoids, with particular reference to their endoplasmic reticulum. Protoplasma 87: 79-90
Hébant C (1976 b) Comparative anatomy of the gametophytes in *Dawsonia* (Polytrichales, Musci) J Hattori Bot Lab 40: 221-246
Hébant C (1977) The conducting tissues of bryophytes. Cramer, Vaduz, Liechtenstein
Hébant C (1979) Conducting tissues in bryophyte systematics. In: Clarke GCS, Duckett JG (eds) Bryophyte systematics. Systematics association special volume 14. Academic Press, New York London, pp 365-383
Hébant C, Schäfer P (1974) Aperçu historique sur les premiers travaux relatifs aux tissus conducteurs des Bryophytes. In: Les Probèmes de la Bryologie, Lille 1972, Bull Soc Bot Fr 121 (Suppl): 49-59
Parthasarathy MV, Pesacreta TC (1980) Microfilaments in plant vascular cells. Can J Bot 58: 807-815
Perry J, Evert RF (1975) Structure and development of the sieve elements in *Psilotum nudum*. Am J Bot 62: 1038-1052
Reinhart D, Thomas R (1981) Sucrose uptake and transport in conducting cells of *Polytrichum commune*. Bryologist 84: 59-64
Scheirer DC (1972) Anatomical studies in the Polytrichaceae. I. The gametophore of *Dendroligotrichum dendroides* (Hedw.) Broth. Bryol 75: 305-314
Scheirer DC (1973) Hydrolysed walls in the water-conducting cells of *Dendroligotrichum* (Bryophyta): histochemistry and ultrastructure. Planta 115: 37-46

Scheirer DC (1974) The anatomy, histochemistry and ultrastructure of the gametophore of *Dendroligotrichum dendroides* (Hedw.) Broth. (Bryopsida: Polytrichaceae). Thesis, Pennsylvania State Univ, University Park

Scheirer DC (1976) Some fine structural observations on the rhizome of *Dendroligotrichum* (Bryophyta). Protoplasma 89: 323-337

Scheirer DC (1977) The thickened leptoid (sieve element) wall of *Dendrologotrichum* (Bryophyta): cytochemistry and fine structure. Am J Bot 64: 369-376

Scheirer DC (1978) Cell wall chemistry and fine structure in leptoids of *Dendroligotrichum* (Bryophyta): the end wall. Am J Bot 65: 1027-1031

Scheirer DC (1980) Differentiation of bryophyte conducting tissue: structure and histochemistry. Bull Torrey Bot Club 107: 298-307

Scheirer DC (1983) Leaf parenchyma with transfer cell-like characteristics in the moss, *Polytrichum commune* Hedw. Am J Bot 70: 987-992

Schofield WB, Hébant C (1984) The morphology and anatomy of the moss gametophore. In: Schuster RM (ed) New manual of bryology, vol 2. Hattori Bot Lab, Nichinan, pp 627-657

Stevenson DW (1974) Ultrastructure of the nacreous leptoids (sieve elements) in the polytrichaceous moss *Atrichum undulatum*. Am J Bot 61: 414-421

Tansley AG, Chick E (1901) Notes on the conducting tissue-system in Bryophyta. Ann Bot (London) 15: 1-38

Thomas RJ, Schiele EM, Scheirer DC (1988) Translocation in *Polytrichum commune* (Bryophyta) I. Conduction and allocation of photoassimilates. Am J Bot 75: 275-281

Wiencke C, Schulz D (1983) The fine structural basis of symplasmic and apoplasmic transport in the "nerve" of the *Funaria* leaflet. Z Pflanzenphysiol 112: 337-350

3 Seedless Vascular Plants

RAY F. EVERT

3.1 Introduction

Prior to research with the electron microscope, little detailed information was available on the ontogeny and structure of the sieve elements in seedless vascular plants. About all that could be said about the mature sieve-element protoplasts in this diverse group of plants was that they generally had a very clear appearance, contained variable numbers of structures called refractive spherules, and apparently lacked nuclei. Little was known about the connections or sieve-area pores between contiguous sieve elements in the seedless vascular plants because their pores are too small (0.5 µm or less in diameter in most plants) to be satisfactorily examined with the light microscope.

Motivated largely by electron microscopy, which enabled them to reveal details of ontogeny and structure unobtainable with traditional light microscope methods, investigators began to turn their attention to the phloem in seedless vascular plants, beginning with the studies by Maxe (1964, 1965, 1966; Liberman-Maxe 1968) on the phloem of the fern *Polypodium vulgare*. Since then the phloem of representatives of every major taxon of seedless vascular plants has been studied with the electron microscope. From these studies, a rather clear picture has emerged of the ontogeny and structure of the sieve elements in the ferns and their so-called allies.

Typically, the sieve elements of seedless vascular plants are quite long, in some species reaching lengths of 30 to 40 mm (Lamoureux 1961). The degree of inclination of their end walls ranges from transverse to very oblique and apparently may vary from plant part to plant part within the same individual. In addition, the degree of inclination of the end walls may be greatly influenced by the growth rate of the plant part (Hébant 1969). The more rapid the growth, the longer the sieve elements and the more greatly inclined the end walls. Branched sieve elements have been found in *Lycopodium lucidulum* (Warmbrodt and Evert 1974a), *Ophioglossum vulgatum* (Warmbrodt and Evert 1979b), and in several species of tropical ferns (Shah et al. 1976).

3.2 The Sieve-Element Protoplast

Ultrastructural details of the very young sieve elements of seedless vascular plants are similar to those of the parenchyma cells bordering them. Both cell types have all of the components commonly associated with meristematic cells.

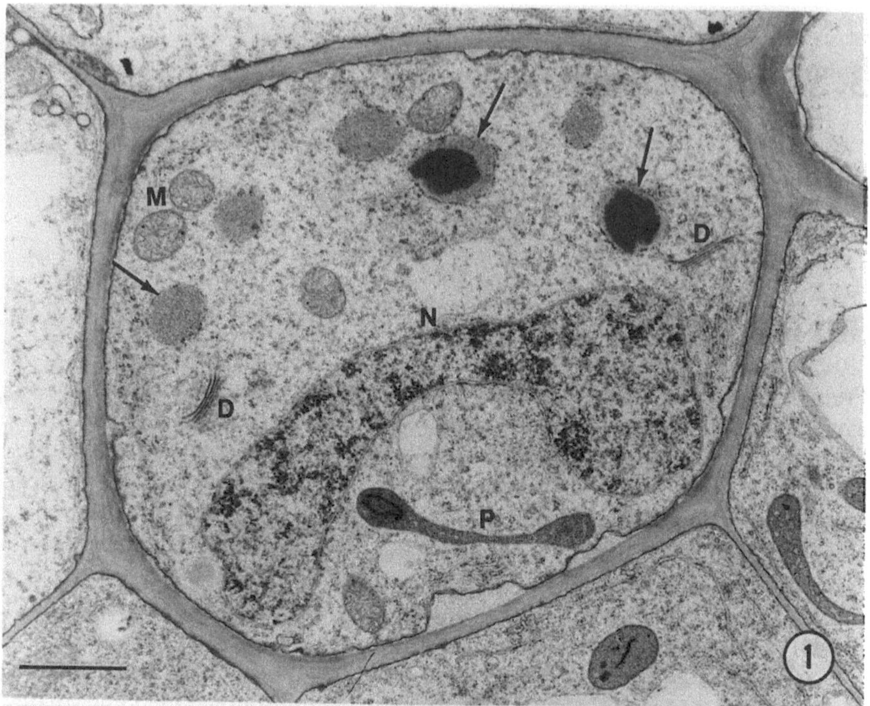
M
D
N
D
P
1

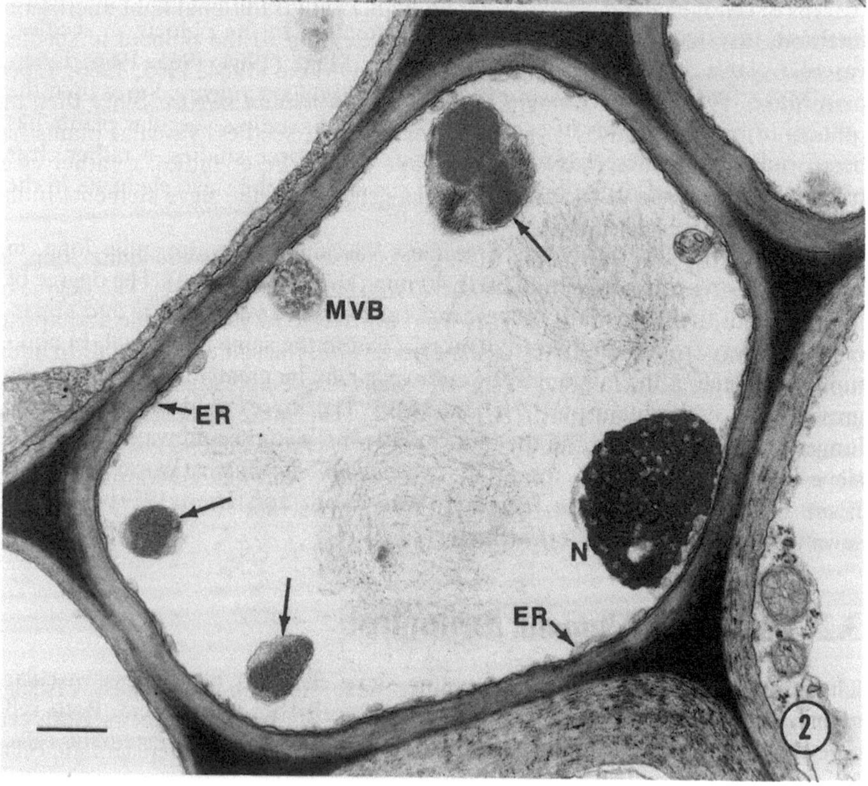
MVB
ER
N
ER
2

As the sieve elements differentiate, however, they eventually undergo a selective autophagy, which results in degeneration of many cellular components and reorganization of others (Figs. 3.1, 3.2 and 3.16–3.19). At maturity, the protoplasts of the sieve elements consist of a plasmalemma, a parietal network of smooth endoplasmic reticulum (ER), and variable numbers of plastids and mitochondria. The mature sieve elements of some species may have occasional aggregates or stacks of ER, in addition to the parietal network. Remnants of nuclei commonly persist in some. With the exception of the lycopods, the most distinguishing feature of the sieve-element protoplast of seedless vascular plants is the presence of the refractive spherules.

3.2.1 Nucleus

Nuclear degeneration, which may be pycnotic or chromatolytic, or by a process intermediate between the two, is a ubiquitous feature of sieve-element differentiation in seedless vascular plants. During pycnotic degeneration, the chromatin increases in quantity or density and accumulates into a more or less continuous mass prior to rupture of the nuclear envelope, whereas chromatolytic degeneration involves disappearance of the stainable contents of the nucleus and retention of the nuclear envelope beyond the disappearance of those contents. As mentioned previously, remnants of nuclei commonly persist in some mature sieve elements.

During a light microscope study of some tropical ferns, Hébant (1969) found nuclear degeneration in protophloem sieve elements to be generally of the pycnotic type. In the metaphloem, however, chromatolysis apparently occurred as often as pycnotic degeneration. Both types of nuclear degeneration occur in the aerial stems of *Equisetum hyemale* (Dute and Evert 1978). Nuclear degeneration in the roots of *E. hyemale* is difficult to classify. There the chromatin does not increase in quantity but forms small, rounded bodies that are released into the cytoplasm when the nuclear envelope begins to rupture (Fig. 3.3). Eventually, both chromatin and nuclear envelope disappear from the cell (Dute and Evert 1977a).

The details of nuclear degeneration may differ markedly from species to species. For example, in *Selaginella kraussiana* (Burr and Evert 1973) tubulate

Figs. 3.1 and 3.2. Transverse sections of sieve elements of *Isoetes muricata*. **Fig. 3.1.** Young sieve element from root prior to initiation of morphological changes in nucleus. Dilated cisternae *(unlabeled arrows)* of endoplasmic reticulum contain a crystalline and/or fibrillar substance. ×29850; *Bar line*=0.5 µm. **Fig. 3.2.** Mature sieve element from leaf. Most of the endoplasmic reticulum *(ER)* occurs along the wall as an anastomosing network. Three dilated ER cisternae *(unlabeled arrows)* can be seen with crystalline and/or fibrillar contents, while some fibrillar material from ruptured cisternae occurs free in the lumen. ×20150; *Bar line*=0.5 µm. *D* dictyosome; *P* plastid; *M* mitochondria; *MVB* multivesicular body; *N* nucleus (Fig. 3.1 from Kruatrachue and Evert 1978; Fig. 3.2 from Kruatrachue and Evert 1974)

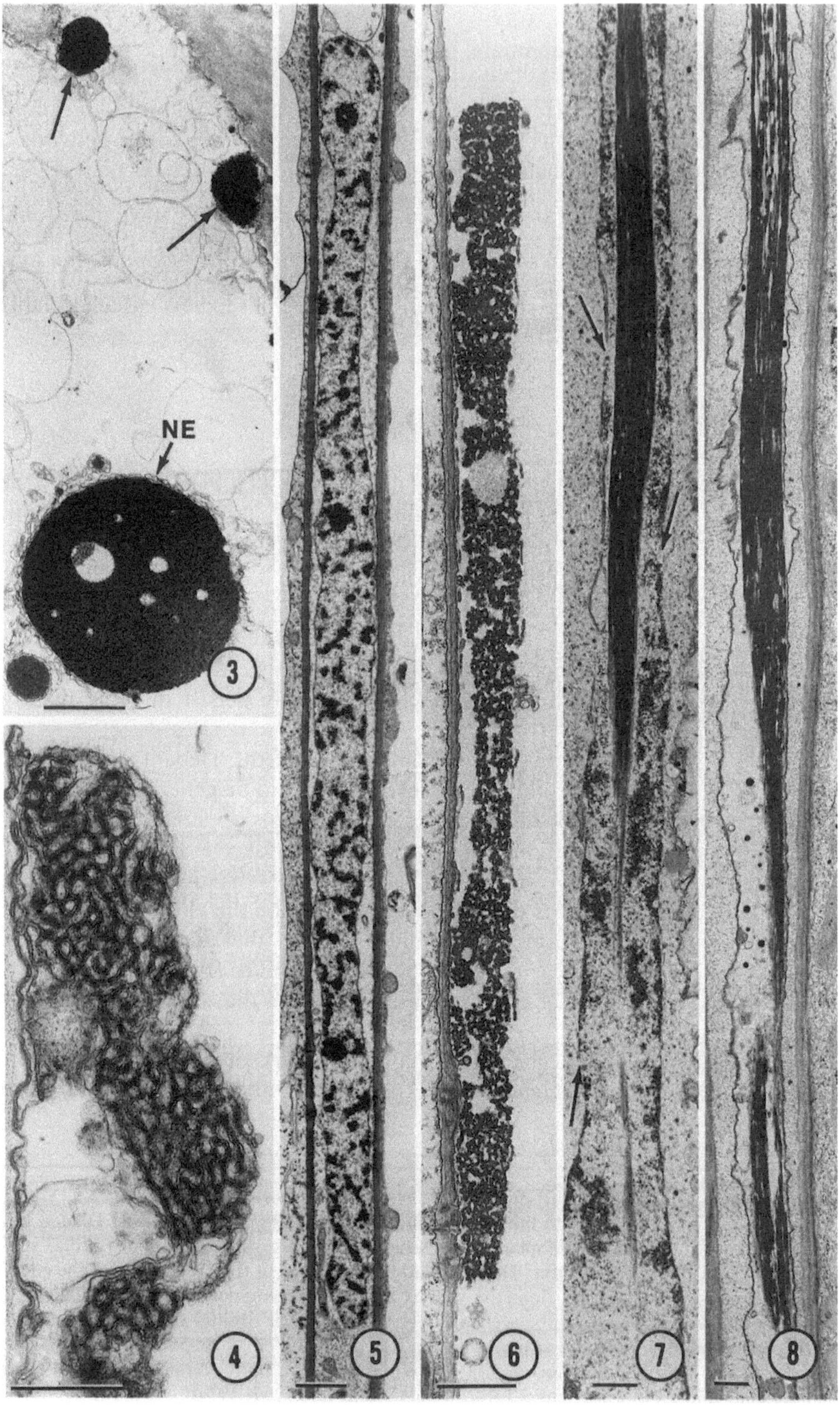
NE
3
4
5
6
7
8

structures interpreted as infoldings of the nuclear envelope begin to appear within the nucleus. These infoldings continue to be produced until ultimately the entire nucleus is a mass of convoluted tubules that persist in the mature sieve element (Fig. 3.4). In sieve elements of the leaves and roots of *Isoetes muricata* (Kruatrachue and Evert 1974, 1978), the very long ("filiform"), persistent nuclei consist of dense chromatin material partly delimited from the rest of the protoplast by remnants of the nuclear envelope (Figs. 3.2, 3.5 and 3.6). Degenerate nuclei also persist in the secondary sieve elements of the corm of *I. muricata* (Kruatrachue and Evert 1977). In the fern *Platycerium bifurcatum* (Evert and Eichhorn 1974a), crystalloids, which may eventually occupy up to a third of the cross-sectional area of the nucleus and extend almost its entire length, arise in the nucleoplasm (Fig. 3.7). At the time of nuclear degeneration, which is intermediate between pycnotic and chromatolytic, the crystalloids are liberated into the lumen of the cell (Fig. 3.8) and eventually degenerate. The very long crystalloids in *Platycerium* are reminiscent of the spindle-shaped crystalline P-protein bodies of the Fabaceae (Wergin and Newcomb 1970; Palevitz and Newcomb 1971; Esau 1978; Lawton 1978a, b). It is pertinent to note, however, that P-protein is lacking in the sieve elements of ferns and other seedless vascular plants. Incidentally, the presence of crystalline inclusions in sieve-element nuclei of certain fern species has long been known (e.g., Poirault 1893).

According to Shah and Nair (1978), nucleoli are sometimes extruded from degenerating nuclei in some fern species. The nucleoli do not persist but degenerate during maturation of the sieve elements. Extruded nucleoli have also been reported for the sieve elements of *Psilotum triquetrum* (Want 1963). Although remnants of nuclei were found in mature *Psilotum nudum* sieve elements, no evidence was found for the presence of extruded nucleoli in that species at any stage of development (Perry and Evert 1975).

◁

Figs. 3.3–3.8. Longitudinal sections showing sieve-element nuclei in various stages of development. **Fig. 3.3.** *Equisetum hyemale.* Late stage of nuclear degeneration in protophloem element. Portions of chromatin *(unlabeled arrows)* have separated from the main nuclear mass. Remnant of the nuclear envelope *(NE)* surrounds the major mass of chromatin. A refractive spherule borders the chromatin mass below and to the left. ×11900; *Bar line*=1 μm. **Fig. 3.4.** *Selaginella kraussiana.* Portion of nucleus with tubulate structure in mature cell. ×32650; *Bar line*=0.5 μm. **Fig. 3.5.** *Isoetes muricata.* Nucleus in very young sieve element. ×3850; *Bar line*=2 μm. **Fig. 3.6.** *I. muricata.* Necrotic nucleus in mature sieve element. ×11600; *Bar line*=1 μm. **Fig. 3.7.** *Platycerium bifurcatum.* Crystalloid-containing nucleus beginning to degenerate. *Arrows* point to discontinuities in nuclear envelope. ×6400; *Bar line*=1 μm. **Fig. 3.8.** *P. bifurcatum.* Crystalloid free in lumen of nearly mature sieve element. ×5000; *Bar line*=1 μm (Fig. 3.3 from Dute and Evert 1978; Fig. 3.4 from Burr and Evert 1973; Figs. 3.5 and 3.6 from Kruatrachue and Evert 1974; Figs. 3.7 and 3.8 from Evert and Eichhorn 1974a)

3.2.2 Endoplasmic Reticulum

During early stages of sieve-element development in seedless vascular plants, the ER is rough-surfaced, mostly cisternal, and fairly uniformly distributed throughout the cytoplasm. As sieve-element development progresses, the ER commonly increases in quantity. Then, in some species (e.g., *Polypodium vulgare,* Maxe 1966; *Selaginella kraussiana,* Burr and Evert 1973; *Platycerium bifurcatum, Phlebodium aureum,* Evert and Eichhorn 1974a; *Psilotum nudum,* Perry and Evert 1975; *Equisetum hyemale,* Dute and Evert 1977a, b, 1978), much of the ER gradually accumulates in stacks, some of which may become closely appressed to the nuclear envelope. Stacking of the ER is initiated at about the same time as the initiation of nuclear degeneration. During stacking of the ER, the ER membranes become agranular. Most of the stacks eventually become parietal in distribution before being reduced in amount or disappearing from the cell altogether. In roots and aerial shoots of *Equisetum hyemale* (Dute and Evert 1977a, b), massive aggregates of ER are formed in the cytoplasm of the sieve elements prior to stacking (Fig. 3.18). In contrast, ER stacking apparently is not a feature of sieve-element differentiation in sieve elements of either the leaf, corm, or root of *Isoetes muricata* (Kruatrachue and Evert 1974, 1977, 1978), nor of that in sieve elements of the stem of *Lycopodium lucidulum* (Warmbrodt and Evert 1974a).

Disappearance of ER from the sieve elements of polypodiaceous ferns may involve different phenomena (Evert and Eichhorn 1976). Regardless of the events leading to maturation, the ER in mature sieve elements of seedless vascular plants commonly consists of a parietal network of smooth, tubular elements (Figs. 3.2, 3.19 and 3.26).

In *Platycerium bifurcatum* (Evert and Eichhorn 1974b) and *Davallia fijiensis* (Fisher and Evert 1979) numerous contacts were observed between the ER and peripheral tubules of the dictyosomes, making it difficult to determine where the ER ends and the dictyosome begins. In *Botrychium virginianum* (Evert 1976), segments of ER frequently were in direct continuity with invaginations of the plasmalemma at the time the cell wall was undergoing a marked increase in thickness. Many of the ER segments contained fibrillar material similar to that found in the invaginations of the plasmalemma.

3.2.3 Plastids and Mitochondria

Plastids and mitochondria are the only cytoplasmic organelles that persist in mature sieve elements of seedless vascular plants. Of the two, the plastids undergo the greatest degree of modification during sieve-element ontogeny. Changes in the plastids commonly begin very early in ontogeny; hence, very young sieve elements often can be distinguished from contiguous parenchyma cells by their plastids. At maturity, relatively little variation exists in the appearance of the sieve-element plastids among the seedless vascular plants and, therefore, plastid structure in this group of plants does not provide useful taxonomic characteristics as in the case of seed plants (Behnke 1981). Typically, the plastids are bounded by two membranes and contain variable numbes of internal membranes, plastoglobuli, and spherical lamellar inclusions. Whereas the young plastids may contain starch grains, the mature plastids lack them. Although chloroplasts have been identified in young sieve elements of *Lycopodium lucidulum* (Warmbrodt and Evert 1974a), only the mature sieve-element plastids of *Microgramma lycopodioides* and *Polypodium schraderi* have been found with internal membranes organized into grana and intergrana lamellae (Evert and Eichhorn 1976).

In most species, the mitochondria undergo no apparent structural modification during maturation of the sieve element. In a few species, the mitochondria of the young sieve element may be relatively long and possibly branched (e. g, in *Platycerium bifurcatum,* Evert and Eichhorn 1976; and *Equisetum hyemale,* Dute and Evert 1977a) or extensively folded, as in *Psilotum nudum* (Perry and Evert 1975). At maturity, however, the mitochondria are similar in appearance from species to species, being round in outline and containing variable numbers of cristae in a relatively dense matrix.

3.2.4 Dictyosomes

Concomitantly with or slightly before the initiation of cell wall thickening in sieve elements of seedless vascular plants, a marked increase occurs in the population of dictyosomes and dictyosome-derived vesicles. Both smooth and coated vesicles are associated with the dictyosomes, and both types of vesicle migrate to the cell wall, where they fuse with the plasmalemma and discharge their contents to the outside. At this stage of development, the plasmalemma has an undulating outline, undoubtedly due in part to this phenomenon.

In the leptosporangiate ferns *Platycerium bifurcatum* and *Phlebodium lycopodioides* (Evert and Eichhorn 1974b) and the eusporangiate fern *Botrychium virginianum* (Evert 1976), the cisternal contents of the dictyosomes become progressively darker from the proximal (forming) to the distal (maturing) faces, due to the presence of increasing amounts of finely granular material. Material of similar appearance occurs in most dictyosome vesicles. The contents of the vesicles often exhibit a granular to fibrillar texture, and they strongly resemble material found in the extracytoplasmic space between plasmalemma and wall. The role of dictyosomes and dictyosome-derived vesicles in cell wall formation in both protophloem and metaphloem sieve elements of *Polypodium vulgare* has been confirmed through the use of Thiéry's method for the detection of polysaccharides (Liberman-Maxe 1978).

As mentioned previously, in the sieve elements of some seedless vascular plants, it is often difficult to distinguish between the ER and peripheral tubules of the dictyosomes because of the many contacts between these two components. The problem is also due in part to the lack of a distinct zone of exclusion about the dictyosomes in some plants. For example, in *Platycerium* and *Phlebodium* (Evert and Eichhorn 1974b) and *Botrychium* (Evert 1976), ribosomes, in particular, are abundant among the peripheral tubules.

As the sieve element approaches maturity, a gradual decrease occurs in the population of cytoplasmic ribosomes and dictyosome vesicles. A concomitant decrease occurs in the dictyosome population itself until, at maturity, all ribosomes and dictyosomes have disappeared from the cell.

3.2.5 Microtubules and Microfilaments

Microtubules (MT) also have been implicated with cell wall formation. Cortical MT, mostly oriented at more or less right angles to the long axis of the cell, have been reported in protophloem and metaphloem sieve elements throughout the period of cell wall formation in a wide variety of seedless vascular plants (e.g., *Isoetes muricata,* Kruatrachue and Evert 1974; *Lycopodium lucidulum,* Warmbrodt and Evert 1974a; *Psilotum nudum,* Perry and Evert 1975; *Platycerium bifurcatum,* Evert and Eichhorn 1976; *Botrychium virginianum,* Evert 1976; *Equisetum hyemale,* Dute and Evert 1977a).

During a quantitative study of MT production over the cell cycle in growing cells of the *Azolla pinnata* root, Hardham and Gunning (1979) found the initiation of differentiation of outer (protophloem) sieve elements to be accompanied by a marked increase in the numbers of cortical MT against the longitudinal walls. The increase in numbers of MT correlated with an increase in the amount of wall thickening. The greatest number of MT were present during the period of greatest cellulose deposition. Specialized activity, in the form of MT, vesicles, and electron-dense, amorphous material, was observed at the cell edges, or corners, of differentiating sieve and xylem elements, indicating that initiation of cortical MT in cells of *A. pinnata* root tips may be correlated at the edge zones; that is, that the specialized edge zones may act as MT-organizing regions. Initiation of wall thickening at the corners of the cells had previously been reported for sieve elements in roots and aerial shoots of *Equisetum hyemale* (Figs. 3.16 and 3.17), although an electron-dense, amorphous material was not found in association with that activity in *E. hyemale* (Dute and Evert 1977a, 1978). As cell wall formation nears completion, the density of MT decreases. At maturity, the sieve element lacks MT.

Microfilaments are common constituents in differentiating sieve elements, where they comprise cytoplasmic fibers oriented mainly parallel to the long axes of the cells. Like the MT, they do not persist beyond maturation of the cell.

3.2.6 Plasmalemma and Tonoplast

In addition to the long-held, general acceptance of the enucleate state of the mature sieve-element protoplast, numerous observations with both light and electron microscopes have led phloem investigators to conclude that mature sieve elements lack a distinct boundary between cytoplasmic and vacuolar contents; that is, that they lack vacuoles, or a tonoplast, at maturity. Typically, the young sieve elements of seedless vascular plants contain numerous vacuoles of varying size.

The behavior of the vacuoles varies considerably from one species to another. In *Selaginella kraussiana*, the young sieve elements can be distinguished from contiguous parenchymatous elements by the presence of a vacuolar protein consisting of fine fibrils which are tightly aggregated into crystalline bundles of varying size (Burr and Evert 1973). The exact fate of the proteinaceous substance and vacuoles could not be determined. Mature sieve elements revealed figures that were interpreted as showing dissociation of the fibrillar aggregates, some of which were and others which were not enclosed by a single membrane, presumably a tonoplast. Although ribosomes, dictyosomes, and rough ER cisternae were plentiful in young sieve elements, no consistent spatial association existed between anyone of them and the vacuoles. Vacuoles containing degenerating membranous material were observed in differentiating sieve elements of *Isoetes muricata* (Kruatrachue and Evert 1977). This probably is a reflection of the role of such vacuoles as lysosomal compartments.

Initiation of wall thickening in sieve elements of *Psilotum nudum* is correlated with the appearance in the cytoplasm of many fibril-containing vacuoles (Perry and Evert 1975). These vacuoles are especially conspicuous during the period of maximal increase in wall thickness. The membranes delimiting many of these vacuoles are bordered internally by many small vesicles which apparently arise from invaginations of the delimiting membranes. Eventually, most of the fibril- and vesicle-containing vacuoles become ensheathed by ER, although at this stage of development most of the ER occurs in parallel arrays. In addition, the delimiting membrane of the vacuoles eventually fuses with the plasmalemma, discharging the vesicles and fibrillar material into the region of the developing wall. Vacuoles lacking fibrillar material are also numerous in differentiating sieve elements of *P. nudum*.

As mentioned previously, mature sieve elements typically lack vacuoles. A notable exception may be found in *Platycerium bifurcatum* in which many of the vacuoles apparently persist after maturation of some sieve elements (Evert and Eichhorn 1976). The plasmalemma maintains its integrity and differentially permeable properties throughout the life of the sieve element.

3.2.7 Refractive Spherules

Refractive spherules have been observed with the electron microscope in sieve elements of all groups of seedless vascular plants, with the exception of the lycopods (*Selaginella*, Burr and Evert 1973; Hébant et al. 1980; *Isoetes*, Kruatrachue and Evert 1974, 1977, 1978; *Lycopodium*, Warmbrodt and Evert 1974a; Hébant et al. 1978). The spherules, which appear highly refractive when viewed in unstained sections with the light microscope, appear as electron-dense

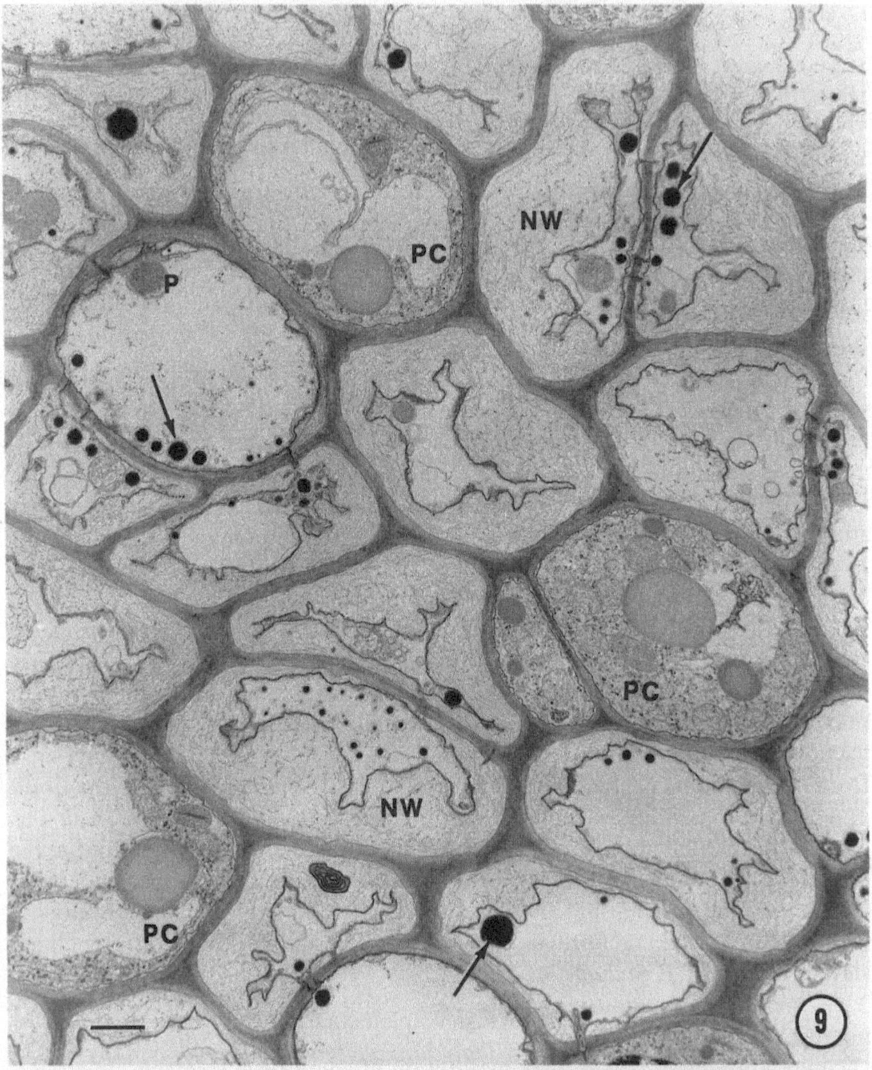
NW
PC
P
PC
NW
PC
9

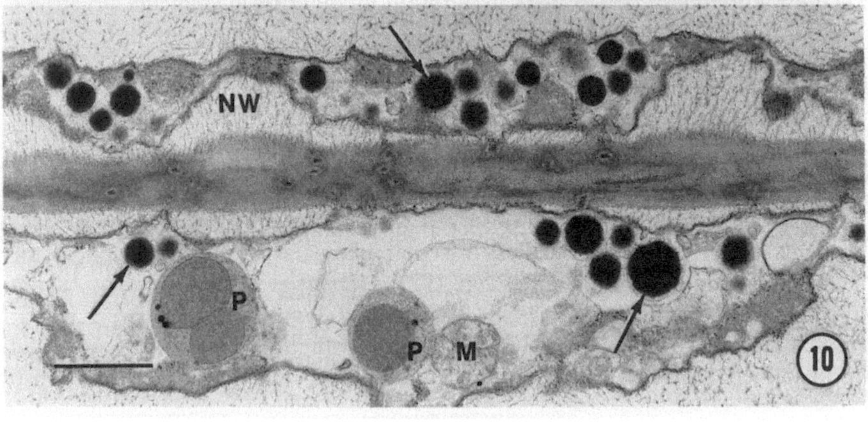
NW
P
P
M
10

bodies, each surrounded by a single membrane, with the electron microscope (Figs. 3.9 and 3.10). Being proteinaceous in nature, refractive spherules often have been compared with the P-type plastids of monocotyledonous sieve-tube members (Behnke 1967). That the plastids and refractive spherules are distinct entities was first demonstrated by Maxe (1964) for the sieve elements in *Polypodium vulgare*.

Among the ferns, refractive spherules have been reported as lacking only in sieve elements of *Pteris longifolia* (Shah and Fotedar 1974b). During that light microscope study, "minute granules" were observed in the parietal cytoplasm but were judged not to be refractive spherules because they did not show a positive reaction with mercuric bromphenol blue, a stain commonly employed for identifying proteins. Since then, however, refractive spherules have been found in sieve elements of other Pteridaceae examined with the electron microscope (Warmbrodt and Evert 1979b). The sieve elements of *P. longifolia* need to be reexamined for the presence of refractive spherules.

Both the ER and the Golgi apparatus have been implicated with the formation of refractive spherules. In *Psilotum nudum* (Perry and Evert 1975), *Botrychium virginianum* (Evert 1976), and *Equisetum hyemale* (Dute and Evert 1977a, 1978), formation of the refractive spherules is preceded by the appearance of granular or flocculent material in dilated portions of predominantly tubular ER. Aggregation or condensation of the material apparently results in formation of the refractive spherules (Figs. 3.11-3.13). Liberman-Maxe (1971) and Sakai and Lamoureux (1973) were the first workers to implicate the ER in the origin of refractive spherules (in *Polypodium vulgare* and *Microsorium scolopendria*, respectively).

Only dictyosomes could be implicated with refractive spherule formation in *Platycerium bifurcatum* and *Phlebodium aureum* (Fig. 3.14; Evert and Eichhorn 1974b). As mentioned previously, however, numerous contacts were encountered between ER and the peripheral tubules of the dictyosomes, often making it difficult to distinguish between membranes of the Golgi apparatus and ER in the vicinity of the dictyosomes. Since then, both the Golgi apparatus and ER have been implicated with refractive spherule formation in sieve elements of *Davallia fijiensis* (Fisher and Evert 1979). There, numerous connections were found between smooth tubular ER and the peripheral tubules of the dictyosomes, indicating that these two cytoplasmic components are indeed parts of a single endomembrane system. Apparently, reports of either ER or dictyosome involvement in refractive spherule formation are not contradictory in fact but "in appearance only," as suggested by Hébant (1976).

Figs. 3.9 and 3.10. Phloem of *Platycerium bifurcatum*. **Fig. 3.9.** Transverse section of portion of vascular bundle with various-aged sieve elements. *Arrows* point to the mostly peripheral refractive spherules, the most conspicuous components of the sieve-element protoplasts. ×7550; *Bar line*=1 μm. **Fig. 3.10.** Longitudinal view of portions of two immature sieve elements. *Arrows* point to refractive spherules. ×13900; *Bar line*=1 μm. *M* mitochondria; *P* plastid; *PC* parenchyma cell; *NW* nacreous wall of sieve element (Evert and Eichhorn 1974b)

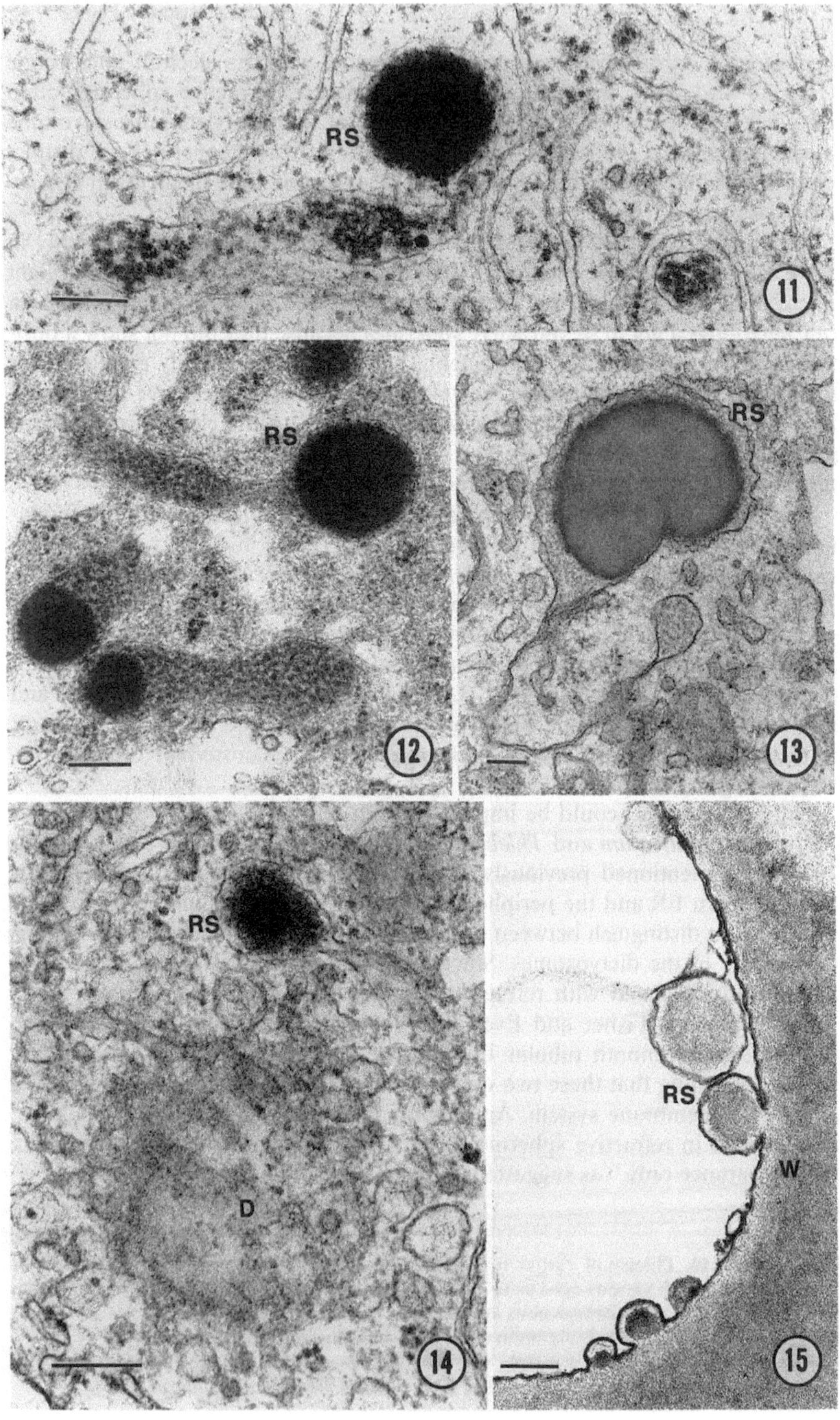
RS
11
RS
12
RS
13
RS
D
14
RS
W
15

In most species, young sieve elements can early be distinguished from neighboring parenchyma cells by the appearance of refractive spherules in the sieve elements. An exception to the use of this criterion alone is found in *Psilotum,* in which refractive spherules may occur in some parenchyma cells (Tarchi and Francalanci 1973) and young tracheary elements (Perry and Evert 1975) as well. Refractive spherules also have been encountered rarely in the procambial cells and occasionally in the so-called "companion cells" of the *Equisetum hyemale* root (Dute and Evert 1977a). Nevertheless, in both *Psilotum* and *Equisetum* the presence of numerous refractive spherules in the young sieve elements can be used to distinguish the young sieve elements from parenchymatous cells.

Many refractive spherules persist in the parietal layer of cytoplasm in mature sieve elements, whereas others are deposited into the region of the wall (Evert and Eichhorn 1974b; Perry and Evert 1975; Evert 1976; Dute and Evert 1977a, 1978; Kruatrachue 1978). The latter phenomenon is accomplished by fusion of the delimiting membrane of the refractive spherule with the plasmalemma of the sieve element (Fig. 3.15). Liberman-Maxe (1978) observed a similar phenomenon in protophloem sieve elements only of *Polypodium vulgare,* but there the pertinent bodies were composed of polysaccharides and not of protein. Further cytochemical study will be necessary to determine whether two types of membrane-bound body - one consisting of polysaccharides and the other of protein - exist side by side in sieve elements of other vascular cryptogams. The exact chemical composition and function of refractive spherules remain to be determined.

As mentioned previously, substantial quantities of crystalline and fibrillar proteinaceous material occur in numerous, relatively small vacuoles of *Selaginella* sieve elements (Burr and Evert 1973). Proteinaceous material also arises in cisternae of rough ER in *Isoetes* sieve elements (Fig. 3.1; Kruatrachue and Evert 1974, 1978). In *Isoetes,* the dilated portions of ER enclosing the proteinaceous material become smooth-surfaced, dissociate from the rest of the ER, and then migrate to the cell wall. Along the way they may form multivesicular bodies,

Figs. 3.11-3.15. Portions of immature (Figs. 3.11-3.14) and mature (Fig. 3.15) sieve elements of some seedless vascular plants showing refractive spherules *(RS).* **Fig. 3.11.** *Botrychium virginianum.* Young sieve element showing anastomosing network of endoplasmic reticulum, parts of which are dilated and filled with granules that apparently form refractive spherules. ×56000; *Bar line*=0.2 μm. **Fig. 3.12.** *B. virginianum.* Later stage of development than shown in Fig. 3.11. Nacreous wall formation was near completion in this cell. Apparently, refractive spherules continue to be formed at this late stage of sieve-element development in this fern. ×45800; *Bar line*=0.2 μm. **Fig. 3.13.** *Psilotum nudum.* Refractive spherule with "tail" extending from dilatation of endoplasmic reticulum. ×27200; *Bar line*=0.2 μm. **Fig. 3.14.** *Platycerium bifurcatum.* Surface view of dictyosome *(D)* and dilated portion of dictyosome tubule with refractive spherule. ×64800; *Bar line*=0.2 μm. **Fig. 3.15.** Final stages in history of refractive spherules in sieve elements of *Microgramma lycopodioides:* fusion of delimiting membranes of refractive spherules with plasmalemma and discharge of spherule material into region of wall *(W).* ×41950; *Bar line*=0.2 μm (Figs. 3.11 and 3.12 from Evert 1976; Fig. 3.13 from Perry and Evert 1975; Figs. 3.14 and 3.15 from Evert and Eichhorn 1974b) which then fuse with the plasmalemma, discharging their contents to the outside (Fig. 3.2). It is quite clear that the crystalline or fibrillar material in *Selaginella* and *Isoetes* is not homologous to the P-protein of angiospermous sieve-tube members. Again, P-protein is lacking in the sieve elements of seedless vascular plants. The absence of refractive spherules in only the lycopods among seedless vascular plants was noted by Hébant et al. (1978) as a condition in agreement with the hypothesis that the lycopods constitutive a relatively isolated group of primitive vascular plants.

3.3 The Wall

Typically, the sieve elements in seedless vascular plants develop walls that are thicker than those of the parenchyma cells bordering them. In some species, the sieve-element wall consists of two morphologically distinct layers, a relatively thin outer layer and a more or less thick inner layer, the so-called nacreous layer, which exhibits a glistening or pearly appearance in fresh sections. Although in other seedless vascular plants the sieve-element walls may become quite thick, they do not consist of morphologically distinct inner and outer layers and, hence, such walls cannot be designated nacreous (Esau and Cheadle 1958). This is true, for example, of the sieve-element walls in *Isoetes muricata* (Figs. 3.1 and 3.2; Kruatrachue and Evert 1974, 1978), *Lycopodium lucidulum* (Warmbrodt and Evert 1974a), and *Equisetum hyemale* (Figs. 3.16 and 3.17; Dute and Evert 1977a).

During his comparative studies of the phloem in seedless vascular plants, Lamoureux (1961) found true nacreous wall layers in only some species of Hymenophyllaceae and Polypodiaceae. In some species, nacreous walls were found in some plant parts and not in others. In contrast, Warmbrodt and Evert (1979b) found nacreous thickenings in sieve elements of all nine families of homosporous leptosporangiate ferns they examined with the electron microscope.

Hébant (1969) found nacreous wall development to be quite variable in tropical ferns and dependent upon the type of organ, the age of the sieve elements,

Figs. 3.16–3.19. Sieve elements from aerial shoot of *Equisetum hyemale*. **Fig. 3.16.** Young protophloem sieve element with wall thickening mostly in corners of cell. ×8000; *Bar line*= 1 µm. **Fig. 3.17.** Protophloem sieve element at a later stage of wall thickening than that of Fig. 3.16. Wall thickening is no longer largely restricted to the corners of the cell. The *light region* bordering the lamellate portion of the wall internally is filled with a fibrillar matrix. ×8200; *Bar line*=1 µm. **Fig. 3.18.** Immature metaphloem sieve element with normal-appearing nucleus *(N)* and cytoplasm filled with endoplasmic reticulum *(ER)*. ×15700; *Bar line*= 1 µm. **Fig. 3.19.** Mature sieve elements are characterized by a clear lumen and peripheral endoplasmic reticulum. ×7100; *Bar line*=1 µm. *D* dictyosome; *M* mitochondrion; *P* plastid; *RS* refractive spherule (Figs. 3.16–3.19 from Dute and Evert 1978).

Fig. 3.20. Transverse section of immature sieve element of *Botrychium virginianum* with fully-formed nacreous thickening bordered internally by electron-dense layer. ×7600; *Bar line*=1 µm (Evert 1976)

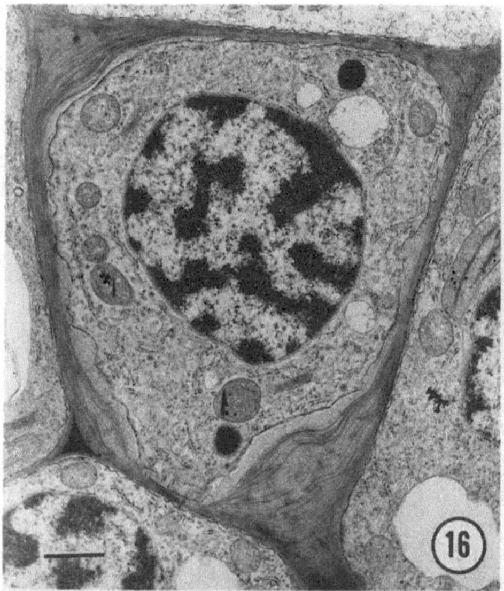
16

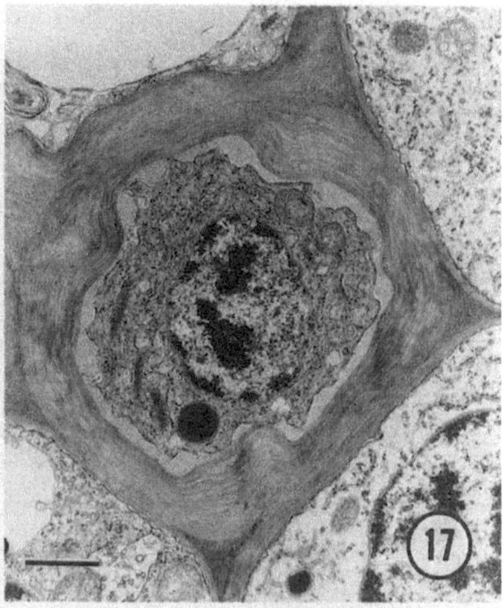
17

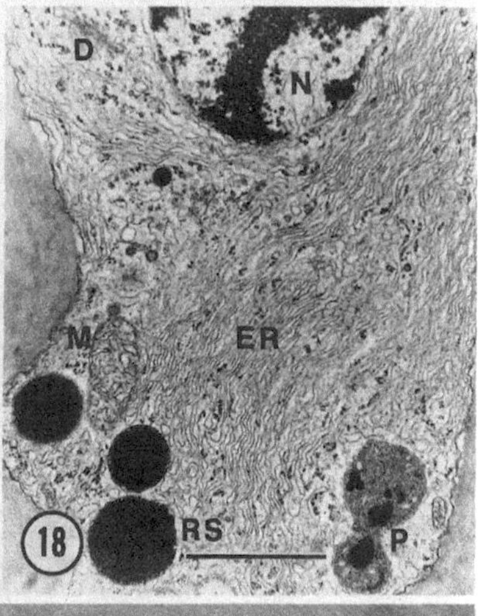
D
N
M
ER
RS
P
18

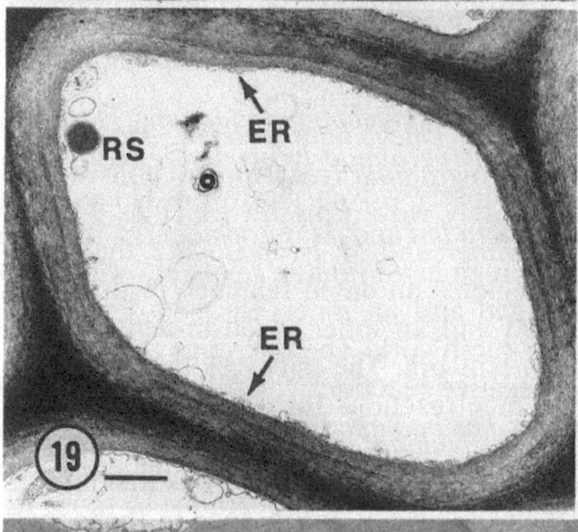
ER
RS
ER
19

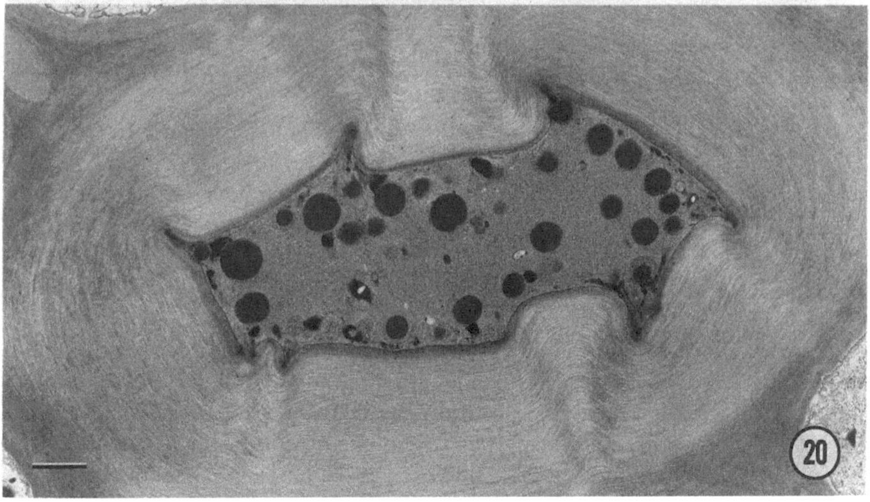
20

and the habitat. He found that ferns such as *Vittaria guineensis,* which go through periods of drought, have well-developed nacreous walls, while aquatic types and those growing in regions of constant humidity lack such wall layers. Nevertheless, Evert and Eichhorn (1976) found nacreous walls in protophloem and metaphloem sieve elements of field-grown *(Polypodium schraderi* and *Microgramma lycopodioides)* and of greenhouse-grown *(Platycerium bifurcatum* and *Phlebodium aureum)* plants. Liberman-Maxe (1978) found nacreous thickenings in only the metaphloem sieve elements of *Polypodium vulgare.*

Considerable variation exists in the structure and development of the so-called nacreous thickenings in vascular cryptogams. The results of two studies, one on the sieve elements of *Botrychium virginianum* (Evert 1976) and the other on four species of polypodiaceous ferns (Evert and Eichhorn 1976), serve to illustrate this point. In *B. virginianum,* the very thick nacreous wall layer has a polylamellate structure, with the cellulose microfibrils arranged parallel to one another and more or less at right angles to the long axis of the cell (Fig. 3.20), and it gives positive reactions for cellulose and pectin. When viewed with polarized light, this nacreous thickening is strongly anisotropic. Other than microtubules, only dictyosome-derived vesicles and possibly ER could be implicated with nacreous wall formation in this eusporangiate fern.

By contrast, the nacreous thickenings in *Phlebodium aureum, Platycerium bifurcatum, Polypodium schraderi,* and *Microgramma lycopodioides* consist of a coarse fibrillar material arranged in a loosely woven meshwork (Figs. 3.9 and 3.10; Evert and Eichhorn 1976). In *P. schraderi* and *M. lycopodioides,* the nacreous thickenings consist of two distinct regions: an outer, relatively wide region in which the fibrils are loosely arranged and an inner, electron-dense region consisting of a more compact fibrillar meshwork. The nacreous thickenings in all four polypodiaceous ferns gave negative reactions for cellulose and pectin and were isotropic. In *Platycerium* and *Phlebodium,* initiation of nacreous wall formation was closely correlated with the appearance of numerous membranes or vesicles in the region of the wall. Walls similar in appearance to those in *P. schraderi* and *M. lycopodioides* are found in *Polypodium vulgare* (Liberman-Maxe 1978). Using Thiéry's test, Liberman-Maxe was able to show the polysaccharide nature of the fibrillar meshwork in *P. vulgare.*

Nacreous walls are also found in *Psilotum nudum* (Perry and Evert 1975). Unlike the sieve-element walls in *B. virginianum,* however, those in *P. nudum* often become lignified. Positive reactions for lignin also have been reported for sieve-element walls in some species of *Equisetum* (Browne 1939; Bohorquez 1970), although lignin was not found in walls of either protophloem or metaphloem sieve elements in a more recent study of *Equisetum hyemale* (Dute and Evert 1978).

The nacreous thickenings in *B. virginianum* (Evert 1976) and some polypodiaceous ferns have at least one thing in common: gradually they decrease in thickness and disappear. The inner, electron-dense portion of the nacreous thickening in the polypodiaceous ferns persists in the mature sieve element. Entire nacreous thickenings may persist, however, in some sieve elements of homosporous leptosporangiate ferns (Liberman-Maxe 1978; Warmbrodt and Evert 1979b).

It should be obvious from this brief discussion of nacreous thickenings in vascular cryptogams that the term nacreous has not been uniformly applied to equivalent wall layers in the sieve elements of this group of vascular plants. The same can be said of the "nacreous" wall layers in angiospermous sieve-tube members. Much systematic ultrastructural research needs to be done on the nacreous wall.

3.4 The Sieve Areas

Considerable variation exists in the distribution, size, and contents of the cytoplasmic connections or pores between sieve elements of seedless vascular plants (Warmbrodt and Evert 1978, 1979a, b). In some species, all of the connections are clustered into sieve areas, whereas in others they occur only singly on both end and lateral walls (e.g., in *Selaginella,* Burr and Evert 1973); in still other species, the pores occur in clusters on the end walls and only singly (e.g., in *Psilotum,* Perry and Evert 1975; in the *Isoetes* root, Kruatrachue and Evert 1978) or not at all (e.g., in the *Isoetes* leaf, Kruatrachue and Evert 1974) on the lateral walls.

Sieve elements commonly are classified as sieve cells or sieve-tube members, depending upon the degree of specialization of their sieve areas and the differences in the distribution of the sieve areas on the walls of a given element (Esau 1969). In most seedless vascular plants, the range in size of the pores on end and lateral walls is generally similar, and the pores or sieve areas are not markedly restricted to certain wall parts in individual elements. By definition, such sieve elements are regarded as sieve cells. In sieve-tube members, sieve areas bearing larger pores than others are localized on the walls to form sieve plates.

Sieve-tube members or sieve-tube member-like sieve elements have been reported in the mid-internodal regions of the aerial stems in several species of *Equisetum* (Lamoureux 1961; Agashe 1968; Bohorquez 1970; Dute and Evert 1978), in the stems of *Cyathea gigantea* (Shah and Fotedar 1974a), in the rhizomes and petioles of *Marsilea quadrifolia* (Loyal and Maheshwari 1979; Miller and Duckett 1979), and in the leaves and roots of *Isoetes muricata* (Kruatrachue and Evert 1974, 1978).

Consistent differences were found between pore sizes in end (average diam 0.4 μm) and lateral (average diam 0.27 μm) walls of metaphloem sieve elements in mid-internodal regions of the aerial stem of *Equisetum hyemale* (Dute and Evert 1978). By definition, therefore, the end walls bear sieve plates and such sieve elements are sieve-tube members. In contrast, internodal sieve elements located near the nodes of the same aerial stems are united across very oblique end walls by "pores" that average only 0.058 μm in diameter, little wider than the average diameter, 0.04 μm, of their plasmodesmata precursors (Dute and

Evert 1977b). The sequence of events in the ontogeny of the protoplast of these primitive-like sieve elements, or sieve cells, is essentially similar to that of the metaphloem sieve elements in the mid-internodal regions. In the root of *E. hyemale* (Dute and Evert 1977a), the pores between contiguous sieve elements are found almost exclusively on the end walls, which range in orientation from transverse to fairly oblique. The pores, which measured from 0.05 μm to 0.32 μm in diameter, are largely restricted to the end walls because most of the sieve elements are laterally separated from one another by parenchyma cells. Inasmuch as the pores of the root sieve elements are largely restricted to the end walls, one might consider calling them sieve-tube members. On the other hand, because of the great variability of pore size in these elements, they might best be considered sieve cells. The authors opted simply to call them sieve elements.

In mature sieve elements of the *Isoetes muricata* root (Kruatrachue and Evert 1978), the occasional solitary pores (0.2 μm in diam) in the lateral walls are considerably smaller than the pores in the end walls (0.4 μm–0.7 μm in diam). In sieve elements of the leaf (Kruatrachue and Evert 1974), the end walls contain either plasmodesmata or pores (0.2 μm–0.57 μm in diam), or both, but only plasmodesmata occur in the lateral walls. In the corm of *I. muricata* (Kruatrachue and Evert 1977), pores of variable size (0.2 μm–1.0 μm in diam) occur in both radial and transverse walls of the secondary sieve elements, but more or less uniformly small pores (0.15 μm–0.2 μm in diam) occur in the tangential walls. Because of the variability of pore size in both radial and transverse walls of the corm sieve elements, these elements might be considered sieve cells. On the other hand, the leaf sieve elements with end-wall pores and the sieve elements of the root could, by definition, be called sieve-tube members. One might question whether the leaf sieve elements with only plasmodesmata in both their end and lateral walls should be considered sieve elements at all. The sequence of events in the ontogeny of their protoplasts is similar, however, to that of the sieve elements with pores in their end walls. Hence, the situation here is similar to that of the primitive-like sieve elements in the aerial stems of *E. hyemale* (Dute and Evert 1977b).

As mentioned previously, sieve-tube members have been reported in the stems of *Cyathea gigantea* (Shah and Fotedar 1974a) and in rhizomes and petioles of *Marsilea quadrifolia* (Loyal and Maheshwari 1979; Miller and Duckett 1979). In *C. gigantea*, the pores in the end walls of the sieve-tube members ranged from 2 μm to 5 μm in diameter, whereas those of the lateral walls measured less than 0.8 μm in diameter. Other sieve elements with small end-wall pores similar in size to those of the lateral walls were also found in the same tissue. Hence, both sieve cells and sieve-tube members are found in the stem of *C. gigantea*. The situation in *M. quadrifolia* is less clear. According to Loyal and Maheshwari (1979), sieve cells far outnumber sieve-tube members in the rhizome and petiole of that aquatic fern. They reported that the sieve-plate pores were small, ranging in diameter from 0.054 μm to 1.0 μm, and located on oblique end walls, and they suggested that these elements represented a transitional stage between typical sieve cells and sieve-tube members. Miller and

Duckett (1979), in an ultrastructural study of *M. quadrifolia,* found what they interpreted as sieve cells in the roots and leaf laminae and sieve-tube members in the rhizomes and petioles. The sieve-tube members were of two types: thick-walled ones with compound sieve plates in oblique end walls and thin-walled ones with simple sieve plates in transverse end walls. Sizes of pores were not mentioned. In her ultrastructural study of the phloem in *M. quadrifolia,* Kruatrachue (1978) reported that the sieve-area pores were quite small in size and referred to the conduits as sieve elements.

The gradient in the nature of the end-wall connections in the *I. muricata* leaf can be interpreted as a reflection of the phylogenetic relationship between sieve elements and parenchyma cells, as can be the presence of the primitive-like sieve elements in the *E. hyemale* internodes. Moreover, the presence, by definition, of sieve-tube members or sieve-tube member-like cells in portions of several vascular cryptogams can be interpreted as a reflection of an early trend in evolutionary specialization toward a greater degree of continuity between protoplasts of contiguous sieve elements - a trend that has begun independently in several, unrelated taxa.

Development of sieve-area pores in *Polypodium vulgare* (Liberman-Maxe 1978) is reminiscent of that in angiosperms (Deshpande 1975; Esau and Thorsch 1985). The pore sites become distinguishable from the rest of the wall by the appearance of paired callose collars or cone-shaped structures, one on either side of the wall, around each plasmodesma. Perforation begins in the region of the middle lamella and then the plasmodesmatal canal widens on both sides of the wall until the pore is fully formed. Callose is not associated with development of the sieve-area pores in *Selaginella* (Burr and Evert 1973), *Lycopodium* (Warmbrodt and Evert 1974a), or *Isoetes* (Kruatrachue and Evert 1974); nor in *Psilotum* (Perry and Evert 1975), *Equisetum* (Dute and Evert 1977a, 1978), or *Botrychium* (Evert 1976). Widening of the plasmodesmatal canals in these sieve elements occurs more or less uniformly along the entire length of the plasmodesmata. In *Psilotum,* the sites of the future pores are clearly indicated by the presence of ER aggregates that arise opposite them early in relation to pore development. In both the root and the aerial stem of *Equisetum,* several tubules of ER early become associated with the developing pore. In general, perforation of the pore sites in the vascular cryptogams occurs late in maturation of the sieve elements, at a time when the nucleus already has a degenerate appearance.

At maturity, the plasmalemma-lined sieve-area pores of *Psilotum* (Perry and Evert 1975) and of the eusporangiate and protoleptosporangiate ferns (Warmbrodt and Evert 1979a) contain numerous electron-dense membranes, apparently tubular elements of ER, which are structurally distinct from the parietal ER in the lumen of the cell (Figs. 3.21 and 3.22). Callose has not been identified in association with the sieve-area pores of either *Psilotum* or the eusporangiate ferns examined with the electron microscope.

The sieve-area pores in the root (Dute and Evert 1977a) and mid-internodal regions of the aerial stems (Dute and Evert 1978) of *Equisetum* are traversed by

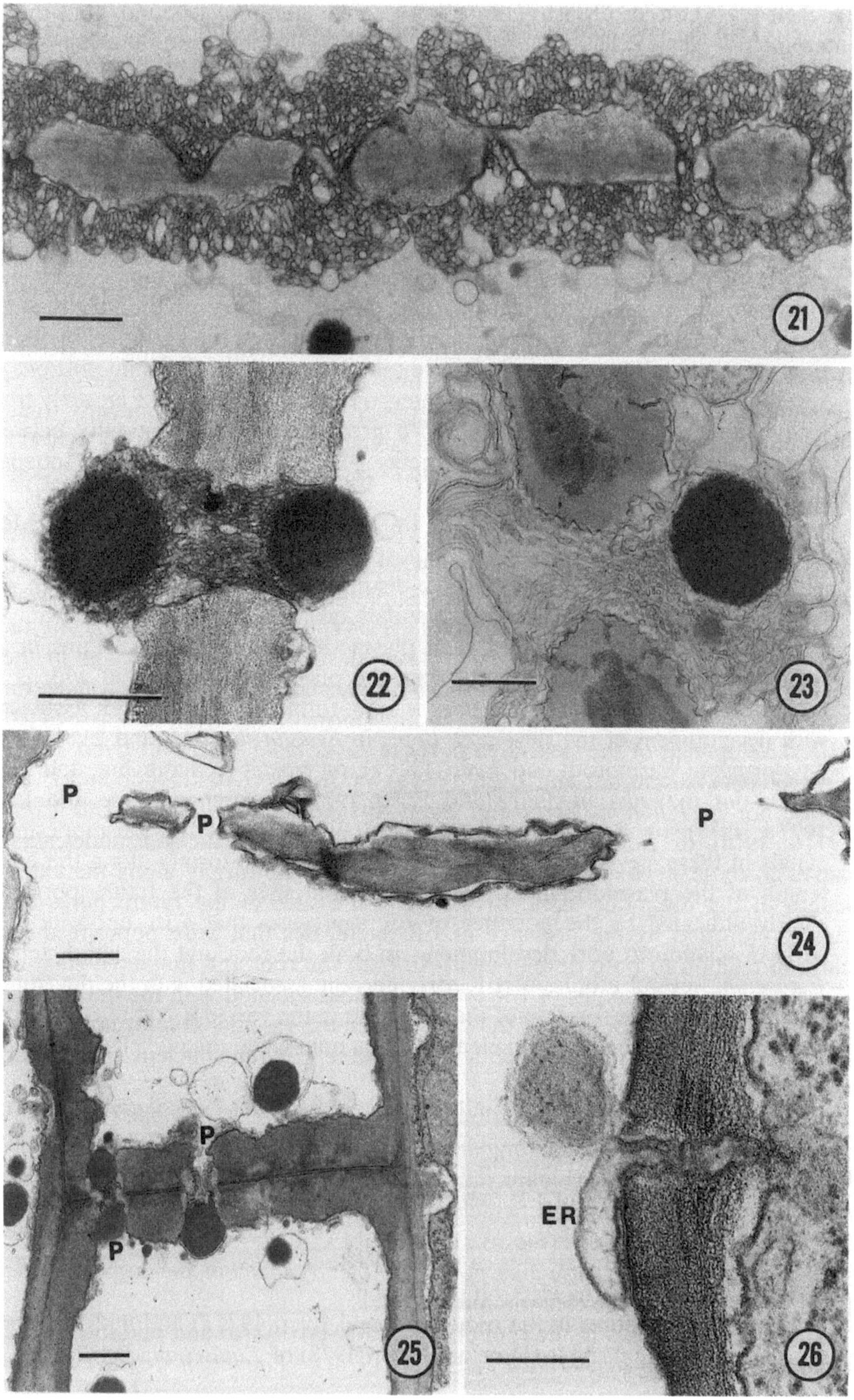
21
22
23
P
P
P
24
P
P
25
ER
26

large numbers of membranes (Fig. 3.23). The nature of the membranes could not be determined in the root largely because of the presence of large quantities of callose (presumably wound callose). In the mid-internodal region, the many membranes apparently are tubules of ER, which are continuous with those in the lumina of the cells.

In heterosporous (Warmbrodt and Evert 1978) and homosporous (Warmbrodt and Evert 1979b) leptosporangiate ferns the sieve-area pores are traversed by variable numbers of ER membranes, which generally do not occlude the pores (Fig. 3.25); nor do these membranes differ structurally from the parietal ER of the lumina. Continuity commonly is found between the pore membranes and those of the parietal network. Considerable variation exists in the quantity of callose associated with the pores in tissues prepared for electron microscopy, ranging from none to massive deposits.

By contrast to those of the other vascular cryptogams, the plasmalemma-lined sieve-area pores of *Selaginella* (Burr and Evert 1973), *Isoetes* (Kruatrachue and Evert 1974, 1977, 1978), and *Lycopodium* (Warmbrodt and Evert 1974a) are virtually wide open, being traversed by only an occasional tubule of ER in apparently undisturbed sieve elements (Fig. 3.24). In most preparations of *Selaginella* and *Isoetes* phloem, some pores commonly were occluded with fibrillar material released from vacuoles (in *Selaginella*) or dilated ER cisternae (in *Isoetes*), which presumably ruptured during manipulation of the tissue. Such views of the pores are reminiscent of the P-protein plugged sieve-plate pores encountered in sieve-tube members in most preparations of angiospermous phloem (Evert 1982).

Although callose has been identified at the mature sieve areas of *Selaginella* and *Isoetes*, none could be detected in sieve elements of *Lycopodium* with the electron microscope. Occasional callose deposits have been found, however, in *Lycopodium* with the light microscope (Lamoureux 1961).

Figs. 3.21-3.26. Sieve-area pores in some seedless vascular plants. **Fig. 3.21.** Endoplasmic reticulum filled pores in wall between sieve elements in aerial shoot of *Psilotum nudum.* ×23200; *Bar line*=0.5 μm. **Fig. 3.22.** In the eusporangiate fern *Botrychium virginianum* the pores are filled with numerous membranes, apparently tubular endoplasmic reticulum. The electron-dense bodies on either side of the pore are refractive spherules. ×34650; *Bar line*=0.5 μm. **Fig. 3.23.** Endoplasmic reticulum filled pore in wall between mature sieve elements in aerial stem of the horsetail *Equisetum hyemale.* The electron-dense body on right is a refractive spherule ×23650; *Bar line*=0.5 μm. **Fig. 3.24.** Unoccluded pores *(P)* in wall between mature sieve elements in corm of the quillwort *Isoetes muricata.* ×17700; *Bar line*=0.5 μm. **Fig. 3.25.** Both pores *(P)* in this transverse end wall between sieve elements in *Marsilea macropoda* are blocked by refractive spherules. ×14250; *Bar line*=0.5 μm. **Fig. 3.26.** Pore-plasmodesma connection between sieve element *(left)* and parenchyma cell *(right)* in *M. macropoda. ER* endoplasmic reticulum; ×72200; *Bar line*=0.2 μm (Fig. 3.21 from Perry and Evert 1975; Fig. 3.22 from Evert 1976; Fig. 3.23 from Dute and Evert 1978; Fig. 3.24 from Kruatrachue and Evert 1977; Figs. 3.25 and 3.26 from Warmbrodt and Evert 1978)

3.5 Parenchymatous Cells Associated with the Sieve Elements

Sieve elements are associated with parenchyma cells in all vascular plants - an association interpreted as a reflection of the close functional relationship between these two cell types and probably a result of the protoplasmic specialization of the sieve element (Fig. 3.9). Although specialized parenchyma cells analogous to companion cells or albuminous (Strasburger) cells are lacking in the seedless vascular plants, cytoplasmic connections (consisting of a small pore on the sieve-element side of the wall and a single, unbranched plasmodesma on the parenchyma-cell side) typically are numerous between sieve elements and parenchyma cells of vascular cryptogams (Fig. 3.26). Despite early reports of the presence of companion cells in the roots of *Equisetum* (Chauveaud 1903), the cells designated there as "companion cells" are not equivalent to the companion cells in angiosperms (Dute and Evert 1977 a).

Sieve-element associated parenchyma cells in vascular cryptogams generally do not differ structurally from other parenchyma cells of the vascular cylinder or bundle. In the petioles of *Polypodium vulgare,* the phloem parenchyma cells, which contain elaborate systems of ER, abundant ribosomes, numerous mitochondria and lipid granules, and some plastids, have numerous cytoplasmic (pore-plasmodesma) connections with the sieve elements (Maxe 1965; Liberman-Maxe 1971). Parenchyma cells reminiscent of those in the *P. vulgare* petioles, but with numerous starch-containing chloroplasts, occur in the vascular bundles of the leaf laminae of *Vittaria guineensis* (Warmbrodt and Evert 1979 a).

Among the seedless vascular plants, only *Lycopodium lucidulum* has been found with apparent specialization among the vascular parenchyma cells (Warmbrodt and Evert 1974b). The vascular cylinder of *L. lucidulum* contains two distinct types of parenchyma cells, one of which is always associated with sieve elements and the other with tracheids. The sieve-element associated parenchyma cells, which have numerous connections with the sieve elements, have very dense protoplasts, due to a high ribosome population and the absence of large vacuoles. The envelopes of the large, ramifying nuclei of these cells have numerous connections with the ER. By contrast, the tracheid-associated parenchyma cells are light in appearance and lack plasmodesmata in their walls. The remaining parenchyma cells have characteristics intermediate between the two extremes.

Most vascular parenchyma cells in the leaves of ferns are in contact with both sieve elements and tracheary elements and, hence, cannot be distinguished as either phloem or xylem parenchyma cells (Warmbrodt and Evert 1978, 1979 a, b). Moreover, the sieve elements usually occur next to the tracheary elements. This close spatial association of the various cell types may facilitate the movement of water and solutes between the tracheary elements and sieve elements.

Transfer cells occur in the vascular bundles of some fern leaves, but they are not the equivalents of the A- and B-type transfer cells in the phloem of some herbaceous dicotyledons (Pate and Gunning 1969). In the ferns, the wall ingrowths occur only opposite the tracheids (Gunning and Pate 1974) or opposite sieve elements, tracheids, and other parenchyma cells (Warmbrodt and Evert 1978, 1979b).

Hébant and co-workers have demonstrated by histochemical methods the presence of intensive phosphatase and respiratory enzyme activities in parenchyma cells associated with sieve elements and tracheary elements in *Lycopodium clavatum* (Hébant et al. 1978), *Selaginella wildenowii* (Hébant et al. 1980), and several species of fern (DeFäy and Hébant 1981). By comparison to the stelar parenchyma cells, the parenchyma cells of the surrounding tissues showed much less intense staining. These results were interpreted by Hébant and co-workers as an indication of a functional relationship between the parenchyma cells and conducting cells of the stele.

3.6 Longevity of the Sieve Elements

Little information is available on the longevity of sieve elements in seedless vascular plants. With the exception of *Isoetes* (and possibly of *Botrychium*), secondary growth is lacking in the vascular cryptogams. Hence, in those lacking secondary vascular tissues, one would expect many of the sieve elements to function for the life of the plant part in which they occur. Parthasarathy (1975) reported finding living sieve elements in the basal part of a 12-year-old tree fern *(Alsophila stipularis)*. Living sieve elements 5 or more years old have been reported in the bracken fern, *Pteridium aquilium* (Hume 1912).

The secondary sieve elements in the corm of *Isoetes muricata* function for more than 1 year (Kruatrachue and Evert 1977). Almost all of those examined in plants that were at least 3 years old were still living. With the onset of winter dormancy, the sieve pores of the sieve elements in *I. muricata* become covered with massive amounts of dormancy callose. In spring, the dormancy callose disappears, beginning with the sieve elements nearest the cambium. Thus, the seasonal cycle of phloem development in *I. muricata* is reminiscent of that in *Vitis* (Esau 1948; Davis and Evert 1970) and *Tilia* (Evert 1962).

3.7 Comments on Terminology

As noted previously, the distinction between sieve cells and sieve-tube members is made largely on the basis of differences in size and distribution of its sieve pores, criteria that have been in general use since they were first proposed by

Cheadle and Whitford in 1941, long before use of the electron microscope in the study of phloem tissue. Accordingly, with few exceptions, the sieve elements in the seedless vascular plants may be classified as sieve cells. With some reservations, sieve-tube members have been identified in a few species of vascular cryptogams. In some instances, pore size and distribution have proved to be inadequate criteria for the classification of the sieve elements in others (e.g., Kruatrachue and Evert 1974, 1977, 1978; Perry and Evert 1975; Dute and Evert 1977b). Moreover, with the electron microscope, it has been demonstrated that considerable variation exists in the development and contents of the sieve pores in this diverse group of vascular plants; and, although refractive spherules characterize the sieve elements of most vascular cryptogams, they are lacking in those of the lycopods.

By contrast, the structure and development of the sieve elements among gymnosperms is remarkably uniform. On the basis of pore size and distribution, the sieve elements in gymnosperms clearly are sieve cells. The sieve areas of all gymnospermous sieve elements develop from a portion of the wall traversed by numerous plasmodesmata and pore formation is accompanied by formation of a median cavity in the middle of the wall. Typically, gymnospermous sieve cells are accompanied by albuminous cells and are united with those parenchymatous elements by cytoplasmic connections consisting of a pore on the sieve-cell side of the wall and branched plasmodesmata on the albuminous-cell side.

With few exceptions (*Austrobaileya scandens,* Srivastava 1970; Behnke 1986), angiospermous sieve elements are characterized by the presence of sieve plates, the development of which is rather uniform from species to species. Hence, on the basis of size and distribution of pores, the sieve elements of the great majority of angiosperms quite clearly are sieve-tube members. In addition to the presence of sieve plates, angiospermous sieve elements typically are associated with companion cells, highly specialized parenchyma cells ontogenetically related to the sieve-tube members and with which they have numerous cytoplasmic connections, consisting of a pore on the sieve-tube member side of the wall and much-branched plasmodesmata on the companion-cell side (Esau and Thorsch 1985). Typically, the sieve-tube members of angiosperms contain P-protein, and their nuclear degeneration is chromatolytic. As one might expect, some angiospermous sieve elements lack one or another character typical of the group. As noted, the sieve elements of *Austrobaileya* lack sieve plates; however, they contain P-protein and are associated with companion cells. Some angiospermous sieve elements possess sieve plates and are associated with companion cells but lack P-protein (e.g., in the grasses and in some palms). Nuclear degeneration often is pycnotic in protophloem sieve elements and some protophloem elements lack companion cells.

Recently, Behnke (1986) proposed that a combination of characters be used in assigning a sieve element to a particular category, especially those of the angiosperms and gymnosperms. While continuing to use "sieve element" as a general term for the conducting elements of the phloem as first proposed by Eames and MacDaniels (1947), and currently in use, he proposes (1) to add a higher taxon name as specification and (2) to restrict the term "sieve cell" to the sieve elements of gymnosperms and the term "sieve tube" to sieve elements of angiosperms, i. e., "sieve-tube member" for the individual sieve element and "sieve tube" for a longitudinal series of "sieve-tube members" (explanation added in consent with Behnke).

This author, while especially emphasizing the retainment of the term sieve-tube member, concurs with most of Behnke's suggestions but for the sake of simplicity proposes that the sieve elements of seedless vascular plants and the assimilate-conducting cells (leptoids) of the leafy gametophytes of certain species of Polytrichales (Hébant 1977) simply be called "sieve elements." Moreover, sieve-tube members would be identified on the basis of sharing any two of the three characters: sieve plates, P-protein, and companion cells. Hence, the sieve elements of *Austrobaileya* would be considered as primitive sieve-tube members (see also Chap. 14.4 for phloem phylogeny).

Acknowledgments. Thanks are expressed to Susan E. Eichhorn for her assistance with preparation of the manuscript, including the figures, and to Kathleen T. Delaney, typist. The preparation of this chapter was supported in part by NSF grant DCB87-01116.

References

Agashe SN (1968) Phloem studies in pteridophytes. pt 1: *Equisetum*. Am Fern J 58: 74-77

Behnke H-D (1967) Über den Aufbau der Siebelement-Plastiden einiger Dioscoreaceen. Z Pflanzenphysiol 57: 243-254

Behnke H-D (1981) Sieve-element characters. Nord J Bot 1: 381-400

Behnke H-D (1986) Sieve-element characters and the systematic position of *Austrobaileya*, Austrobaileyaceae - with comments to the distinction and definition of sieve cells and sieve-tube members. Plant Syst Evol 152: 101-121

Bohorquez LZ (1970) Phloem anatomy of *Equisetum telmateia* and *E. hyemale*. MS Thesis, Univ Ill, Chicago

Browne IMP (1939) Anatomy of the aerial axes of *Equisetum kansanum*. Bot Gaz 101: 35-50

Burr FA, Evert RF (1973) Some aspects of sieve-element structure and development in *Selaginella kraussiana*. Protoplasma 78: 81-97

Chauveaud G (1903) Recherches sur le mode de formation des tubes criblées dans la racine des Cryptogames vasculaires et des Gymnospermes. Ann Sci Nat Bot Ser 8, 18: 165-277

Cheadle VI, Whitford NB (1941) Observations on the phloem in the Monocotyledoneae. I. The occurrence and phylogenetic specialization in structure of the sieve tubes in the metaphloem. Am J Bot 28: 623-627

Davis JD, Evert RF (1970) Seasonal cycle of phloem development in woody vines. Bot Gaz 131: 128-138

DeFäy E, Hébant C (1981) Respiratory and phosphatase activities in parenchyma cells associated with conducting elements of ferns. Ann Bot (London) 47: 703-707

Deshpande BP (1975) Differentiation of the sieve plate of *Cucurbita:* a further view. Ann Bot (London) 39: 1015-1022

Dute RR, Evert RF (1977a) Sieve-element ontogeny in the root of *Equisetum hyemale*. Am J Bot 64: 421-438

Dute RR, Evert RF (1977b) Primitive-like metaphloem sieve elements in the aerial stem of *Equisetum hyemale*. Protoplasma 91: 257-266
Dute RR, Evert RF (1978) Sieve-element ontogeny in the aerial shoot of *Equisetum hyemale* L. Ann Bot (London) 42: 23-32
Eames AJ, MacDaniels LH (1947) An introduction to plant anatomy. McGraw-Hill, New York
Esau K (1948) Phloem structure in the grapevine, and its seasonal changes. Hilgardia 18: 217-296
Esau K (1969) The phloem. In: Zimmerman W, Ozenda P, Wulff HD (eds) Encyclopedia of plant anatomy, vol 5, pt 2. Borntraeger, Berlin Stuttgart, 505 pp
Esau K (1978) Developmental features of the primary phloem in *Phaseolus vulgaris* L. Ann Bot (London) 42: 1-13
Esau K, Cheadle VI (1958) Wall thickening in sieve elements. Proc Natl Acad Sci USA 44: 546-553
Esau K, Thorsch J (1985) Sieve plate pores and plasmodesmata, the communication channels of the symplast: ultrastructural aspects and developmental relations. Am J Bot 72: 1641-1653
Evert RF (1962) Some aspects of phloem development in *Tilia americana*. Am J Bot 49: 659
Evert RF (1976) Some aspects of sieve-element structure and development in *Botrychium virginianum*. Isr J Bot 25: 101-126
Evert RF (1982) Sieve-tube structure in relation to function. BioScience 32: 789-795
Evert RF, Eichhorn SE (1974a) Sieve-element ultrastructure in *Platycerium bifurcatum* and some other polypodiaceous ferns: the nucleus. Planta 119: 301-318
Evert RF, Eichhorn SE (1974b) Sieve-element ultrastructure in *Platycerium bifurcatum* and some other polypodiaceous ferns: the refractive spherules. Planta 119: 319-334
Evert RF, Eichhorn SE (1976) Sieve-element ultrastructure in *Platycerium bifurcatum* and some other polypodiaceous ferns: the nacreous wall thickening and maturation of the protoplast. Am J Bot 63: 30-48
Fisher DG, Evert RF (1979) Endoplasmic reticulum-dictyosome involvement in the origin of refractive spherules in sieve elements of *Davallia fijiensis* Hook. Ann Bot (London) 43: 255-258
Gunning BES, Pate JS (1974) Transfer cells. In: Robards AW (ed) Dynamic aspects of plant ultrastructure. McGraw-Hill, New York, pp 441-448
Hardham AR, Gunning BES (1979) Interpolation of microtubules into cortical arrays during cell elongation and differentiation in roots of *Azolla pinnata*. J Cell Sci 37: 411-442
Hébant C (1969) Observations sur le phloème de quelques Filicinées tropicales. Nat Monspel Ser Bot 20: 135-196
Hébant C (1976) Studies on the development of the conducting tissue-system in gametophytes of some Polytrichales. III. Further observations on leptoids, with particular reference to their endoplasmic reticulum. Protoplasma 87: 79-90
Hébant C (1977) The conducting tissues of bryophytes. Cramer, Vaduz, Liechtenstein
Hébant C, Guiraud R, Barthonnet J, Ba AT (1978) Le phloème de *Lycopodium clavatum:* organisation, ultrastructure et histochimie. Can J Bot 23: 2973-2980
Hébant C, Guiraud R, Martin MG (1980) Le phloème de *Selaginella willdenowii:* histophysiologie comparée. Plant Syst Evol 135: 159-169
Hume EMM (1912) The histology of the sieve tubes of *Pteridium aquilinum*, with some notes on *Marsilea quadrifolia* and *Lygodium dichotomum*. Ann Bot (London) 26: 573-587
Kruatrachue M (1978) Some aspects of phloem structure and development in *Marsilea quadrifolia* L. J Sci Soc Thailand 4: 127-138
Kruatrachue M, Evert RF (1974) Structure and development of sieve elements in the leaf of *Isoetes muricata*. Am J Bot 61: 253-266
Kruatrachue M, Evert RF (1977) The lateral meristem and its derivatives in the corm of *Isoetes muricata*. Am J Bot 64: 310-325
Kruatrachue M, Evert RF (1978) Structure and development of sieve elements in the root of *Isoetes muricata* Dur. Ann Bot (London) 42: 15-21

Lamoureux CH (1961) Comparative studies on phloem of vascular cryptogams. Ph D Thesis, Univ Cal, Davis
Lawton DM (1978a) P-protein crystals do not disperse in uninjured sieve elements in roots of runner bean *(Phaseolus multiflorus)* fixed with glutaraldehyde. Ann Bot (London) 42: 353-361
Lawton DM (1978b) Ultrastructural comparison of the tailed and tailless P-protein crystals respectively of runner bean *(Phaseolus multiflorus)* and garden pea *(Pisum sativum)* with tilting stage electron microscopy. Protoplasma 97: 1-11
Liberman-Maxe M (1968) Différenciation des pores dans les cellules criblées de *Polypodium vulgare* (Polypodiacée). C R Acad Sci Paris Ser D 266: 767-769
Liberman-Maxe M (1971) Etude cytologique de la différenciation des cellules criblées de *Polypodium vulgare* (Polypodiacée). J Microsc (Paris) 12: 271-288
Liberman-Maxe M (1978) La paroi des cellules criblées dans le phloème d'une Fougère, le Polypode. Biol Cell 31: 201-210
Loyal DS, Maheshwari AK (1979) Sieve cells vs sieve tube members in *Marsilea* - a reevaluation. Phytomorphology 29: 68-70
Maxe M (1964) Aspects infrastructuraux des cellules criblées de *Polypodium vulgare* (Polypodiacée). C R Acad Sci Paris Ser D 258: 5701-5704
Maxe M (1965) Aspects infrastructuraux des cellules du parenchyme phloémien de *Polypodium vulgare* (Polypodiacée). C R Acad Sci Paris Ser D 260: 5609-5612
Maxe M (1966) Etude de la dégénérescence nucléaire dans les cellules criblées de *Polypodium vulgare* (Polypodiacée). C R Acad Sci Paris Ser D 262: 2211-2214
Miller DJ, Duckett JG (1979) A study of stelar ultrastructure in the heterosporous water fern *Marsilea quadrifolia* L. Ann Bot (London) 44: 231-238
Palevitz BA, Newcomb EH (1971) The ultrastructure and development of tubular and crystalline P-protein in the sieve elements of certain papilionaceous legumes. Protoplasma 72: 399-426
Parthasarathy MV (1975) Sieve-element structure. In: Zimmermann MH, Milburn JA (eds) Encyclopedia of plant physiology, vol 1. Transport in plants. I. Phloem transport. Springer, Berlin Heidelberg New York, pp 1-38
Pate JS, Gunning BES (1969) Vascular transfer cells in angiosperm leaves. A taxonomic and morphological survey. Protoplasma 68: 135-156
Perry JW, Evert RF (1975) Structure and development of the sieve elements in *Psilotum nudum*. Am J Bot 62: 1038-1052
Poirault G (1893) Recherches anatomiques sur les Cryptogames vasculaires. Ann Sci Nat Bot Ser 7 18: 113-256
Sakai WS, Lamoureux CH (1973) Refractive spherules in phloem of *Microsorium scolopendria* and *Psilotum nudum*. Protoplasma 77: 221-229
Shah JJ, Fotedar RL (1974a) Sieve-tube members in the stem of *Cyathea gigantea*. Am Fern J 64: 27-28
Shah JJ, Fotedar RL (1974b) Structure and development of phloem in the rachis of *Pteris longifolia*. Phytomorphology 24: 107-113
Shah JJ, Nair MNB (1978) Nuclear autolysis in the young sieve cells of some ferns. Nucleus (Calcutta) 21: 161-168
Shah JJ, Fotedar RL, Nair MNB, Kothari IL (1976) Occurrence of branched sieve elements. Acta Bot Indica 4: 53-55
Srivastava LM (1970) The secondary phloem of *Austrobaileya scandens*. Can J Bot 48: 341-359
Tarchi AMF, Francalanci C (1973) Osservazioni sull'ultrastruttura delle cellule cribose di *Psilotum nudum* (L.) Beauv. Caryologia 26: 425-456
Want F (1963) The phloem of *Psilotum triquetrum*. Aust J Sci 25: 469
Warmbrodt RD, Evert RF (1974a) Structure and development of the sieve element in the stem of *Lycopodium lucidulum*. Am J Bot 61: 267-277
Warmbrodt RD, Evert RF (1974b) Structure of the vascular parenchyma in the stem of *Lycopodium lucidulum*. Am J Bot 61: 437-443

Warmbrodt RD, Evert RF (1978) Comparative leaf structure of six species of heterosporous ferns. Bot Gaz 139: 393-429
Warmbrodt RD, Evert RF (1979a) Comparative leaf structure of six species of eusporangiate and protoleptosporangiate ferns. Bot Gaz 140: 153-167
Warmbrodt RD, Evert RF (1979b) Comparative leaf structure of several species of homosporous leptosporangiate ferns. Am J Bot 66: 412-440
Wergin WP, Newcomb EH (1970) Formation and dispersal of crystalline P-protein in sieve elements of soybean (*Glycine max* L.). Protoplasma 71: 365-388

4 Conifers

ALEXANDER SCHULZ

4.1 Introduction

The Coniferophytina represent a very successful subdivision of the plant kingdom. While most of its families are spread over the entire northern hemisphere and often establish extremely large populations, some also occur in the southern hemisphere. Moreover, since the conifers include individuals of the bristlecone pine *(Pinus longaeva)* from the Californian White Mountains which are almost 5000 years old (according to tree-ring counts, cf. Ferguson 1968) and the giant Californian Redwood trees *(Sequoia sempervirens)* - which measure more than 110 m in height and over 11 m in diameter - this group also holds the longevity and size records among seed plants. Rapid growth and high yield are other factors specifically attributed to them which have led to the preference of the conifer softwood in commercial wood production.

All these features - but longevity and prodigious growth in particular - are largely dependent upon the existence of a permanent embryonic tissue, the vascular cambium, which for 100 and more years provides an annual increment of cells ready to specialize for water and food conduct. This developmental pattern establishes an unbroken continuum of less than 1-year-old vascular cells between the autotrophic, assimilate-producing green parts of a tree and its heterotrophic, water-absorbing roots.

The same developmental pattern likewise applies to *Ginkgo* (in this chapter included into the term "conifers" if not further specified), cycads (see Chap. 5) and dicotyledons (see Chap. 6), but palaeobotanic investigations showed that secondary (i. e., cambium-borne) sieve elements with sieve areas comparable to those of recent gymnosperms already occurred in the Pteridosperm *Medullosa* during the Carboniferous period (Smoot 1984), a long time before angiosperms appeared at the end of the Jurassic time (see also Chap. 14).

Compared to lower vascular plants, the phloem of the three subdivisions of seed plants is distinguished by an intimate relation between sieve elements and certain parenchyma cells (Esau 1969, p. 386), - a specialization which might have likewise contributed to the success of seed plants in evolution.

4.2 General Description

4.2.1 Primary and Secondary Phloem

During primary growth of seed plants the sieve elements originate in procambial strands and are often found to be the first cell type to specialize. If the organ is still elongating, the first developed (protophloem) sieve elements are obliterated in the course of extension growth. They are replaced by parallel-running cell files of the procambium which are successively induced to sieve-element differentiation, thus constituting the metaphloem. In plants with secondary growth the metaphloem is supplemented and later replaced by secondary, i. e., cambium-borne sieve elements.

Due to the early start of cambial activity in conifers there is a gradual transition of primary to secondary phloem, the latter being recognized in transverse sections by its regular scale-like pattern (Fig. 4.1; cf. Gourret and Strullu 1974; Neuberger and Evert 1974).

The surprisingly high regularity applies not only to the shape of cells, but also to the determination of cambial derivatives. Formed as a rhythmic succession of uniseriate bands, the alternation of cell types (e. g., fibers, sieve cells, parenchyma cells, sieve cells, fibers and so on) is particularly obvious in transverse sections of Taxaceae, Taxodiaceae, Cupressaceae and Podocarpaceae (Esau 1969). This predictable differentiation of cambial derivatives has been used to advantage in a physiologically oriented electron microscopic analysis of the ontogeny of secondary phloem in *Metasequoia glyptostroboides* (Kollmann and Schumacher 1961, 1962a, b, 1963, 1964).

Most of the ultrastructural studies reviewed in this article were concerned with the differentiation of secondary phloem. The few investigations that dealt with primary phloem avoided interference with early developed secondary phloem by using only 2-week-old pine hypocotyls (Neuberger and Evert 1976), vascular traces of sprouting needles (Singh 1984 a) or slowly growing, mycorrhizal rootlets (Warmbrodt and Eschrich 1985).

Conifers have a special tissue which is commonly related to the primary phloem but is not described for other vascular plants. This *"precursory phloem"* ("liber précurseur" after Chauveaud 1902 a, b) is found near the root apex and together with the first xylary elements makes up the stele. It is described by light microscopy as consisting of indistinctly differentiated cells which, in radial succession, exhibit increasingly more developed sieve areas (Esau 1969). The precursory phloem of *Ginkgo* contains, in addition, secretory cells (Chauveaud 1903). Ultrastructural studies of young roots of Pinidae revealed two cell types at the phloem poles: one located just inside the pericycle and commonly designated as phloem parenchyma, and the other representing immature sieve cells. The cells of the *"peripheral phloem parenchyma"* contain a high number of actin-filament bundles and have prominent wall thickenings where groups of plasmodesmata cross the wall (Pesacreta and Parthasarathy 1984; Warmbrodt and

Eschrich 1985). The immature *sieve cells,* confusingly named "sieve cell precursors" by Warmbrodt and Eschrich (1985), exhibit similar thickenings at the site of future sieve areas. After maturing all features of these sieve cells - e.g., pycnotic nucleus, plastids with starch grains and protein filaments, ER complexes and size of sieve pores (Pesacreta and Parthasarathy 1984; Warmbrodt and Eschrich 1985) - resemble those of primary sieve cells of pine hypocotyls (cf. Neuberger and Evert 1976), and clearly contrast with the peripheral phloem parenchyma.

A possible function of peripheral phloem parenchyma, as accessory to the first sieve cells, might be derived from its close spatial association, from the slightly increased diameter of the plasmodesmata between the individual parenchyma cells (Warmbrodt and Eschrich 1985), and the simultaneous obliteration of both cell types during later stages of stelar development. The relation between the two cell types is, however, much less pronounced than that between Strasburger cells and sieve elements, the latter being structurally characterized by specific plasmatic connections (see Sect. 4.4).

Since protophloem and peripheral phloem parenchyma are easily discriminated on the basis of their ultrastructure, it might be proper to either omit the term "precursory phloem" entirely, or - in agreement with its literal meaning - restrict it to the undifferentiated tissue portion of phloem poles in the stele of conifer roots. Judged from its ultrastructure, however, this tissue could as well be labeled as procambium (cf. Fig. 1 in Pesacreta and Parthasarathy 1984).

4.2.2 Shape and Size of Sieve Cells

The shape of the conifer sieve cells is very uniform. The narrow, spindle-like cells may reach a length of up to 5000 μm (Esau 1969), while their diameter is in the range of 5 to 50 μm. Due to the increments of cambial activity, secondary sieve cells seen in cross-sections generally show a rectangular outline (Fig. 4.1). Their radial walls bear a series of pit-like sieve areas (Figs. 4.2, 4.20), each consisting of groups of sieve pores (Fig. 4.18). Although the longitudinal contour of sieve elements is very straight, it is difficult to discriminate between lateral walls and the long-tapering end walls. The latter can be recognized, however, by the higher density of sieve areas (Fig. 4.20; cf. Esau 1969). Sieve areas in lateral and end walls have the same degree of specialization; this is one criterion for the definition of gymnospermous sieve cells, and contrasts with angiospermous sieve elements where lateral sieve areas are comparatively less specialized (Esau 1969; cf. Behnke 1986).

Within the Coniferophytina the Pinaceae is the only family which has sieve cells with conspicuous, nacré-like secondary walls (Fig. 4.17; cf. Esau 1969; Srivastava 1969) and lacks phloem fibers. The physical support given by fibers present in the other families (Cupressaceae, Podocarpaceae, Taxaceae, Taxodiaceae and *Ginkgo;* Esau 1969) might, in Pinaceae, be taken over by the development of secondary walls in the sieve cells.

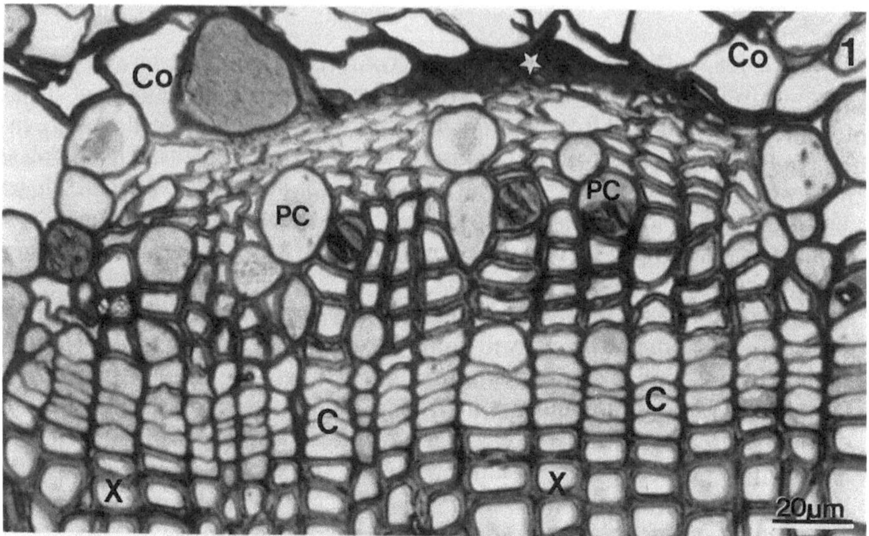
1
Co
Co
PC
PC
C
C
X
X
20μm

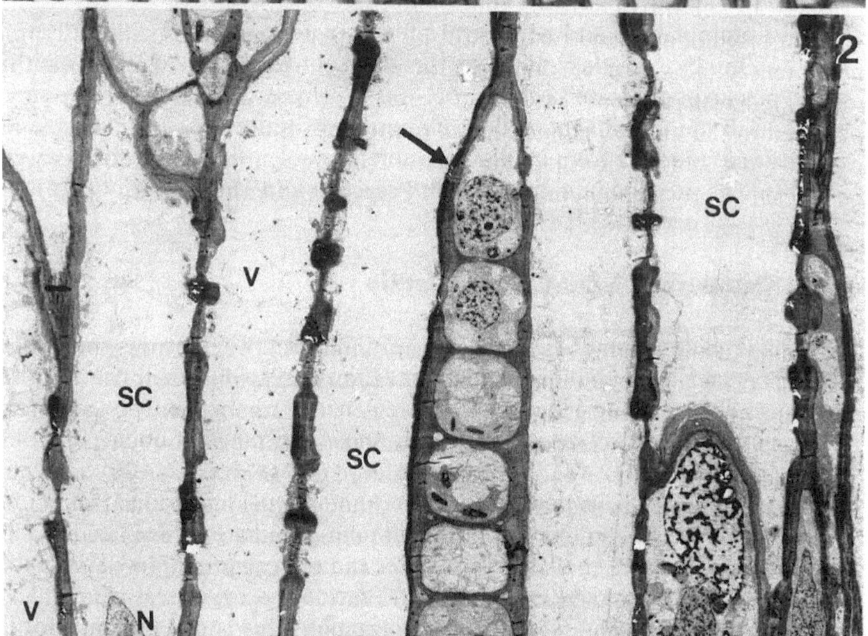
2
SC
SC
SC
V
V
N
TC
20μm

In the Pinaceae an increase in wall thickness may be taken as the first character for the recognition of future secondary sieve elements (Figs. 4.1, 4.6). This is also valid for primary phloem in hypocotyls (Neuberger and Evert 1976) and young needle traces (Singh 1984 a). The sieve cells of mycorrhizal rootlets, however, had only slightly thickened walls which were not reminiscent of nacreous walls at any developmental stage (Warmbrodt and Eschrich 1985).

Correspondingly, the appearance of sieve areas in Pinaceae depends upon the relative thickness of the wall they are part of: a sieve area may be a prominent structure as in primary phloem of roots, hypocotyls and needle traces (Neuberger and Evert 1976; Warmbrodt and Eschrich 1985; Singh 1984 a; Schulz and Behnke 1987), or may form pit-like depressions in the secondary phloem (Figs. 4.2, 4.3). In the latter, the secondary wall develops borders around each sieve area (Esau 1969; Srivastava 1969), resembling bordered pits.

4.3 Development of the Sieve Cell

The life of an individual sieve cell - i. e., its differentiation, function and death - starts with the completion of mitosis of its procambial or cambial precursor. While the sister cell may or may not continue dividing, the fate of the future sieve element, the most specialized cell type of conifers, is definitely determined. A succession of structural changes, i. e., the *differentiation*, characterized by a drastic reorganization of the cytoplasm and by the formation of intimate symplastic connections to neighboring cells [sieve cells and Strasburger (= albuminous) cells], leads to a stage that most investigators describe as maturity and assume that it represents the state of *function*, i. e., the capability of sieve cells to conduct assimilates.

However, the inadequate methods used for electron microscopic fixation and preparation do not provide an answer to the question of the in-vivo structure of translocating sieve cells, although this has been the subject of continuous efforts.

The final cessation of sieve-element function is clearly marked by the *death* of the protoplast, visible in the breakdown of the plasmalemma, obliteration and final autolysis of the cells. The crushed outlines of closely layered cell walls are the only remnants of nonfunctional sieve cells (Fig. 4.1; cf. Esau 1969).

◁

Fig. 4.1. Cross-section through the vascular bundle of a half-year-old branch with regular secondary phloem above the thin-walled cambium *(C)*. Obliterated primary phloem at *. *PC* parenchyma cells; *Co* cortex; *X* xylem. *Picea abies*, collected Dec. 85, semithin section. × 540.
Fig. 4.2. Tangential view of immature phloem from the main root of *Picea abies*. Sieve cells *(SC)* still contain a nucleus *(N)* and numerous spherical vacuoles *(V)*; thin wall portions are developing sieve areas. A Strasburger cell *(arrow)* is located at the margin of a uniseriate, cross-sectioned ray. *TC* Parenchyma cell with phenolic contents in the vacuole. × 470

The variation in the differentiation of sieve cells is small. This applies equally to the comparison of primary and secondary sieve cells, to sieve cells which overwinter in an intermediate stage (Kollmann and Schumacher 1964; Alfieri and Evert 1968 b; Esau 1969), and to sieve cells in different species of the Coniferophytina. This review deals, therefore, with changes of cytoplasm and cell walls which can be generalized from various ultrastructural investigations. Aberrations from the general way of differentiation will be indicated.

4.3.1 The Nucleus

During sieve-cell differentiation the nucleus changes quite remarkably from an appearance common to meristematic cells (Fig. 4.6) to an often ovoid, electron-dense mass of condensed chromatin (Figs. 4.4, 4.17) persisting within the mature sieve cell.

There are two types of nuclear disintegration in sieve elements, referred to as *pycnosis* and *chromatolysis*. Both terms are related to the *final* fate of chromatin which is either aggregated into a dense mass or disappears completely. In gymnosperms only pycnosis has been recorded (cf. Behnke 1986). It is confusing that the chromatolysis of angiosperms includes a first step of chromatin condensation to which the term pycnosis was originally assigned (Salmon 1946). In contrast to this primary event, in gymnosperms the condensation of chromatin is accompanied by a reduction of the karyoplasm (Wooding 1966; Behnke 1986) and the persistence of pycnotic nuclei in mature sieve cells (Wooding 1966; Neuberger and Evert 1974, 1976), the latter being described also with the light microscope (Evert et al. 1970). It is significant that the pycnotic chromatin of mature sieve cells still gives strongly positive reaction with the DNA-specific DAPI-dye (Schulz and Behnke 1987), while, due to successive chromatolysis, this dye shows decreasing nuclear fluorescence in a developing protophloem-sieve tube of the angiosperm *Pisum* (Schulz 1987). There is, however, no doubt that transcriptional activity stops at sieve-cell maturity, since tritiated uridine is not incorporated into pycnotic nuclei (Hébant 1975).

In early stages of sieve-cell differentiation the nucleus shows a wavy outline (Kollmann and Schumacher 1961) and lobing (Fig. 4.3; Srivastava and O'Brien

Fig. 4.3. Sieve cell of the trunk of *Abies alba* with a lobed *(arrow)* nucleus *(N)* that shows an even distribution of granular chromatin. ER is aggregated among spherical vacuoles *(V)*. The pit-like wall depressions are future sieve areas. *P* sieve-element plastids. × 3500. **Fig. 4.4.** Mature sieve cell of a branch of *Picea abies* with pycnotic nucleus *(PC)* and a plastid *(P)* within the electron-lucent lumen. × 3500. **Fig. 4.5.** Nuclear pores *(arrows)* indicate persistence of the nuclear envelope which covers the condensed chromatin of a pycnotic nucleus. *Abies alba*, × 40000. **Fig. 4.6.** Cross-section through narrow sieve cells of a branch of *Abies alba* with various stages of maturing sieve cells. Young sieve cells *(upper left)* have nuclei *(N)* with small chromatin clumps, the mature one shows a pycnotic nucleus *(PN)*. *W* thick secondary walls, *arrow* points to a nucleolus. × 3500

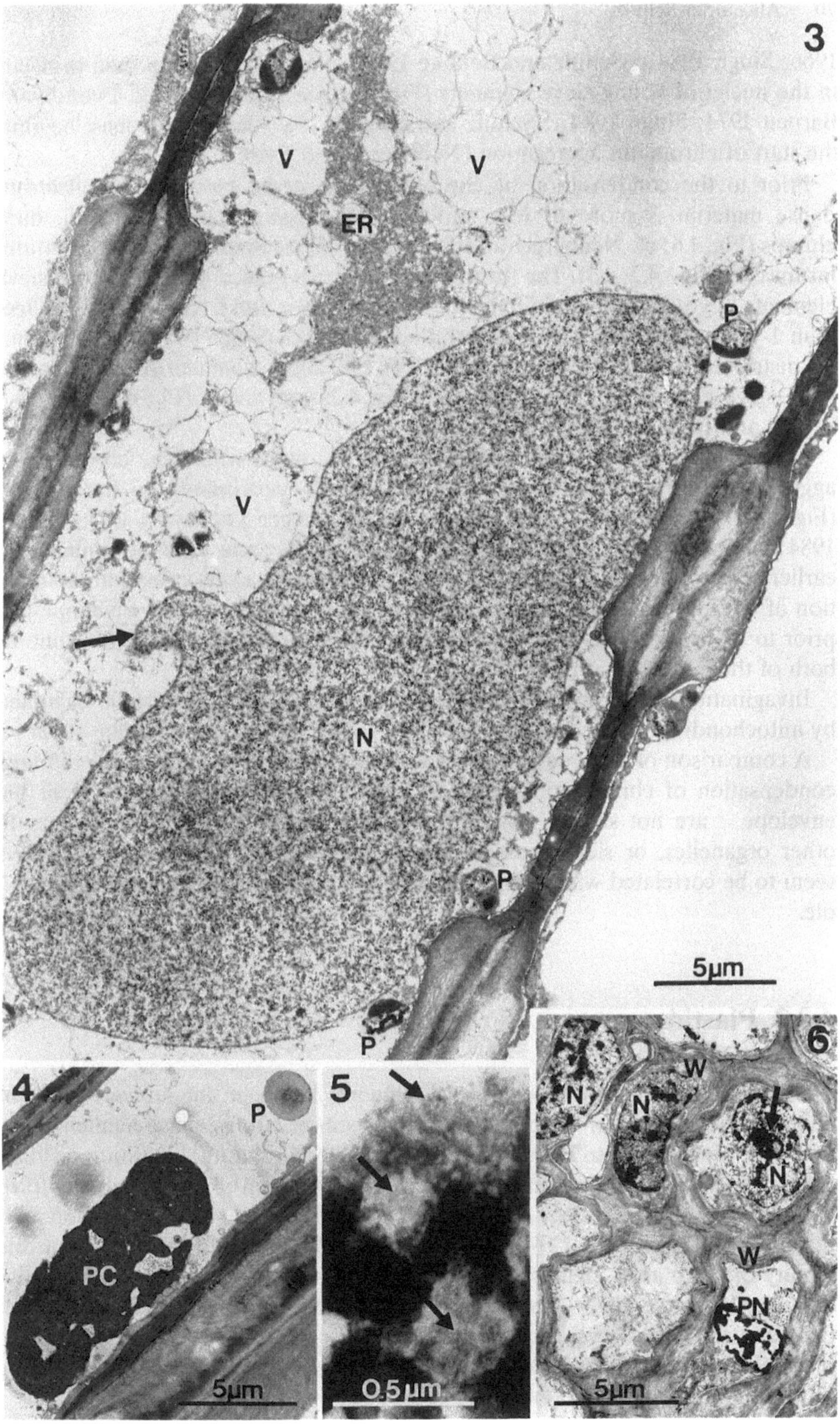
3
V
V
ER
V
P
N
P
P
5µm
4
P
PC
5µm
5
0.5µm
6
N
N
W
N
W
PN
5µm

1966; Singh 1984 a; Schulz and Behnke 1987). Nucleoli were reported to occur in the nuclei of young sieve elements (Fig. 4.6; see Murmanis and Evert 1966; Barnett 1974; Singh 1984 a; Schulz and Behnke 1987) and may persist beyond the start of chromatin aggregation (Neuberger and Evert 1976).

Prior to the condensation of chromatin into larger portions this electron-dense material is more or less uniformly distributed, either in small, dark clumps (Fig. 4.6; cf. Neuberger and Evert 1974, 1976; Singh 1984 a) or as granular masses (Figs. 4.2, 4.3). The former appearance is typical for the narrow sieve elements of young phloem in hypocotyls (Neuberger and Evert 1974, 1976), less than 1-year-old branches (Fig. 4.6) and needle traces (Singh 1984 a). The granular distribution seems to be correlated with the larger diameter of nuclei in the wide sieve elements of the main root (Fig. 4.2) and trunk (Fig. 4.3) (see also Schulz and Behnke 1987).

The nuclear envelope may persist even to the stage where the chromatin is aggregated into a single mass and only partly penetrated by karyoplasm (Fig. 4.5 shows typical nuclear pores; cf. Neuberger and Evert 1976; Singh 1984 a; Behnke 1986), but it was also reported to become discontinuous at an earlier stage (Neuberger and Evert 1974). Singh (1984 a) described an association of ER tubules and noncortical microtubules with the nuclear envelope just prior to its breakdown, which is suggestive of a participation of either one or both of these in its disruption.

Invaginations of the pycnotic nuclear mass were often found to be occupied by mitochondria or other organelles (Singh 1984 a; Schulz and Behnke 1987).

A comparison of different reports reveals that the stages of pycnosis - lobing, condensation of chromatin, disappearance of nucleoli and breakdown of the envelope - are not strictly synchronized with each other or with changes in other organelles, or sieve areas. Fully widened sieve pores, however, always seem to be correlated with completed pycnosis and disappearance of the vacuole.

4.3.2 Plastids

The fate of plastids in maturing sieve elements has been the subject of major controversies. Since, in electron microscopic observations, sieve-element plastids were often found to be broken, it was concluded that the liberation of their specific inclusions is a natural event during maturation of sieve elements (Srivastava and O'Brien 1966; Parameswaran 1971; Harris 1972; Gourret and Strullu 1974; Murmanis 1974). It is now generally accepted, however, that the lability of the plastid membranes during fixation is the reason for their breakdown (cf. Behnke 1974; Neuberger and Evert 1974). The detection of intact plastids with typical inclusions in obliterating sieve cells already devoid of a plasmalemma (Schulz and Behnke 1987) further corroborates that the bursting of plastids is not a natural event.

The inclusions of sieve-element plastids are of diagnostic value for the classification of conifers. *Ginkgo,* Cephalotaxaceae, Cupressaceae, Taxaceae and Taxodiaceae contain S-type plastids (e.g., Figs. 4.7, 4.8; *Thuja plicata*), while Pinaceae contain P-type plastids (Fig. 4.9: *Pseudotsuga menziesii,* Figs. 4.10, 4.13: *Abies alba,* Figs. 4.14–4.16: *Picea abies*) (cf. Behnke 1974).

The differentiation of sieve-element plastids includes changes in shape, gradual dissolution of the matrix, and synthesis of starch grains and, in Pinaceae, of two forms of protein inclusions.

Derived from procambial or cambial proplastids with plastoglobuli, thylakoids and sometimes starch grains (Neuberger and Evert 1974; Barnett 1974), the plastids of very young sieve cells are still variable in shape, often elongated (Fig. 4.14), amoeboid (Fig. 4.16) or ovoid (Fig. 4.13) (see also Wooding 1968; Barnett 1974; Neuberger and Evert 1974). In Pinaceae, protein filaments and crystals are the first protoplasmic characters which allow the recognition of sieve cells, otherwise only detected by an early increase of wall thickness (Neuberger and Evert 1974; Singh 1984a). The matrix of these young plastids is still rather dense and may contain vesiculated thylakoids (Fig. 4.14; cf. Srivastava and O'Brien 1966; Schulz and Behnke 1987). During further differentiation the plastids become round (Figs. 4.7, 4.8), loose their thylakoids (= "lamellae" of Neuberger and Evert 1974) and obtain an electron-lucent matrix (Fig. 4.15). In all conifers, starch grains - the synthesis of which occurs before the plastids get a spherical shape - tend to change from a club-shaped form into small granules (Figs. 4.9–4.11; cf. Behnke 1974; Neuberger and Evert 1974); they typically stain red (Esau 1969, p. 335), or reddish blue (Wooding 1968) with iodine. In well-preserved, mature sieve elements all plastid inclusions remain enclosed by the intact double membrane (Figs. 4.9, 4.15; see also Behnke 1974; Neuberger and Evert 1974; Warmbrodt and Eschrich 1985; Schulz and Behnke 1987).

The protein inclusions of the P-type plastids of Pinaceae consists of filaments and crystals, both found in all Pinaceae except for *Tsuga canadensis* and *Larix decidua* where only filaments have been depicted so far (Figs. 4.10, 4.11, 4.14; cf. Behnke 1974; Murmanis and Evert 1966; Srivastava and O'Brien 1966; Parameswaran 1971; Timell 1973; Barnett 1974; Neuberger and Evert 1974; Schulz and Behnke 1987).

In both, young and mature, sieve elements the plastid filaments are arranged parallel to each other and form a ring-like structure, which, depending upon the angle of sectioning, are depicted as incomplete annulae (Fig. 4.10) or two transversely cut bundles (Figs. 4.9, 4.15). Diameter measurements of single filaments are reported to be 4 to 8 nm (Parameswaran 1971) or 5.5 to 6.5 nm (Neuberger and Evert 1974) for different pine species.

As a result of a detailed analysis of the differentiation of sieve-cell plastids in *Pseudotsuga menziesii,* Gourret and Strullu (1974) described the formation of filament bundles - in mature plastids similarly recognized as bi-dissected or arc-like parts of rings - from rigid, rod-like and in cross-sections hexagonal protein inclusions. These rods, never detected by the present author in young plas-

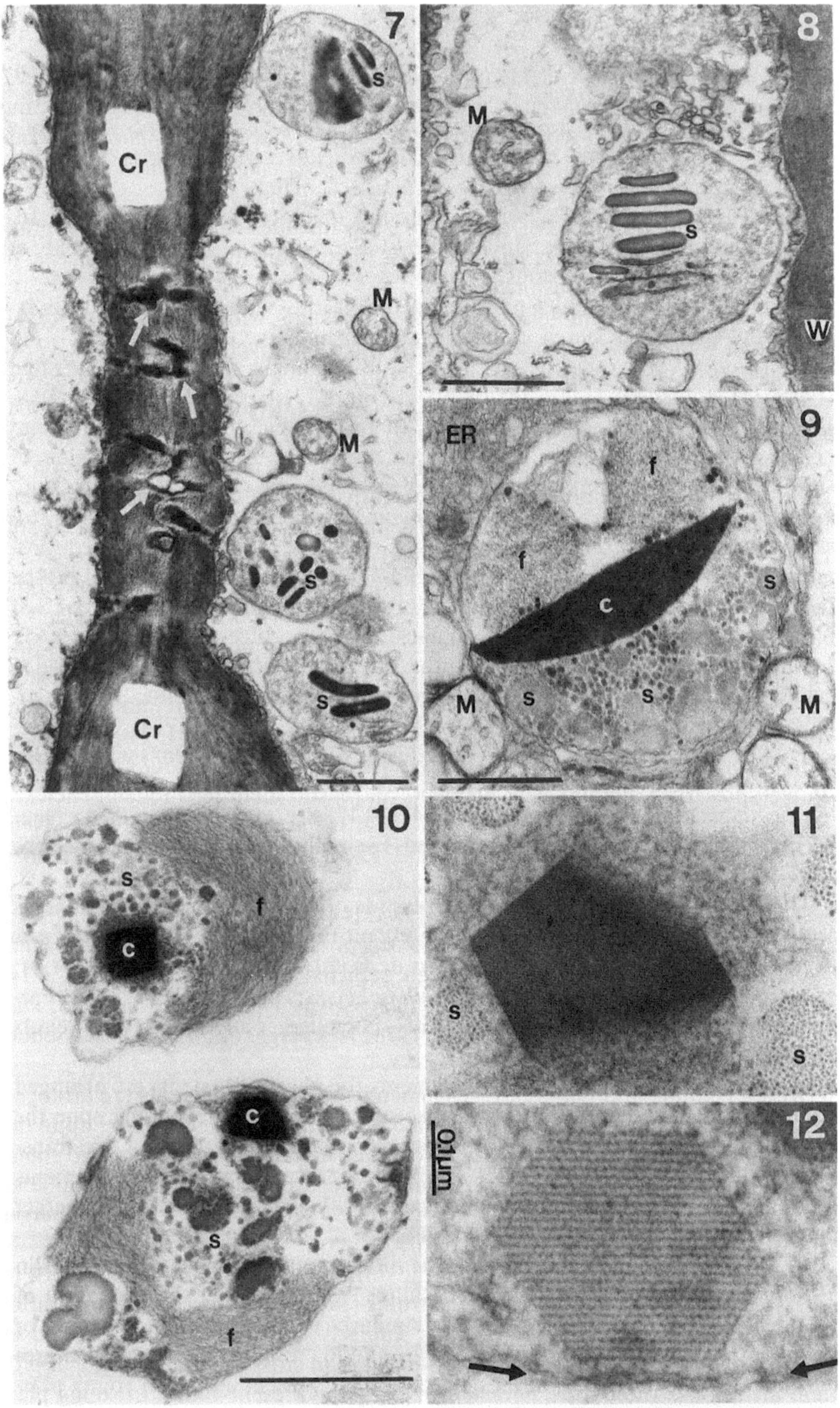
7
Cr
s
M
M
s
s
Cr
8
M
s
W
9
ER
f
f
c
s
M
s
s
M
10
s
f
c
c
s
f
11
s
s
12
0.1µm

tids of *Abies* and *Picea* nor described for *Pinus* species (Srivastava and O'Brien 1966; Wooding 1966, 1968; Neuberger and Evert 1974, 1976), might be peculiar to *Pseudotsuga menziesii*.

Electron-dense crystals occur in plastids of primary and secondary sieve cells except for primary phloem of hypocotyls (Neuberger and Evert 1976) and mycorrhizal roots (Warmbrodt and Eschrich 1985). The number of crystals per plastid and the density of their typical striations (Figs. 4.11, 4.12), however, seem to be different in different species. If a three-dimensional reconstruction is attempted from crystal outlines depicted in various ultrathin-section, the protein inclusions in *Abies, Pinus, Picea* and *Cedrus* appear to be polygonal bodies (cf. Figs. 4.10–4.14, 4.16 and literature cited above), while in *Pseudotsuga* they form an ellipsoid to lens-shaped body (Fig. 4.9; see also Gourret and Strullu 1974). Center-to-center measurements of the distances between the electron-dense striations within protein crystals of *Abies* (Fig. 4.11) are reported to be 3.5 to 4 nm, for *Cedrus* 10 to 12 nm (Behnke 1974), for *Pinus* 6 to 6.5 nm (Neuberger and Evert 1974) and for *Pseudotsuga* 6 nm (Gourret and Strullu 1974). In addition to a crystal with dense striations, sieve-element plastids of *Abies* contain a second crystal with a spacing of 12–15 nm (Fig. 4.12).

A biochemical comparison of the proteins from phloem and xylem extracts in *Pinus sabiana* revealed a phloem-specific protein composed of two 24 to 25 kdalton polypeptides (Alosi and Park 1983). Immuno-cytochemical experiments using antibodies against these proteins and sections of different species of Pinaceae indicated their localization in the plastid crystal (Schulz et al. 1989).

4.3.3 Mitochondria

Mitochondria do not show conspicuous changes during sieve-cell differentiation. In young sieve cells they are elongated, contain a dense matrix and several cristae and are distributed over the entire protoplast (Fig. 4.21; cf. Neuberger and Evert 1974). During cell differentiation they gradually lose their matrix (cf.

◁

Fig. 4.7. Sieve area of *Thuja plicata* with widening sieve pores interlinked by small median cavities *(arrows)*. The spherical S-type plastids contain club-shaped starch grains *(s)*. *M* mitochondria; *Cr* wall crystals. × 13000. (From Behnke 1974.) **Fig. 4.8.** Sieve-element plastid with starch grains *(s)* in granular matrix. *M* mitochondrion; *W* wall. *Thuja plicata*, × 18000. (Courtesy of H-D Behnke.) **Fig. 4.9.** P-type plastid of *Pseudotsuga menziesii* with dense, lens-shaped protein crystal *(c)*, bi-dissected filament bundles *(f)* and particulated starch grains *(s)* in contact with ER complex and mitochondria *(M)*. × 18000. (From Behnke 1974.) **Fig. 4.10.** P-type plastids of *Abies alba* with polygonal protein crystal *(c)*, partly sectioned rings of filaments *(f)* and particulated starch grains *(s)* in mature sieve cell. Remnants of the plastidal matrix around crystal. × 25000. **Fig. 4.11.** Dense crystal of *Abies alba* with fine striations among small starch particles *(s)*. × 100000 (bar line in Fig. 4.12). (From Schulz and Behnke 1987.) **Fig. 4.12.** Less densely striated crystal of *Abies alba* close to the double plastid membrane *(arrows)*. × 100000. (From Schulz and Behnke 1987.)

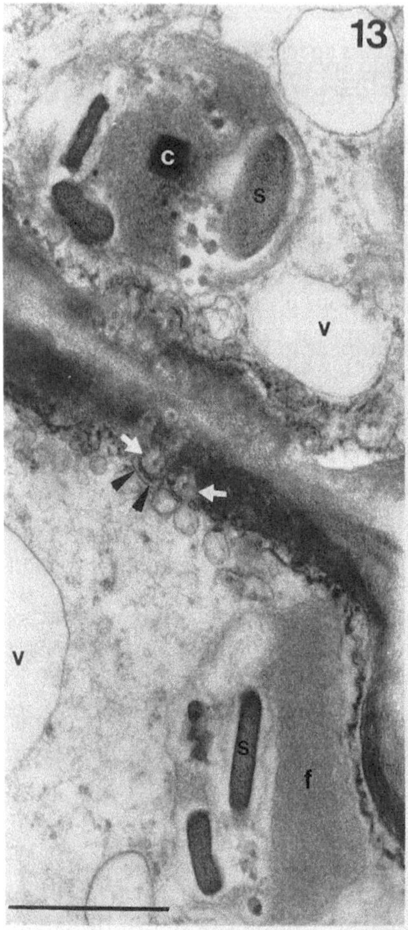
13
c
s
v
v
s
f

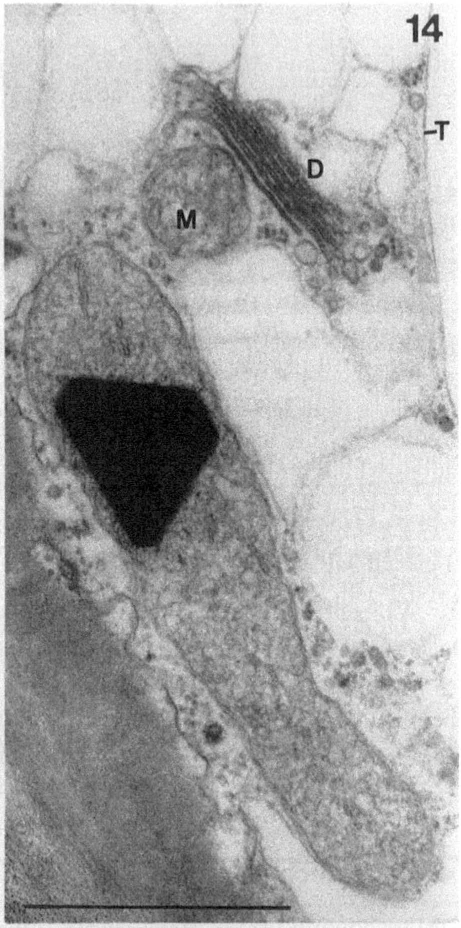
14
T
D
M

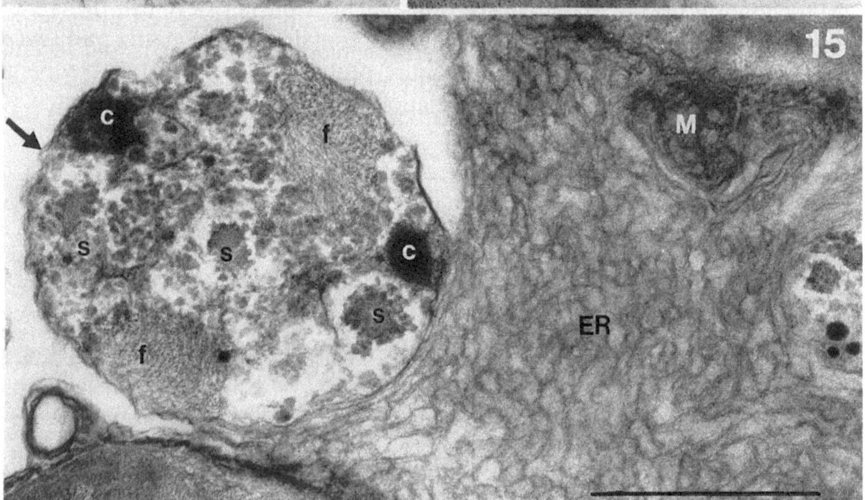
15
c
f
M
s
s
c
s
f
ER

Fig. 4.21 with Fig. 4.9) and start to become round much as the plastids do. Mitochondria of mature elements are spherical and obtain a parietal distribution (Fig. 4.15).

4.3.4 Dictyosomes

Young sieve elements contain numerous dictyosomes; the number of vesicles around them indicate a high degree of activity (Figs. 4.14, 4.21; see also Neuberger and Evert 1976; Singh 1984 a). The undulating contour of the plasmalemma in young sieve elements (Singh 1984 a) as well as numerous profiles of vesicles fusing with the plasmalemma (Neuberger and Evert 1974) support the view that the main activity of the dictyosomes is concerned with wall deposition.

In later stages of differentiation the entire Golgi apparatus disappears, including dictyosomes and coated and uncoated vesicles (Singh 1984 a). This disappearance apparently coincides with the end of the increase in wall thickness and the breakdown of the vacuole (Kollmann and Schumacher 1964). According to Singh (1984 a) the disappearance of dictyosomes might be effected by a loosening of intercisternal bonds.

4.3.5 Endoplasmic Reticulum (ER) and Ribosomes

Besides nuclear pycnosis and plastid differentiation, changes of the ER are the most impressive feature within the maturing sieve cell. At the start of this process rough ER is depicted in electron micrographs as single cisternae (Fig. 4.26; Neuberger and Evert 1974, 1976; Singh 1984 a). Coincident with the loss of ribosomes, the ER cisternae change to a branched (Figs. 4.16, 4.21; cf. Wooding 1966; Singh 1984 a; Schulz and Behnke 1987) or tubular form (Fig. 4.22; Kollmann and Schumacher 1962 b). Subsequently, these tubules are often found to be oriented parallel to the long axis of sieve cells (Fig. 4.23; see Kollmann and Schumacher 1962 b). The lability of the ER during preparation and fixation may lead to the vesiculation of tubules (Kollmann and Schumacher 1964). This was also discussed by Neuberger and Evert (1974) who concluded from their

◁

Fig. 4.13. Spheroid and ovoid plastids of *Abies alba* with starch *(s)*, filaments *(f)* and crystal *(c)* at future sieve area of two immature sieve cells. *Arrowheads* ER close to callose deposits *(arrows)* that surround plasmodesmata; *V* vacuoles. × 22000. **Fig. 4.14.** Parietal cytoplasm of a very young sieve cell of *Picea abies* exhibiting a dictyosome *(D)* with vesicles, a mitochondrion *(M)* and an elongated plastid with a conspicuous protein crystal. The thylakoidal membranes appear vesiculated. *T* tonoplast. × 37000. **Fig. 4.15.** Parietal cytoplasm of a mature sieve element of *Picea abies* containing a complex network of smooth ER cisternae and a plastid with particulated starch grains *(s)*, crystals *(c)* and two profiles of filament bundles *(f)*. *Arrow* points to the intact double membrane of the plastid; *M* mitochondrion. × 32000

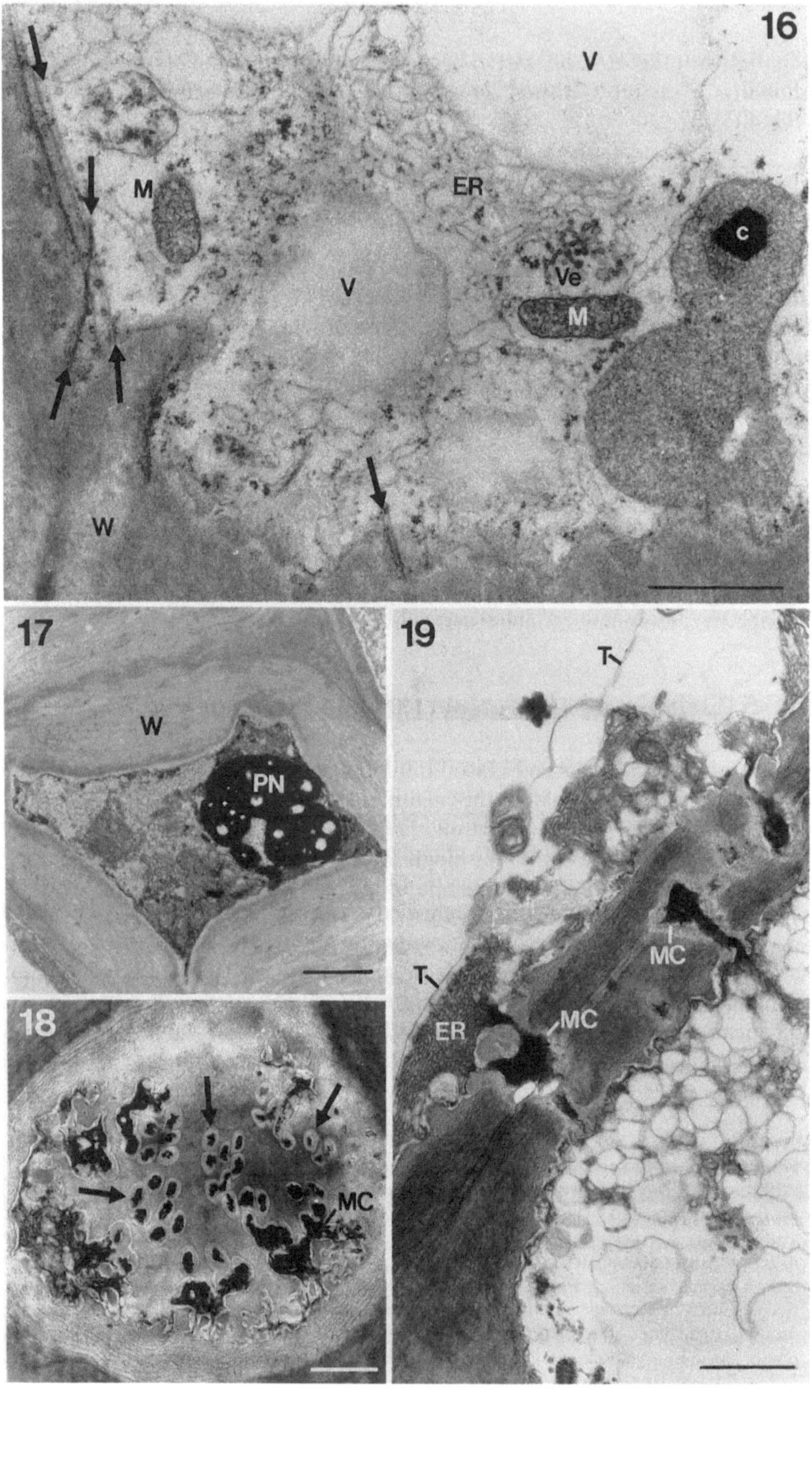
16
V
M
ER
c
V
Ve
M
W
17
W
PN
18
MC
19
T
MC
T
MC
ER

results that "the original, rough ER of the young sieve cells is incorporated into the extensive, newly-formed system of smooth, tubular ER".

Smooth ER-tubules tend to contact each other (Figs. 4.22, 4.23), forming complex networks at sieve areas (Fig. 4.25; see Wooding 1966; Neuberger and Evert 1974, 1976) and also at other regions (Figs. 4.3, 4.24; cf. Neuberger and Evert 1976; Schulz and Behnke 1987). In the mature sieve cell most ER tubules were found in groups and at a parietal position, rarely as separate units (Wooding 1966; Schulz and Behnke 1987). A peculiar feature of petiolar sieve cells of *Ginkgo* seems to be a stacking of the ER in concentric layers (Dute 1983). A role of the ER in the breakdown of the nuclear envelope was discussed by Singh (1984 a).

ER complexes appear at sieve areas prior to the dissolution of the tonoplast and are coincident with the first indications of a widening of the sieve pores (Figs. 4.19, 4.27; see also Fig. 7 in Neuberger and Evert 1975). They persist at the sieve areas and at parietal positions throughout the life span of a mature sieve cell (Figs. 4.25, 4.28; see also Kollmann and Schumacher 1962 a, 1963; Wooding 1968; Neuberger and Evert 1974, 1975, 1976). Some tubules of the ER run through the sieve-pore lumina into the median cavity (see below). In this region the ER is in contact with that of the neighboring sieve cell (Fig. 4.28; cf. Kollmann and Schumacher 1962 a, 1963; Wooding 1968; Neuberger and Evert 1975). On the sieve-cell side, the sieve areas linked to the Strasburger cells are also provided with extensive ER complexes (Figs. 4.31, 4.32; cf. Neuberger and Evert 1975; Sauter et al. 1976).

The in-vivo location of the ER complexes at sieve areas of functional sieve cells has often been questioned. Due to the loss of osmotic pressure during EM preparation, ER profiles could be shifted onto the sieve areas, thus blocking the pores (cf. Wooding 1966; Warmbrodt and Eschrich 1985). The investigation of *living* sieve cells with an ER-specific, vital dye and with a newly developed confocal laser-scanning microscope (see Schulz 1988) revealed evidence that ER is symmetrically distributed on either side of sieve areas. This is true for sieve areas in end walls as for those in lateral walls (Fig. 4.20). In contrast, a wound-induced transfer should place ER complexes toward one side of the sieve areas only.

Fig. 4.16. Microtubules *(arrows)* are associated with a portion of the secondary wall which borders the pit-like depression of a sieve area *(below this section)*. The cytoplasm includes an amoeboid plastid with a protein crystal *(c)*, elongated mitochondria *(M)* and branching ER cisternae in contact with few ribosomes. Dictyosome vesicles at *Ve*. *V* vacuoles; *W* wall. *Picea abies*, main root. × 20000. **Fig. 4.17.** Cross-section through a sieve cell of a needle trace of *Picea abies* with a pycnotic nucleus *(PN)* and a conspicuous secondary wall *(W)*. × 10000. **Fig. 4.18.** Glancing section through a sieve area with numerous, callose-lined sieve pores *(arrows)* and portions of the median cavity *(MC)*. *Picea abies*, × 10000. **Fig. 4.19.** Developing sieve area with widened median cavities *(MC)* that are connected by sieve pores to two sieve cells. The smooth ER complex of the left sieve cell is bordered by a tonoplast *(T)*, while the vesiculation of the ER in the right cell may already be a wound response. *Picea abies*, × 14000

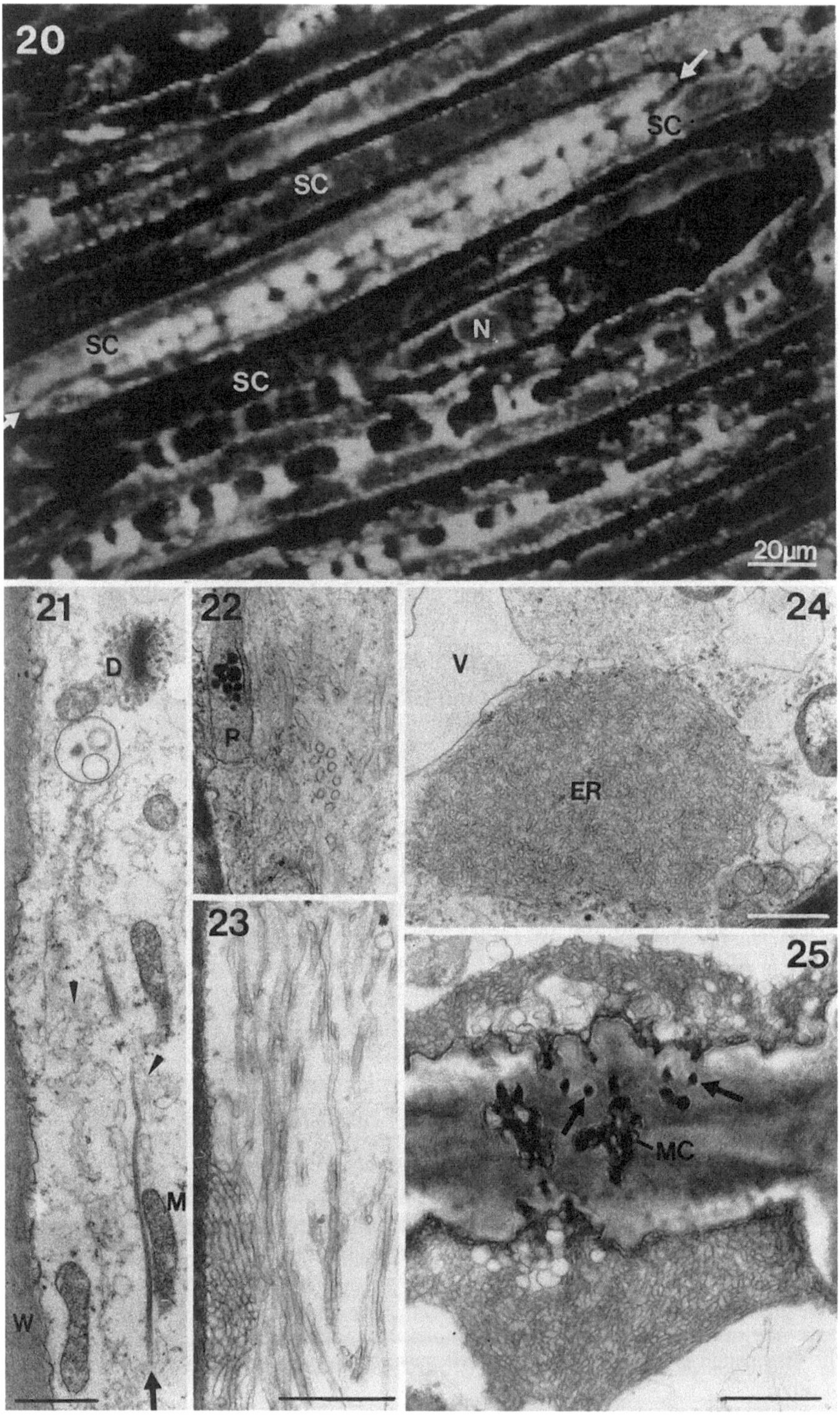
20
SC
SC
SC
SC
SC
N
20µm
21
D
M
W
22
P
23
24
V
ER
25
MC

Ribosomes occur bound to the rough ER cisternae of young sieve cells and as free helical polysomes (Neuberger and Evert 1974). Liberation of ribosomes from the ER was described to be accompanied by their aggregation in the ground plasm (Wooding 1966). Subsequently, the number of ribosomes diminishes until eventually they can no longer be detected in evacuolate, mature sieve cells (e.g., Figs. 4.15, 4.23, 4.28).

4.3.6 Vacuole and Ground Plasm

During the early stages of sieve-cell differentiation the vacuoles occupy a considerable part of the cytoplasm (Fig. 4.3; cf. Singh 1984a; Neuberger and Evert 1974). In the wide and elongated sieve elements found in the trunk and main roots, the vacuoles occur in the form of densely packed spheres about 10 to 20 μm in diameter and are surrounded by ground plasm and organelles (Figs. 4.2, 4.3). Although relations between the degeneration of the tonoplast and other events in differentiation are not known, it may be pertinent that this event takes place *after* the condensation of chromatin, appearance of ER-complexes and the start of sieve-pore widening (see also Kollmann and Schumacher 1964).

Since dictyosomes, vesicles, ribosomes and cytoskeletal proteins disappear during differentiation, the ground plasm of the mature sieve cells is electron-translucent and their cytoplasm is devoid of any structures except for remnants of pycnotic nuclei, plastids, mitochondria and ER. In addition, lipid droplets occur in the sieve-cell cytoplasm in cotyledons of *Pinus radiata* (Singh 1984b) and hypocotyls of *Pinus resinosa* (Neuberger and Evert 1974).

Fig. 4.20. *Picea abies*, trunk phloem. Tangential, optical section through living sieve cells obtained by confocal laser scanning microscopy. The vital, ER-specific fluorochrom DiOC is able to delineate the symmetrical distribution of ER complexes attached to sieve areas of end walls (e.g., *between arrows*) as well as lateral wall parts. The cell walls proper remain unstained. *N* nucleus in a Strasburger cell; *SC* sieve cells. × 500. **Fig. 4.21.** Microfilament bundle *(arrow)* in close contact with an elongated mitochondrion *(M)*. The dictyosome *(D)* is surrounded by numerous vesicles. Branching of the ER is indicated by *arrowheads*. *W* wall. *Picea abies*, × 12000. **Fig. 4.22.** Longitudinally and transversely cut smooth ER tubules of an immature sieve cell in contact with each other and with a still elongated plastid *(P)* with plastoglobuli. *Picea abies*, × 16000. (From Schulz and Behnke 1987.) **Fig. 4.23.** Mature sieve cell of *Picea abies* with smooth ER tubules, partly parallel arranged to the cell axis, partly forming a network. × 16000. (From Schulz and Behnke 1987.) **Fig. 4.24.** Immature sieve cell of *Picea abies* with extensive ER network. *V* vacuole. × 12000. **Fig. 4.25.** Mature sieve area with smooth ER complexes apposed to both sides, callose-lined sieve pores *(arrows)* and a compound median cavity *(MC)*. *Picea abies*, × 14000. (From Schulz and Behnke 1987.)

4.3.7 Structural Proteins

Cortical microtubules were recorded in younger stages of sieve cells (Fig. 4.16; cf. Srivastava and O'Brien 1966; Parameswaran 1971; Neuberger and Evert 1974; Singh 1984 a). They were often found around the border of a developing sieve area (Fig. 4.16; Schulz and Behnke 1987). This situation suggests a participation of microtubules in delimiting the deposition of secondary wall material upon sieve areas. Microtubules disappear at about the same time as the vacuole, but they were found just prior to this event deeper in the cytoplasm and close to the pycnotic nucleus (Singh 1984 a). A role of microtubules in the breakdown of the nuclear envelope was discussed by Singh (1984 a).

The large size of coniferous sieve cells (with lengths of up to 5 mm) requires a system which facilitates intracellular transport and coordinates metabolism and wall deposition. Accordingly, during the phase of high Golgi activity the sieve cells show bundles of microfilaments alongside the axis of the cell (Fig. 4.21; cf. Parthasarathy and Pesacreta 1980; Pesacreta and Parthasarathy 1984; Singh 1984 b). Associations between these microfilament bundles and organelles are frequent (Fig. 4.21: mitochondrion; Singh 1984 b; see also Fig. 5 in Wooding 1966).

During the first decade of ultrastructural description of gymnosperm sieve cells, the presence and nature of P-protein (or slime) were a source of major disputes. Pycnotic nuclei, plastid filaments as well as badly preserved ER were all considered to be forms of P-protein. It has been known, however, since the early 1970's, that the *lack of P-protein* is a consistent feature in all gymnosperms (cf. Behnke and Paliwal 1973; Neuberger and Evert 1974; Behnke 1983; Evert 1984). The absence of P-protein is of classificatory value for the delimitation of gymnosperm sieve cells from angiosperm sieve-tube members (Behnke 1986).

4.3.8 The Wall

Unlike those in other conifers, the sieve elements of Pinaceae have conspicuous secondary walls. The composition and the arrangement of secondary wall materials were intensively investigated in *Pinus strobus* by Srivastava (1969) who used different staining techniques and analyzed the orientation of cellulose fibrils. The orderly arrangement of these fibrils can easily be demonstrated by the birefrigence of the wall in polarized light (cf. Srivastava 1969; Parameswaran 1971). According to Srivastava (1969) the secondary walls are composed predominantly of cellulose with lesser amounts of polyuronides and pectins. Cellulose fibrils occur in parallel arrays and at two different angles. The lamellate appearance of the secondary wall in electron microscopy is explained by the higher density of cellulose at the plane where the arrays are crossing. In polarization microscopy the borders around sieve areas are visible as birefrigent rims, thus indicating a concentric deposition of cellulose (Srivastava 1969).

4.3.9 Intercellular Communication

EM studies of coniferous sieve cells have comprehensively dealt with the fine structure of sieve areas. Narrow pores and the attachment of ER complexes - both attributes of mature sieve areas and agreed upon by most authors - would be expected to slow down translocation considerably, if mass flow of assimilates is proposed as the mechanism for translocation.

Sieve areas are circular to oval wall portions which are penetrated by numerous sieve pores (Fig. 4.18). In contrast to angiosperms, however, an individual sieve pore runs only half-way through the wall common to two sieve elements. In the middle of this wall, the pore channel approaches an extensive median cavity or median nodule (Esau 1969) which is lined by plasmalemma (Figs. 4.25, 4.28). The plasmalemma is continuous with that of the pore channel. From the cavity the symplastic continuity to the neighboring cell is established by another sieve pore. The conduit through the sieve area is therefore achieved via a sieve pore-median cavity-sieve pore connection. It is furthermore influenced by the presence of ER membranes that are found all the way through the wall, i. e., in the pore lumen as well as in the median cavity (Figs. 4.19, 4.28).

The structure of a sieve area in conifers resembles that of a primary pit field, but the plasmodesmata as well as the median cavity are extended. The basic structure of sieve areas and their similarity to primary pit fields have been described by Kollmann and Schumacher (1962 a, 1963). These, and subsequent investigations showed callose rings lining the pore channel (Fig. 4.18; cf. Wooding 1966; Neuberger and Evert 1975; Schulz and Behnke 1987). At the same time, however, mature sieve pores are occasionally found without any callose, indicating that in conifers the synthesis of this wall substance might be only a wound response (Singh 1984 a; Schulz and Behnke 1987).

The differentiation of sieve pores in conifers starts from plasmodesmata which cross the wall at the sites of future sieve areas (Fig. 4.26; cf. Neuberger and Evert 1975; Singh 1984 a; Schulz and Behnke 1987). Individual cisternae of the ER may be in contact with plasmodesmata (Figs. 4.13, 4.26: arrowheads; cf. Neuberger and Evert 1975). The widening of the symplastic channel starts in the region of the middle lamella. Here, small cavities arise (Fig. 4.27; cf. Neuberger and Evert 1975; Singh 1984 a) while the plasmodesmata become less straight and develop "kinks" (Fig. 4.7; cf. Singh 1984 a). The formation of smooth ER complexes more or less coincides with a lateral extension of the small cavities (Figs. 4.19, 4.27; cf. Neuberger and Evert 1975). Finally, these cavities fuse, thus giving rise to a compound median nodule which interlinks a large number of sieve pores (Figs. 4.25, 4.28; cf. Harris 1972; Neuberger and Evert 1975; Singh 1984 a), now widened to a diameter of 100 to 800 nm. Throughout the differentiation of sieve pores, desmotubules and other membranes of the ER run through the symplastic channels (Kollmann and Schumacher 1962 a, 1963; Neuberger and Evert 1975; Singh 1984 a). At maturity the ER network is found attached to the sieve areas and is continuous through the

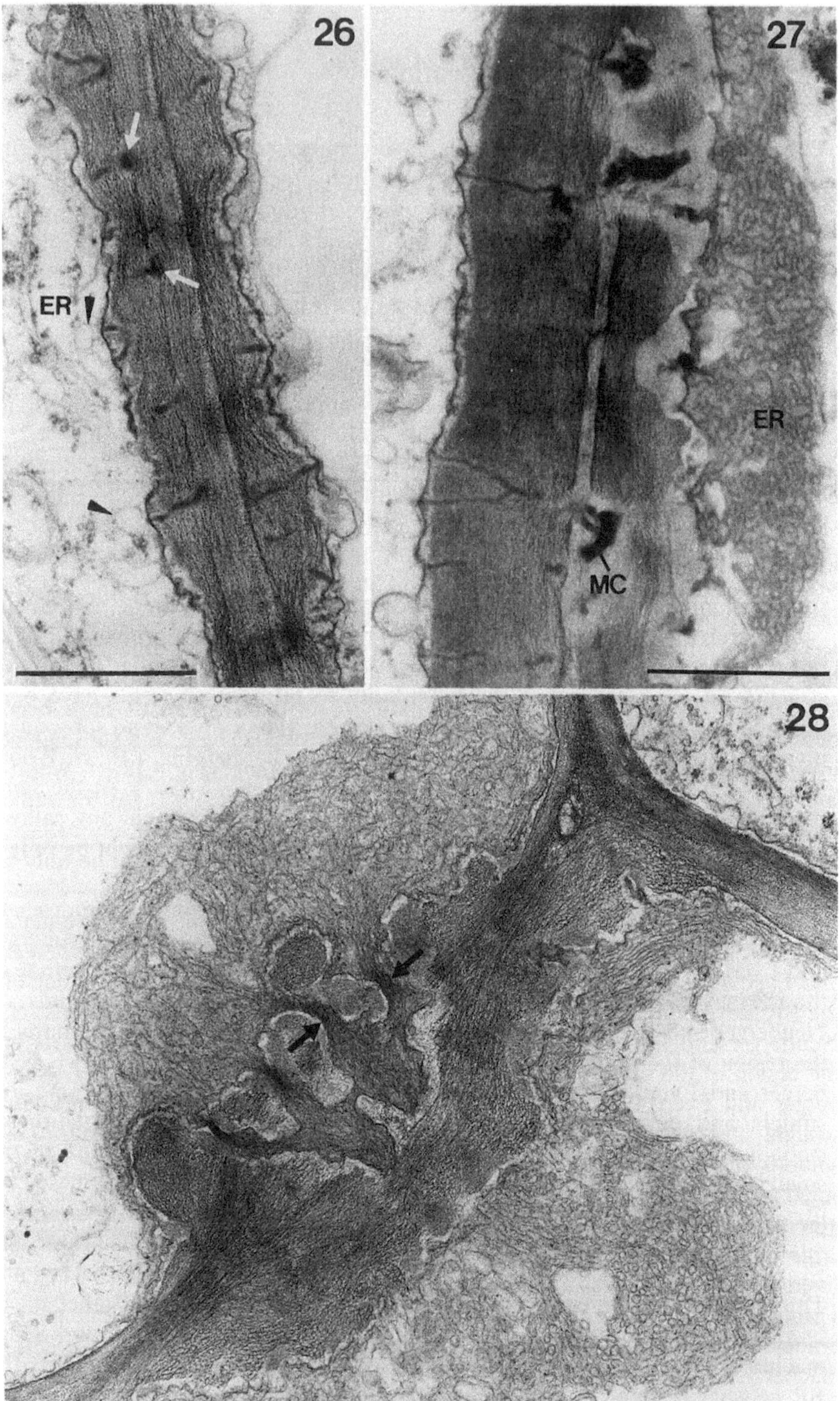

26
ER
27
ER
MC
28

pore lumen, via the median cavity, to the ER network of the neighboring cell (Fig. 4.28; cf. Neuberger and Evert 1975).

In contrast to the sieve-pore development of angiosperms there is no obvious participation of callose platelets or of small ER cisternae which would indicate the site and dimension of the future pore (Neuberger and Evert 1975). Only occasionally small callose deposits were found around future sieve pores still of plasmodesmatal size (Fig. 4.13). Mature sieve pores are often constricted by callose collars (Figs. 4.18, 4.25; cf. Srivastava and O'Brien 1966; Wooding 1966; Neuberger and Evert 1975), thus indicating a mechanism of wound response as effective as that in angiosperms. Comparable to the plugging of angiosperm sieve pores with P-protein during wounding, the ER of conifer sieve cells may change its conformation. In badly preserved material the ER tubules seem to swell, thereby blocking the sieve pores (compare Schulz and Behnke 1987).

Cessation of sieve-element function leads to the development of heavy deposits of callose, entirely covering the sieve area (Esau 1969: "definitive callose"). Together with those sieve-cell organelles which persist throughout the mature stage, definitive callose disappears eventually when the cell is obliterated.

4.4 Strasburger Cells

Sieve elements of gymnosperms and angiosperms have a specialized relation to certain parenchyma cells. This relation is structurally expressed by symplastic connections which consist of typical sieve pores at the sieve-element side, and of numerous plasmodesmata at the parenchyma-cell side (Figs. 4.31, 4.32).

In contrast to the companion cells in angiosperms, however, the *Strasburger cells* of gymnosperms (= albuminous cells - see discussion about this term in Esau 1969) have no ontogenetic relation to the sieve elements with which they are linked. In Pinaceae the origin of Strasburger cells in declining fusiform initials and marginal ray initials - and only extremely rarely as sister cells of sieve cells - was demonstrated by Srivastava (1963), using serial sections. A tier of successively more differentiated Strasburger cells at the margin of a ray is shown in Fig. 4.29 (cf. Sauter et al. 1976).

◁

Fig. 4.26. Future sieve area with several plasmodesmata which are slightly extended in the region of the middle lamella *(arrows)*. The cytoplasm of the left cell shows separately running, rough ER cisternae, some of which are in contact to plasmodesmata *(arrowheads)*. *Picea abies*, × 26000. **Fig. 4.27.** Future sieve area, on the right side more advanced in development than on the left. Plasmodesmata of the left side are in contact to extending, small cavities *(MC)*, where the wall is less contrasted. An ER complex consisting of smooth tubules appears on the right sieve-cell side. *Picea abies*, × 16000. (From Schulz and Behnke 1987.) **Fig. 4.28.** Mature sieve area of *Pinus resinosa* overlaid on both sides with an extensive ER network. Tubules of this complex are also found within the widened pores *(arrows)* and the extended median cavity. × 27000. (From Neuberger and Evert 1976.)

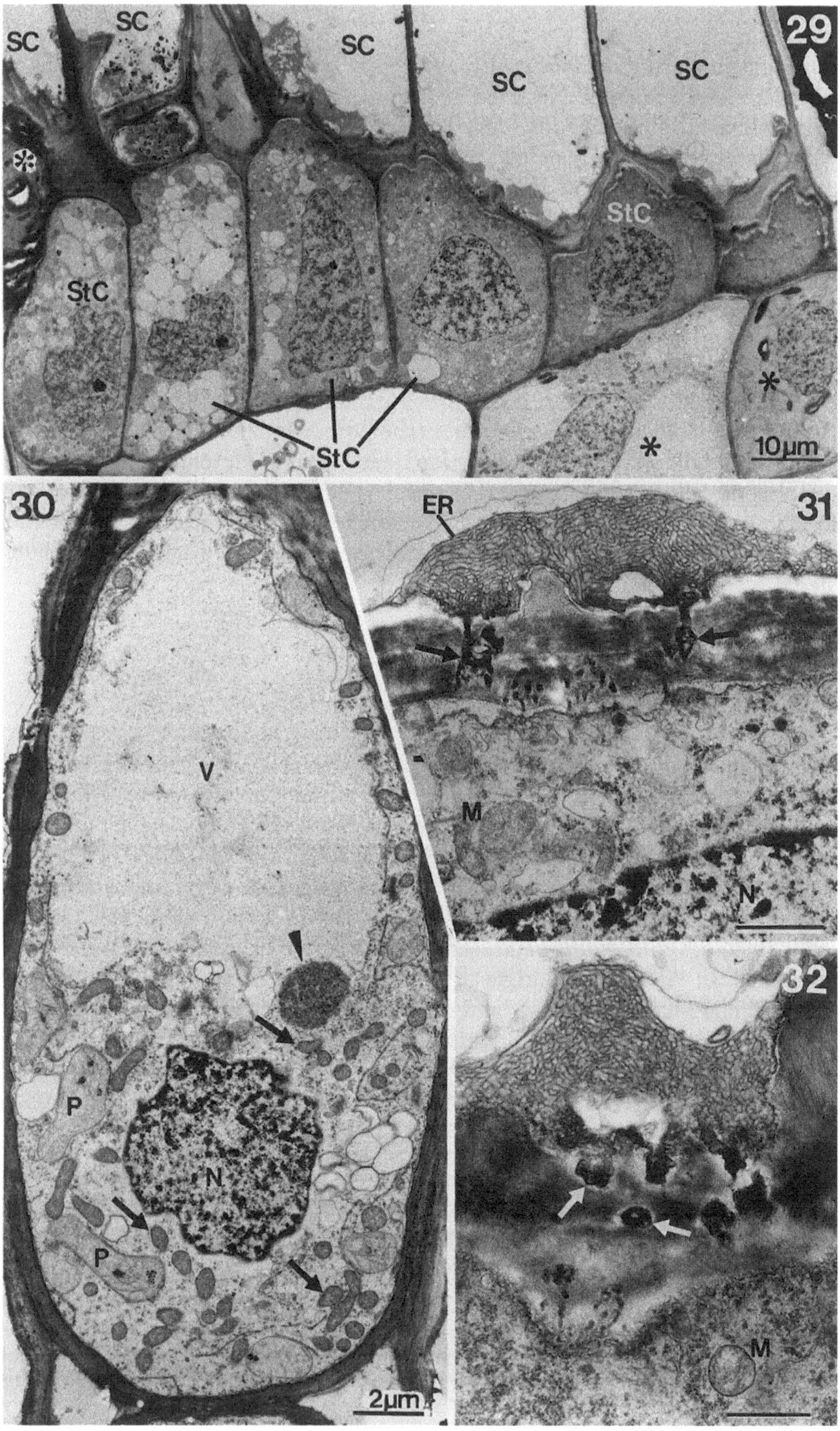
29
SC
SC
SC
SC
SC
StC
StC
StC
10µm
30
V
P
N
P
2µm
31
ER
M
N
32
M

The role of mature Strasburger cells for sieve elements is reflected by their dense cytoplasm, numerous mitochondria, a lobed nucleus (Fig. 4.30; cf. Carde 1973; Sauter et al. 1976) as well as by a high activity in different enzymes (see Sauter 1974). The density of the protoplast in partly due to the high number of free ribosomes which often form clusters (Figs. 4.31, 4.32; and Carde 1973; Neuberger and Evert 1975; Sauter et al. 1976; Schulz and Behnke 1987). Bodies of granular material were also described in mature Strasburger cells (Fig. 4.30: arrowhead; cf. Carde 1973 "formations morulaires", Sauter et al. 1976; Schulz and Behnke 1987). The shape and size of these structures is in accordance with those interpreted as "slime bodies" after light microscopical staining of Strasburger cells (Alfieri and Evert 1968a). The granular appearance in the EM, however, resembles aggregated ribosome-like particles (Schulz and Behnke 1987).

In addition to the normal content of parenchymatous cells - nucleus, plastids, mitochondria, rough ER and ribosomes - microbodies are also found in Strasburger cells (Neuberger and Evert 1975; Schulz and Behnke 1987). Starch grains, most abundant in neighboring ray cells, are only occasionally found in the plastids of Strasburger cells (Esau 1969; Sauter et al. 1976).

The differentiation of the protoplast of Strasburger cells involves changes of the vacuom. Close to the cambium, Strasburger cells show only a few small vacuoles (Fig. 4.29). Subsequently, the number of small vacuoles increases and a polar organization becomes visible in tangential sections; the nucleus occupies a proximal position to the ray, while a large vacuole is found in the distal, conical part of the cell (Figs. 4.2, 4.30; see Schulz and Behnke 1987). Vacuolar changes during differentiation were also described by Sauter et al. (1976).

In the needle, there seem to be three cells which function together as an accessory complex to sieve elements. According to Carde (1973) the first cell, adjacent to a sieve cell, is a transition cell, the second, well-linked by plasmodesmata, has all the characters of the Strasburger cells described above, and the third contains chloroplasts ("chlorophyll cells"; Carde 1973).

The structure of the half-sided sieve area is essentially the same as described above for double-sided sieve areas between two sieve elements. However, from

◁

Fig. 4.29. Margin of a ray with a series of differentiating Strasburger cells *(StC)*, starch-containing cells (*) and mature sieve cells *(SC)*. In the course of differentiation, the nuclei of the Strasburger cells show increasing lobing and, the amount of small vacuoles rises (cambium *outside the right margin*). *Abies alba*, × 1000. **Fig. 4.30.** Tangential view of a mature Strasburger cell of *Picea abies* with a vacuole *(V)* distal and the lobed nucleus *(N)* proximal to the ray. Numerous mitochondria *(arrows)*, several plastids *(P)* and a dense, granular body *(arrowhead)* are embedded in the cytoplasm. × 5000. **Fig. 4.31.** Specific plasmatic connection between a mature sieve cell characterized by an ER complex, and a Strasburger cell with dense cytoplasm, mitochondria *(M)* and nucleus *(N)*. The connection consists of a half-sided sieve area which is linked by the median cavity *(arrows)* and numerous plasmodesmata to the Strasburger cell. *Picea abies*, × 12000. **Fig. 4.32.** Glancing section through a half-sided sieve area with ER complex, conspicuous pores *(arrows)* and numerous plasmodesmata. The cytoplasm of the Strasburger cell contains a large amount of free ribosomes. *M* mitochondrion. *Abies alba*, × 12000

an extensive median cavity numerous plasmodesmata form the other side of this specific connection (Figs. 4.31, 4.32; cf. Wooding 1966, 1968; Srivastava and O'Brien 1966; Carde 1974; Neuberger and Evert 1975; Sauter et al. 1976, and others). It is important to note that the Strasburger cells are also linked by well-developed primary pit fields to neighboring ray parenchyma cells (Timell 1973; Carde 1974; Sauter et al. 1976). Thus, the particularly well-developed symplastic compartment of sieve cells in conifers - with sieve pores of 500 nm and more in width - is well linked via Strasburger cells and plasmodesmata to the parenchymatous regions of rays and the cortex.

Acknowledgment. Supported in part by the Projekt Europäisches Forschungszentrum für Maßnahmen zur Luftreinhaltung.

References

Alfieri FJ, Evert RF (1968 a) Observations on albuminous cells in *Pinus*. Planta 78: 93-97

Alfieri FJ, Evert RF (1968 b) Seasonal development of the secondary phloem in *Pinus*. Am J Bot 55: 518-528

Alosi MC, Park RB (1983) Fractionation and polypeptide analysis of phloem tissue of *Pinus sabiniana* Dougl. Planta 157: 298-306

Barnett JR (1974) Secondary phloem in *Pinus radiata* D. Don. I. Structure of differentiating sieve cells. N Z J Bot 12: 245-260

Behnke H-D (1974) Sieve-element plastids of gymnospermae: their ultrastructure in relation to systematics. Plant Syst Evol 123: 1-12

Behnke H-D (1983) Cytology and morphogenesis of higher plant cells - phloem. In: Ellenberg H, Esser K, Kubitzki K, Schnepf E, Ziegler H (eds) Progress in botany, vol 45. Springer, Berlin Heidelberg, pp 18-35

Behnke H-D (1986) Sieve element characters and the systematic position of *Austrobaileya*, Austrobaileyaceae - with comments to the distinction and definition of sieve cells and sieve-tube members. Plant Syst Evol 152: 101-121

Behnke H-D, Paliwal GS (1973) Ultrastructure of phloem and its development in *Gnetum gnemon*, with some observations on *Ephedra campylopoda*. Protoplasma 78: 305-319

Carde J-P (1973) Le tissu de transfert (= cellules de Strasburger) dans les aiguilles du Pin maritime (*Pinus pinaster* Ait.) I. Etude histologique et infrastructurale du tissu adulte. J Microsc 17: 65-88

Carde J-P (1974) Le tissu de transfert (= cellules de Strasburger) dans les aiguilles du Pin maritime (*Pinus pinaster* Ait.) II. Caractères cytochimiques et infrastructuraux de la paroi et des plasmodesmes. J Microsc 20: 51-72

Chauveaud G (1902 a) De l'existence d'éléments précurseurs des tubes criblés chez les Gymnospermes. C R Acad Sci 134: 1605-1606

Chauveaud G (1902 b) Développement des éléments précurseurs des tubes criblés dans le *Thuja orientalis*. Bull Mus Hist Nat 8: 447-454

Chauveaud G (1903) Recherches sur le mode de formation des tubes criblés dans la racine des Cryptogames vasculaires et des Gymnospermes. Ann Sci Nat Bot 18: 165-277

Dute RR (1983) Features of sieve-element ontogeny in *Ginkgo biloba*. Am J Bot 70 (5, 2): 19

Esau K (1969) The phloem. In: Zimmermann W, Ozenda P, Wulff HD (eds) Encyclopedia of plant anatomy, vol 5, pt 2. Borntraeger, Berlin Stuttgart, 505 pp

Evert RF (1984) Comparative structure of phloem. In: White RA, Dickison WC (eds) Contemporary problems in plant anatomy. Academic Press, New York London, pp 145-234

Evert RF, Davis JD, Tucker CM, Alfieri FJ (1970) On the occurrence of nuclei in mature sieve elements. Planta 95: 281-296
Ferguson (1968) Bristlecone pine: science and esthetics. Science 159: 839-846
Gourret J-P, Strullu D-G (1974) Evolution des protéoplastes du phloème de *Pseudotsuga menziesii* Mirb. (Abietacées) au cours de la différenciation des éléments criblés. J Microsc 20: 73-82
Harris WM (1972) Ultrastructural observations in Pinaceae leaf phloem. I. The spring condition. New Phytol 71: 169-173
Hébant C (1975) Lack of incorporation of tritiated uridine by nuclei of mature sieve elements in *Metasequoia glyptostroboides* and *Sequoiadendron giganteum*. Planta 126: 161-163
Kollmann R, Schumacher W (1961) Über die Feinstruktur des Phloems von *Metasequoia glyptostroboides* und seine jahreszeitlichen Veränderungen, I. Mitteilung: Das Ruhephloem. Planta 57: 583-607
Kollmann R, Schumacher W (1962a) Über die Feinstruktur des Phloems von *Metasequoia glyptostroboides* und seine jahreszeitlichen Veränderungen, II. Mitteilung: Vergleichende Untersuchungen der plasmatischen Verbindungsbrücken in Phloemparenchymzellen und Siebzellen. Planta 58: 366-386
Kollmann R, Schumacher W (1962b) Über die Feinstruktur des Phloems von *Metasequoia glyptostroboides* und seine jahreszeitlichen Veränderungen, III. Mitteilung: Die Reaktivierung der Phloemzellen im Frühjahr. Planta 59: 195-221
Kollmann R, Schumacher W (1963) Über die Feinstruktur des Phloems von *Metasequoia glyptostroboides* und seine jahreszeitlichen Veränderungen, IV. Mitteilung: Weitere Beobachtungen zum Feinbau der Plasmabrücken in den Siebzellen. Planta 60: 360-389
Kollmann R, Schumacher W (1964) Über die Feinstruktur des Phloems von *Metasequoia glyptostroboides* und seine jahreszeitlichen Veränderungen, V. Mitteilung: Die Differenzierung der Siebzellen in Verlaufe einer Vegetationsperiode. Planta 63: 155-190
Murmanis L (1974) Filamentous component of secondary phloem sieve elements in *Pinus strobus* L. Ann Bot (London) 38: 859-863
Murmanis L, Evert RF (1966) Some aspects of sieve cell ultrastructure in *Pinus strobus*. Am J Bot 53: 1065-1078
Neuberger DS, Evert RF (1974) Structure and development of the sieve element protoplast in the hypocotyl of *Pinus resinosa*. Am J Bot 61: 360-374
Neuberger DS, Evert RF (1975) Structure and development of sieve areas in the hypocotyl of *Pinus resinosa*. Protoplasma 84: 109-125
Neuberger DS, Evert RF (1976) Structure and development of sieve cells in the primary phloem of *Pinus resinosa*. Protoplasma 87: 27-37
Parameswaran N (1971) Zur Feinstruktur der Assimilatleitbahnen in der Nadel von *Pinus silvestris*. Cytobiologie 3: 70-88
Parthasarathy MV, Pesacreta TC (1980) Microfilaments in plant vascular cells. Can J Bot 58: 807-815
Pesacreta TC, Parthasarathy MV (1984) Microfilament bundles in the roots of a conifer, *Chamaecyparis obtusa*. Protoplasma 121: 54-64
Salmon J (1946) Différenciation des tubes criblés chez les Angiospermes: Recherches cytologiques. Rev Cytol Cytophysiol Veget 9: 55-168
Sauter JJ (1974) Structure and physiology of Strasburger cells. Ber Dtsch Bot Ges 87: 327-336
Sauter JJ, Dörr I, Kollmann R (1976) The ultrastructure of Strasburger cells (=albuminous cells) in the secondary phloem of *Pinus nigra* var. *austriaca* (Hoess) Badoux. Protoplasma 88: 31-49
Schulz A (1987) Sieve-element differentiation and fluoresceine translocation in wound-phloem of pea roots after complete severance of the stele. Planta 170: 289-299
Schulz A (1988) Vascular differentiation in the root cortex of peas: premitotic stages of cytoplasmic reactivation. Protoplasma 143: 176-187
Schulz A, Behnke H-D (1987) Feinbau und Differenzierung des Phloems von Buchen, Fichten und Tannen aus Waldschadensgebieten. PEF-Ber 16, Kernforschungszentrum Karlsruhe, 95 pp

Schulz A, Alosi AC, Sabnis DD, Park RB (1989) A phloem-specific, lectin-like protein is located in pine sieve-element plastids by immunocytochemistry. Planta 179: 506-515
Singh AP (1984 a) *Pinus radiata* needle trace studies: fine structure of immature sieve cells in the primary phloem. Cytologia 49: 359-384
Singh AP (1984 b) Microfilaments in the phloem of *Pinus radiata* cotyledons. Cytologia 49: 385-393
Smoot EL (1984) Phloem anatomy of the carboniferous pteridosperm *Medullosa* and evolutionary trends in gymnosperm phloem. Bot Gaz 145: 550-564
Srivastava LM (1963) Secondary phloem in the Pinaceae. Univ Cal Publ Bot 36, 142 pp
Srivastava LM (1969) On the ultrastructure of cambium and its derivatives. III. The secondary walls of the sieve elements of *Pinus strobus*. Am J Bot 56: 354-361
Srivastava LM, O'Brien TP (1966) On the ultrastructure of cambium and its vascular derivatives. II. Secondary phloem of *Pinus strobus* L. Protoplasma 61: 277-293
Timell TE (1973) Ultrastructure of the dormant and active cambial zones and the dormant phloem associated with formation of normal and compression woods in *Picea abies* (L.) Karst. Tech Publ 96: 3-94
Warmbrodt RD, Eschrich W (1985) Studies on the mycorrhizas of *Pinus sylvestris* L. produced in vitro with the basidiomycete *Suillus variegatus* (SW. ex FR.) O. Kuntze. II. Ultrastructural aspects of the endodermis and vascular cylinder of the mycorrhizal rootlets. New Phytol 100: 403-418
Wooding FBP (1966) The development of the sieve elements of *Pinus pinea*. Planta 69: 230-243
Wooding FBP (1968) Fine structure of callus phloem in *Pinus pinea*. Planta 83. 99-110

5 Cycads and Gnetophytes

H.-Dietmar Behnke

5.1 Introduction

The cycads and gnetophytes, which make up the subdivision Cycadophytina, represent relicts ("living fossils") of gymnospermous groups which had their main expansion in Jurassic and Cretaceous times. Consequently, their extant species (some 100 in the cycads and about 70 in the gnetophytes) are regarded as the specialized remnant of a once very successful line which in the Devonian has the same origin as the Coniferophytina, i. e., in the so-called Progymnospermae. Since this connection is mediated via the seed ferns in upper Carbonian/lower Permian times, a group to which the Angiospermae are also linked, the Cycadophytina have often been regarded as connecting links between gymnosperms and angiosperms. *Gnetum,* in particular, seems to be well suited for this role since, unlike all other gymnosperms, this genus has been shown to include vessels in its xylem and at some time was also thought to contain companion cells in its phloem. The few ultrastructural investigations of the phloem in this group available, however, corroborate the view presently held by taxonomists that the gnetophytes represent a highly specialized but distinct gymnospermous taxon.

Despite its importance as a separate line of the seed plants little phloem research has been done on the Cycadophytina. Light microscopic studies of this group date back to the second half of the nineteenth century (see review by Esau 1969); cycads, *Gnetum, Ephedra,* and *Welwitschia* were included in Strasburger's (1891) monograph of vascular tissues in plants. Ultrastructural investigations of this group, however, are rare and there are no thorough, ontogenetic studies of many taxa; the entire cycads is left almost untouched. This chapter will review the few data available and will also demonstrate that the Cycadophytina represent a largely neglected but very promising field for future phloem research.

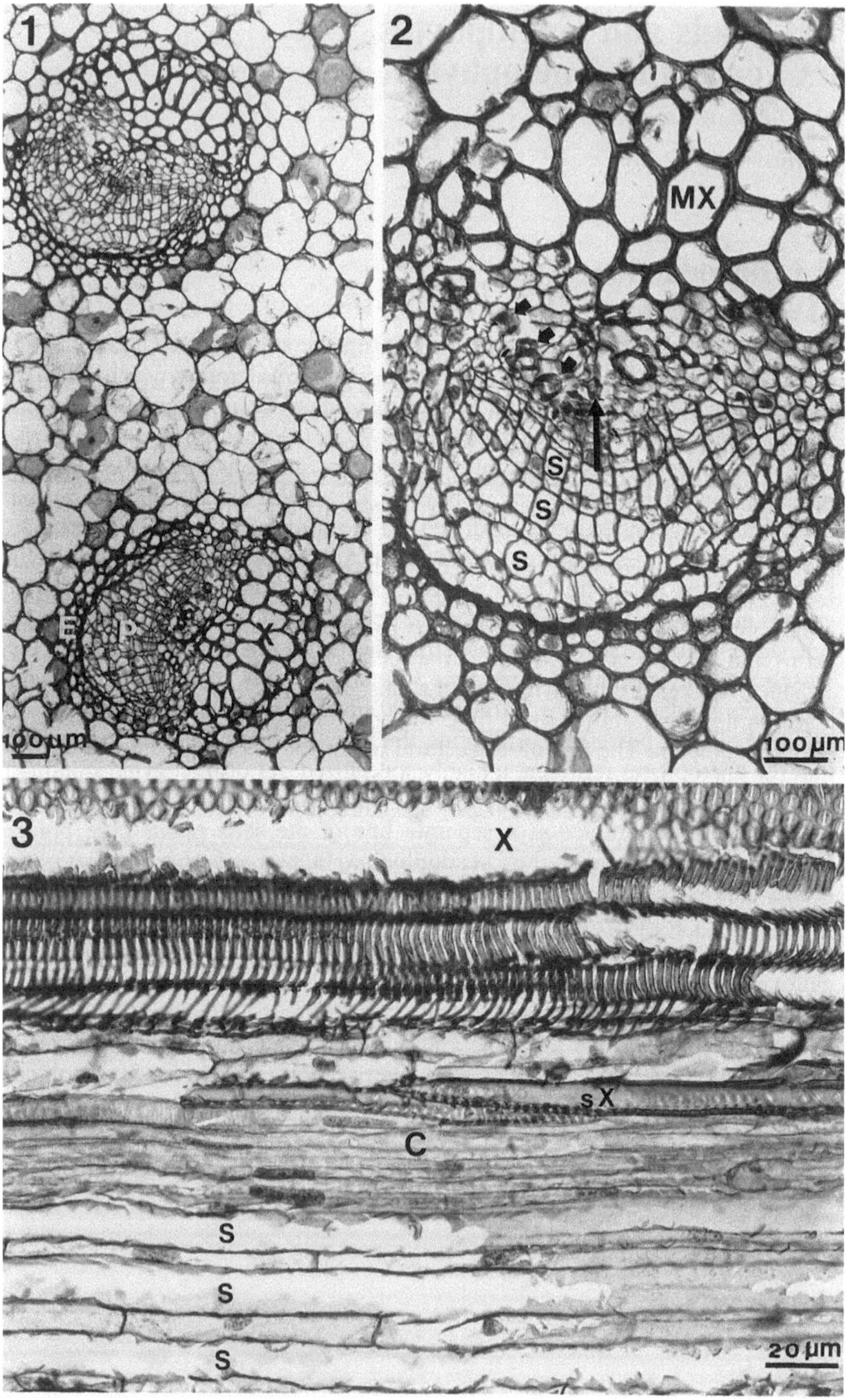
1
E
P
X
100 µm
2
MX
S
S
S
100 µm
3
X
sX
C
S
S
S
20 µm

5.2 Cycads

5.2.1 Organization and Composition of the Phloem

Since the scanty ultrastructural details of phloem available derive from the leaf, the following description of vascular bundles is confined to studies of this organ (for phloem organization in other parts see Esau 1969). Numerous discrete vascular bundles innervate the midrib of a *Cycas* leaf, part of which is shown in a cross-section in Fig. 5.1. Following the almost circular outline of the midrib, the bundles are arranged in a peripheral, ring-like pattern with their xylem oriented to the adaxial side of the leaf. A single bundle (Figs. 5.2, 5.3) is surrounded by a sclerenchymatic sheath (endodermis) and contains an unusual distribution of its xylem, composed of a small protoxylem strand, a broad endarch (i. e., developing toward the adaxial side) metaxylem and a few secondary xylem elements derived from the cambium and situated next to the phloem.

The phloem consists of wide sieve elements and narrower parenchyma cells; phloem fibers are missing in the leaf. The protophloem is crushed and situated at the far abaxial side of the bundle. In transverse-sectioned secondary phloem and part of the metaphloem, radial files can be recognized which consist of almost regularly alternating bands of sieve elements and parenchyma cells (Fig. 5.2). In radial longitudinal sections through a leaf bundle, comparatively long sieve elements (about 0.5 mm) alternate with a series of rather short parenchyma cells (Figs. 5.3, 5.4).

5.2.2 Ultrastructure of the Sieve Elements

Young developing sieve elements of a cycad leaf were first depicted by Parthasarathy and Pesacreta (1980) in their study of microfilaments in vascular cells. An elongate nucleus, a plastid, many mitochondria, tubular ER and small vacuoles were shown by these authors in the young sieve elements of *Zamia*, corresponding to what is known from other gymnosperms. However, the regular presence of microfilament bundles which were in contact with different organelles represented an outstanding feature. One of their figures also showed a

◁

Fig. 5.1. Part of a cross-section through the midrib of a leaf of *Cycas revoluta* showing two vascular bundles composed of phloem *(P)* and xylem *(X)* and surrounded by a sclerenchymatic sheath *(E)*; x 50. **Fig. 5.2.** Cross-section through a leaf vascular bundle of *C. revoluta* composed of a small protoxylem strand *(at large arrow)*, a broad endarch metaxylem *(MX)*, a few secondary xylem elements *(small arrows)* adjacent to the cambium, and a broad area of metaphloem which is aligned in radial rows. Sieve elements *(S)* occur in tangential bands, regularly alternating with parenchyma cells. × 100. **Fig. 5.3.** Radial longitudinal section through a leaf vascular bundle of *C. revoluta* showing primary *(X)* and secondary *(sX)* xylem, cambium and metaphloem with alternating sieve elements *(S)* and parenchyma cells. × 500

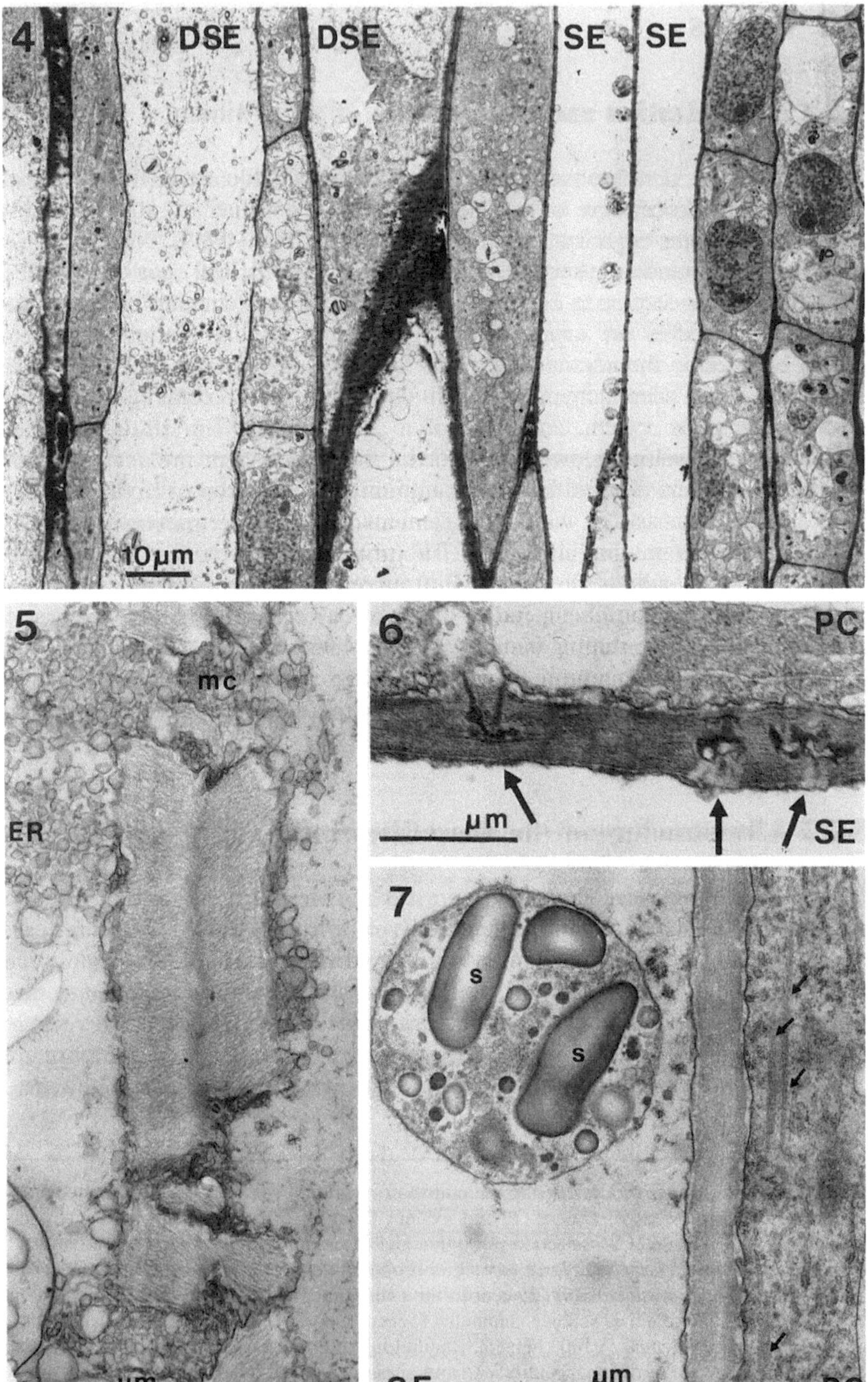
4
DSE
DSE
SE
SE
10µm
5
mc
ER
µm
6
PC
µm
SE
7
s
s
SE
µm
PC

small piece of a mature sieve element with a part of a pycnotic nuclear mass and a starch containing sieve-element plastid.

The following short description of the sieve-element ultrastructure refers to the present author's investigations of the four cycads *Cycas revoluta, Encephalartos horridus, Macrozamia secunda,* and *Stangeria paradoxa* (Behnke 1974, 1986 and unpublished results) which demonstrated that in the midrib of leaves the differentiation of the cycad sieve elements includes the same drastic structural changes as are known from the intensive studies of conifers (see Chap. 4). These changes result in the degeneration of the nucleus, the loss of the vacuole(s), and the depletion of almost all the sieve-element contents, except for the plasmalemma, plastids, mitochondria, some tubular ER, and remnants of the nuclei (cf. Figs. 5.4, 5.9).

The nuclear degeneration, which is of the pycnotic type found in all other gymnosperms, is recognized by an initial coarsening and subsequent increasing condensation of the chromatin, while at the same time its matrix is depleted of other material (Fig. 5.11). For some time the nuclear envelope follows the shrinking matrix and gives the nucleus an extremely lobed appearance (Fig. 5.9), until it eventually dissolves and leaves some patches of pycnotic material attached to the plasmalemma (Fig. 5.10) or free in the cell lumen (Fig. 5.8). These nuclear remnants may remain after all other lytic changes have occurred within the sieve element. The only other organelles present in mature sieve elements are plastids (S-type, with ovoid to disc-shaped starch grains, cf. Fig. 5.7), mitochondria, and agranular ER.

The cell walls of cycad sieve elements are not differentiated into inner and outer parts (i. e., do not contain nacreous parts). Sieve areas are chiefly located in the oblique to transverse end walls but may also be found along the entire radial longitudial walls (see Fig. 5.9 and Behnke 1986). The composition of the sieve areas is the same as in the conifers, i. e., several narrow sieve pores penetrate the wall from both sides and in the region of the middle lamella connect to form median cavities. Among the species investigated the size of the sieve pores may vary from narrow (about 0.1 μm in diameter) to wide openings (up to 0.5 μm). Profiles of ER are always found in the vicinity of the sieve areas and the membranes also penetrate the sieve pores and median cavities (see Fig. 5.5 and Behnke 1986).

◁

Fig. 5.4. Longitudinal section through part of metaphloem from a leaf of *Cycas revoluta* demonstrating developing *(DSE)* and mature *(SE)* sieve elements, irregularly alternating with parenchymatic cells. × 800. **Fig. 5.5.** Symplastic connections between two sieve elements of *C. revoluta* composed of comparatively broad sieve-area pores on either side of the connection joined in the middle of the cell wall by median cavities *(mc)*. ER in form of vesicles (artefacts of processing) is accumulated in front of and penetrates into the sieve area. × 18000. **Fig. 5.6.** Sieve pore/plasmodesmata connections *(arrows)* between a sieve element *(SE)* and a parenchyma cell *(PC)* in leaf metaphloem of *C. revoluta*. × 18000. **Fig. 5.7.** Sieve element *(SE)* with S-type plastid containing numerous starch grains *(s)* and parenchyma cell *(PC)* with microfilament bundles in leaf phloem of *Encephalartos horridus*. × 18000

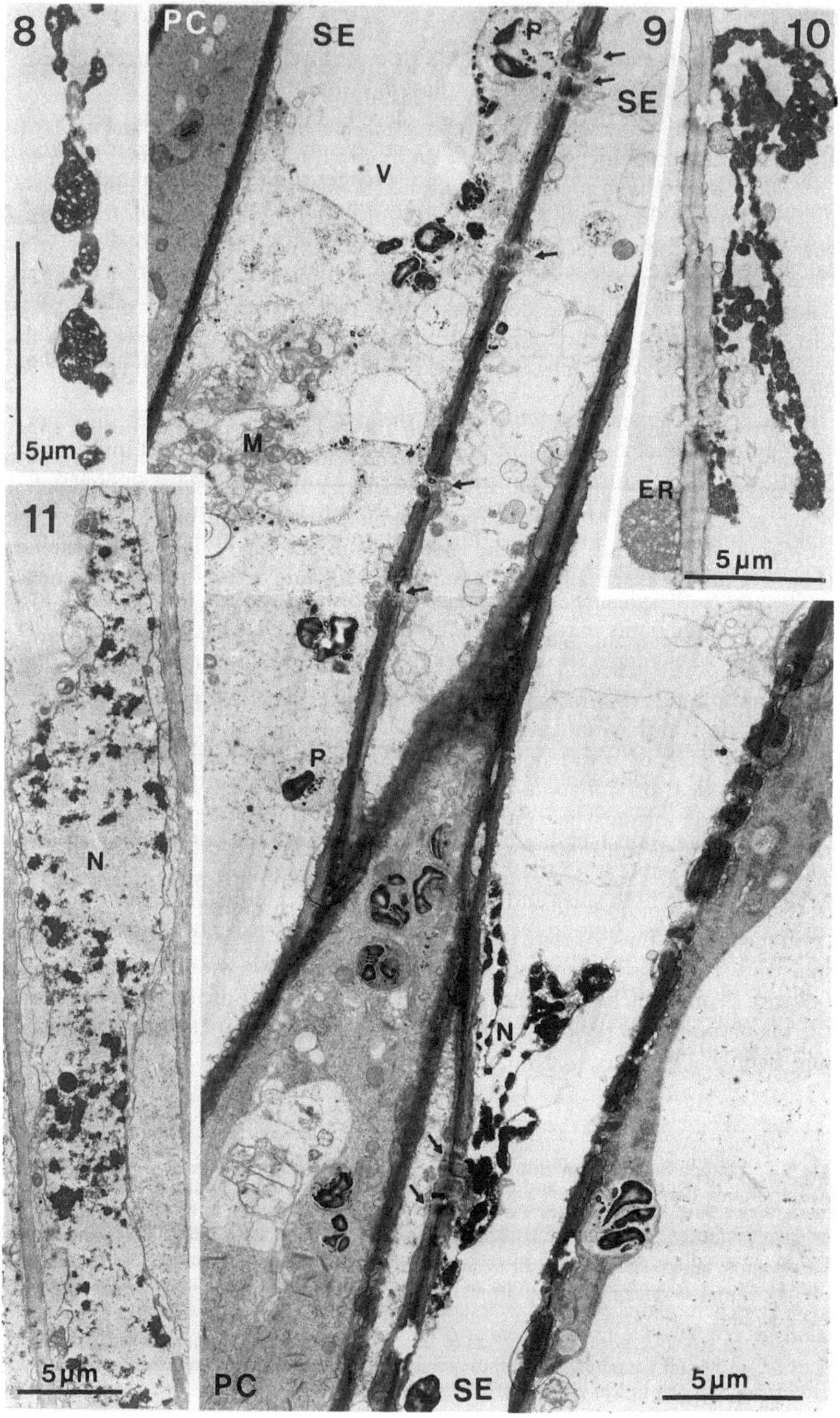
8
5µm
PC
SE
P
9
SE
V
M
P
N
PC
SE
5µm
10
ER
5µm
11
N
5µm

5.2.3 Parenchymatous Cells Associated with the Sieve Elements

In the stem of *Cycas circinalis* Strasburger (1891) distinguished between starch-containing parenchyma cells and albuminous cells (i. e., Strasburger cells) without starch, the latter collapsing when the associate sieve elements cease to function. Our studies of the phloem in leaves of *Cycas revoluta* did not permit a clear distinction of the two different types of parenchyma cells which laterally accompany the sieve elements in long files of short cells (Figs. 5.4, 5.9). These cells are characterized by large nuclei, starch-containing plastids, and a comparatively dense cytoplasm which occasionally also contains bundles of microfilaments (Fig. 5.7). In the absence of an exact tier-by-tier analysis, the limited light microscopic studies revealed only one kind of phloem-parenchyma cell (Figs. 5.2, 5.3), and the few markers available from ultrathin sections could not unambiguously distinguish putative Strasburger cells from other parenchyma cells. However, in the few examples where typical sieve area/plasmodesmata connections were found between sieve elements and parenchyma cells (cf. Fig. 5.6), the plastids of the latter were devoid of starch.

5.3 Gnetophytes

5.3.1 Organization and Composition of the Phloem

The few studies performed on the phloem of stem and leaves in the gnetophytes describe open collateral vascular bundles which are arranged in the axis to form a hollow cylinder and are separated from each other by multiseriate primary rays (Fig. 5.12). Regular, secondary growth derived from the vascular cambium is reported for all three genera, including the leaves of *Welwitschia* (Evert et al. 1973a). The delimitation of the primary and the initiation of secondary tissues is somewhat obscured by the rather strict radial filing of cells observed already in the metaphloem and metaxylem (see Paliwal and Behnke 1973; Alfieri and

◁

Fig. 5.8. Nuclear remnant of a sieve element of *Stangeria paradoxa*. × 6000. **Fig. 5.9.** Longitudinal section through leaf phloem of *Cycas revoluta* showing sieve elements *(SE)* in different stages of differentiation next to parenchyma cells *(PC)*. A mature sieve element contains a highly lobed nuclear remnant *(N)*. Sieve elements are connected via sieve areas in their radial longitudinal walls *(at arrows)*. *P* plastids; *M* mitochondria; *V* vacuole. × 4000 (Behnke 1986). **Fig. 5.10.** Nuclear remnant in a mature sieve element of *C. revoluta*. × 4000. **Fig. 5.11.** The nucleus *(N)* in a differentiating sieve element of *Macrozamia secunda* contains irregularly distributed patches of deeply staining chromatic material, indicating early degenerative changes. × 3000

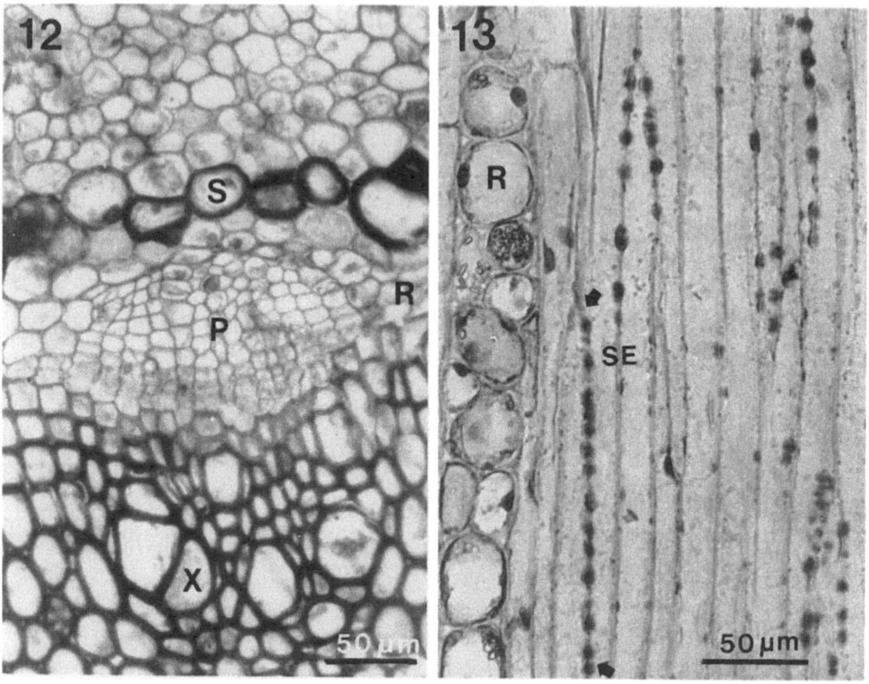

Fig. 5.12. Cross-section through secondary stem phloem of *Gnetum gnemon* showing xylem with large vessels *(X)*, phloem *(P)*, rays *(R)* and sclerenchyma *(S)*. × 250. (From Paliwal and Behnke 1973). **Fig. 5.13.** Longitudinal section through stem phloem of *G. gnemon* with sieve elements *(SE)* connected by sieve areas concentrated in the tapering end of radial longitudinal walls *(between arrows)*. *R* ray. × 290 (Paliwal and Behnke 1973)

Mottola 1983 for *Gnetum* and *Ephedra*). The vascular cylinder is delimited from the outer bark by one to several layers of sclereids (Fig. 5.12).

Strasburger (1891) described the phloem of *Ephedra* and *Gnetum* as being composed of sieve elements and two types of axial parenchyma cells, i. .e., the more abundant starch-free Strasburger cells (albuminous cells) connected to the sieve elements and the less common starch-containing parenchyma cells. Both sieve elements and Strasburger cells occurred in alternating radial files, while parenchyma cells were less regularly distributed. Thompson (1919) named the Strasburger cells companion cells, although he mentioned their origin from different cambial intials. Among the more recent studies, Evert et al. (1973b) followed Strasburger (1891) and used the term albuminous cells for *Welwitschia* phloem. Alosi and Alfieri (1972) in their ontogenetic study of *Ephedra* secondary phloem, pointed to the unusual distribution and arrangement of albuminous cells having characters intergrading with parenchyma cells, while for *Gnetum* Paliwal and Behnke (1973) used the more general term (related) parenchyma cells.

The sieve elements have extremely tapered walls, the end parts of which contain many sieve areas (between 7 and 19 were counted in radial sections by Paliwal and Behnke 1973). The size of the sieve elements has been measured in *Ephedra* and *Gnetum*. While there are no differences recorded for their diameter, 15-24 μm, two length classes, 400 and 220 μm (the latter derived by transverse division from fusiform initials) were found in *Ephedra californica* (Alosi and Alfieri 1972) and lengths from 440 to 840 μm were recorded in *Gnetum* (Paliwal and Behnke 1973). Parenchyma cells had a length of 100-220 μm and their diameter was 10-20 μm.

5.3.2 Ontogeny and Structure of the Mature Sieve Element

The Sieve Element Protoplast. Only two ultrastructural studies have been carried out which cover the leaf veins of *Welwitschia* (Evert et al. 1973a) and the young stem of *Gnetum* (Behnke and Paliwal 1973). The protoplast of young sieve elements (just derived from procambial or cambial tissue) differs little from that of developing parenchyma cells but rather soon it undergoes the drastic changes characteristic of all sieve elements of vascular plants, transforming the cell into a more or less open conduit for assimilate transport.

The nucleus of young sieve elements contains much heterochromatic material in clumps which are scattered over the matrix or attached to the envelope. In transverse sections the nuclei are shown to extend over more than half of the cross-sectional area of a sieve element (Evert et al. 1973a), longitudinal sections demonstrate their extreme length (Fig. 5.14; cf. Behnke and Paliwal 1973). An increasing condensation of the chromatin material and a gradual depletion of their matrix contents combined with a gradual shrinkage of their volume characterizes some of the degenerative changes of the nucleus as revealed by the electron microscope (see Figs. 5.14, 5.15; cf. Evert et al. 1973a; Behnke and Paliwal 1973). Nuclear remnants composed of dark-stained and condensed chromatin material still surrounded by the nuclear envelope persist within the mature sieve element (cf. Evert et al. 1973a). These details of nuclear degeneration conform to the pycnotic mode and to what is known from the conifers (see Chap. 4).

In *Welwitschia* Evert et al. (1973a) describe a close association between degenerating nuclei and mitochondria which were frequently shown to be attached to the nuclear envelope (cf. also Fig. 5.14) or even partly surrounded by the latter. This association was shown to continue beyond the depletion of the protoplast in maturing sieve elements. While there is no structural modification visible in mitochondria, the persisting plastids undergo considerable structural changes: their dense matrix is depleted, their internal membranes are lost and their shape changed from elongate to spherical. Plastids of mature sieve elements are characterized by their starch grains often exhibiting discoid forms.

More decisive changes occur in the endoplasmic reticulum (ER). The young sieve element contains granular, cisternal ER, but early in development aggre-

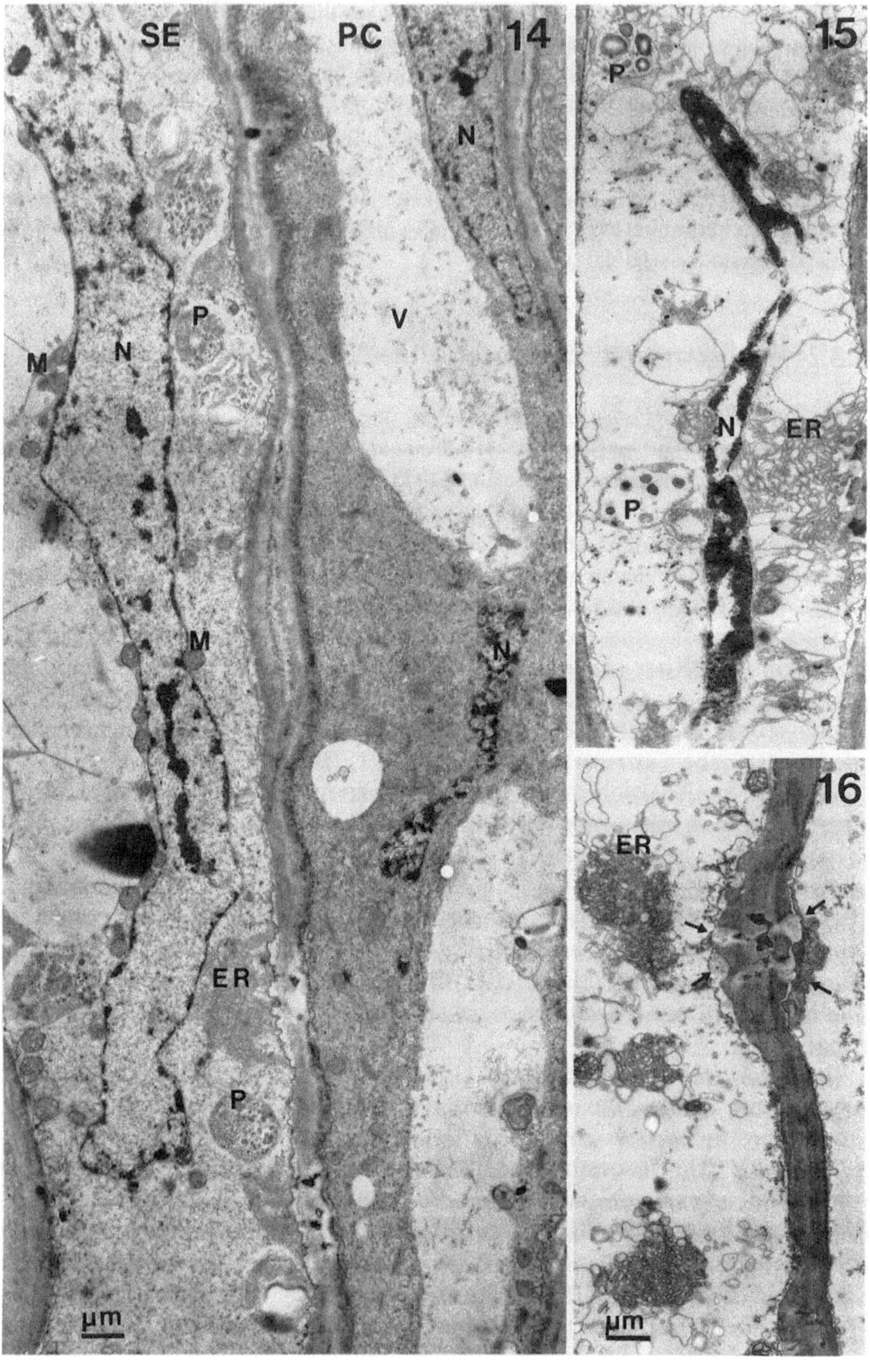
SE
PC
14
N
P
V
M
N
M
N
ER
P
µm
15
P
N
ER
P
16
ER
µm

gates of agranular, tubular ER appear at different places within the differentiating sieve element (Evert et al. 1973 a). While the granular ER disappears, these tubular aggregates enlarge (cf. Figs. 5.14, 5.15: ER) and persist in the mature sieve element (Fig. 5.16). Massive amounts of tubular ER may occasionally be found in contact with plastids and the nuclear remnants (see Behnke and Paliwal 1973), but it is always associated with sieve areas, a situation also found in the conifer sieve elements (cf. Chap. 4).

Other cytoplasmic organelles, such as dictyosomes, microtubules, microfilaments, ribosomes, and the vacuole, disintegrate during sieve element differentiation. The mature sieve element is delimited by the plasmalemma and contains only plastids, mitochrondria, nuclear remnants and tubular ER aggregates. P-protein has never been found. Apparently, aggregates of ER, difficult to identify with the light microscope, were mistaken as "slime bodies" and were described (Alosi and Alfieri 1972) in *Ephedra* (but see Behnke and Paliwal 1973).

The Cell Wall and Sieve Areas. The cell walls of the sieve elements in the gnetophytes are rather thin-walled and do not appear to be secondarily thickened. Sieve areas are confined to the radial walls. Up to 19 contiguous sieve areas were counted within the end parts of these walls in *Gnetum* (see Fig. 5.13; cf. Paliwal and Behnke 1973), a concentration which led previous investigators to compare them to a compound sieve plate (Boodle and Worsdell 1894). In *Ephedra,* Alosi and Alfieri (1972) describe a subdivision of sieve areas into several, distinct groups of sieve pores which they name sieve fields.

Ultrastructural studies by Evert et al. (1973 b) and Behnke and Paliwal (1973) clearly confirmed that the sieve areas of the Gnetales are of the type found in other gymnosperms and that they are not comparable to those in angiosperm sieve plates. A sieve area is composed of several small sieve pores which, in the region of the middle lamella, join to form a large median cavity. The sieve pores and the median cavity are bordered by plasmalemma and are penetrated by profiles of agranular ER which, in addition, often aggregate in front of a sieve area. The size of the sieve pores varies but rarely approaches or exceeds 0.1 μm. The pores of mature sieve elements are often found to be narrowed by or even occluded by callose (Figs. 5.16; cf. Evert et al. 1973 b; Behnke and Paliwal

◁

Fig. 5.14. Longitudinal section through phloem of *Gnetum gnemon*. A differentiating sieve element *(SE)* is distinguished from its neighboring parenchyma cell *(PC)* by its very long nucleus *(N)* which contains darkly stained patches of chromatin, partly attached to the inside of the nuclear envelope, by mitochondria *(M)* often in close vicinity to the outside of the nuclear envelope, by starch-containing plastids *(P)* and a less dense cytoplasmic matrix containing some ER aggregates. The nucleus of the parenchyma cells extends partly into a cytoplasmic bridge passing through the vacuole *(V)*. × 6000. **Fig. 5.15.** Mature sieve element of *G. gnemon* with nuclear remnant *(N)*, S-type plastids *(P)* and aggregates of ER. × 6000 (Behnke and Paliwal 1973). **Fig. 5.16.** Sieve area between two mature sieve elements of *G. gnemon* with their pores *(arrows)* occluded by callose and a median cavity in the middle of the wall. Aggregates of ER are present in the sieve element lumen. × 6000 (Behnke and Paliwal 1973)

1973), but callose has not been found to take part in the development of the pores (see also Chap. 4).

Sieve areas develop from aggregates of plasmodesmata, which in *Welwitschia* are described as not occurring in primary pit-fields (see Evert et al. 1973b). The differentiation of a sieve area starts with the formation of the median cavity and continues with the widening of the plasmodesmata, a process which may not be synchronized on both sides of the cavity if the respective sieve elements are of different developmental stages (cf. Evert et al. 1973b; see also Chap. 4, Fig. 4.27). During the differentiation of a sieve area portions of the cell wall increase in thickness, causing the sieve areas to protrude from the other parts of the wall and the developed sieve pores to become longer than the plasmodesmata from which they were derived.

5.3.3 Parenchymatous Cells Associated with the Sieve Elements

Two types of parenchymatous elements are described in *Ephedra* by Alfieri and Mottola (1983), densely staining albuminous (= Strasburger) cells occurring in regular radial files of cells alternating with the sieve element files, and considerably smaller phloem parenchyma cells less frequently found in tangential bands and occasionally alternating with the sieve elements. Paliwal and Behnke (1973) did not detect the latter type in *Gnetum* and simply called the former (related) parenchyma cells. In their ultrastructural study of *Welwitschia* Evert et al. (1973b) found the two types to be distinct on the basis of their plastids and their connections to neighboring sieve elements. Phloem parenchyma cells contained chloroplasts with internal thylakoids well organized into grana and with starch grains, while Strasburger cells develop less elaborate plastids with only some parallel thylakoids (Evert et al. 1973b).

The Strasburger cells (as well as related parenchyma cells) of all three genera were found to be interconnected by plasmodesmata occurring in groups and united in the region of the middle lamella by median cavities. Their connections to the sieve elements are intermediate between a sieve area and a plasmodesmatal field. A cluster of sieve pores penetrates the wall on the sieve element side and joins in a median cavity with the plasmodesmata penetrating through the wall from the Strasburger cell. The sieve pores and the median cavity are shown to contain elements of agranular ER. The contents of the plasmodesmata remains obscure, although ER cisternae are often found in their vicinity (cf. Evert et al. 1973b; Behnke and Paliwal 1973).

These characteristics, and the fact that Strasburger cells of the gnetophytes also degenerate at the time when the associated sieve element dies, indicate that the specialized parenchyma cells of the gnetophytes are similar to the Strasburger cells of other gymnosperms (cf. Chap. 4).

Acknowledgment. Supported in part by the Deutsche Forschungsgemeinschaft.

References

Alfieri FJ, Mottola PM (1983) Seasonal changes in the phloem of *Ephedra californica* Wats. Bot Gaz 144: 240-246

Alosi MC, Alfieri FJ (1972) Ontogeny and structure of the secondary phloem of *Ephedra*. Am J Bot 59: 818-827

Behnke H-D (1974) Sieve-element plastids of gymnospermae: their ultrastructure in relation to systematics. Plant Syst Evol 123: 1-12

Behnke H-D (1986) Sieve element characters and the systematic position of *Austrobaileya* (Austrobaileyaceae) - with comments to the distinction and definition of sieve cells and sieve-tube members. Plant Syst Evol 152: 101-121

Behnke H-D, Paliwal GS (1973) Ultrastructure of phloem and its development in *Gnetum gnemon*, with some observations on *Ephedra campylopoda*. Protoplasma 78: 305-319

Boodle LA, Worsdell WC (1894) On the comparative anatomy of the Casuarinaceae, with special reference to the Gnetaceae and Cupuliferae. Ann Bot (London) 8: 231-264

Esau K (1969) The phloem. In: Zimmermann W, Ozenda P, Wulff HD (eds) Encyclopedia of plant anatomy, vol 5, pt. 2. Borntraeger, Berlin Stuttgart, 505 pp

Evert RF, Bornman CH, Butler V, Gilliland MG (1973a) Structure and development of the sieve cell protoplast in leaf veins of *Welwitschia*. Protoplasma 76: 1-21

Evert RF, Bornman CH, Butler V, Gilliland MG (1973b) Structure and development of sieve areas in leaf veins of *Welwitschia*. Protoplasma 76: 23-34

Paliwal GS, Behnke H-D (1973) Light microscopic study of the organization of phloem in the stem of *Gnetum gnemon*. Phytomorphology 23: 183-193

Parthasarathy MV, Pesacreta TC (1980) Microfilaments in plant vascular cells. Can J Bot 58: 807-815

Strasburger E (1891) Über den Bau und die Verrichtungen der Leitungsbahnen in den Pflanzen. Histologische Beiträge, Vol 3. Fischer, Jena, XXXII + 1000 pp

Thompson WP (1919) Companion cells in bast of *Gnetum* and angiosperms. Bot Gaz 68: 451-459

6 Dicotyledons

RAY F. EVERT

6.1 Introduction

The sieve elements of angiosperms are regarded as the most highly evolved among vascular plants, and those of the dicotyledons have received the greatest attention in the literature on phloem. By definition, the sieve elements of dicotyledons are sieve-tube members; that is, sieve elements in which some of the sieve areas are more highly specialized (having larger pores) than others and are localized on the walls to form sieve plates (Esau 1969). Typically, the sieve plates occur on the end walls, and the sieve-tube members are arranged end-on-end to form sieve tubes, the sieve-plate pores providing a high degree of protoplasmic continuity between the superimposed cells (Figs. 6.1–6.3). The protoplasts of dicotyledonous sieve-tube members typically contain P-protein (phloem protein). In addition to the presence of sieve plates and P-protein, the sieve-tube members of dicotyledons typically are associated with companion cells, specialized parenchyma cells closely related to the sieve-tube members both ontogenetically and functionally.

6.2 The Sieve-Tube Member Protoplast

The young sieve-tube member contains all of the cellular components characteristic of young plant cells and initially is indistinguishable from other young cells of the phloem or identifiable only by its close spatial association with its companion cell(s). As the sieve-tube member differentiates it undergoes profound changes, the major ones being the breakdown of the nucleus and of the tonoplast. A selective autophagy also results in loss of the ribosomes, dictyosomes, microtubules, and microfilaments. At maturity, the sieve-tube member retains a plasmalemma, endoplasmic reticulum (ER), plastids, and mitochondria, all of which occupy a parietal position within the cell. In addition, with few exceptions (see below), the mature sieve-tube members of dicotyledons contain some form of P-protein.

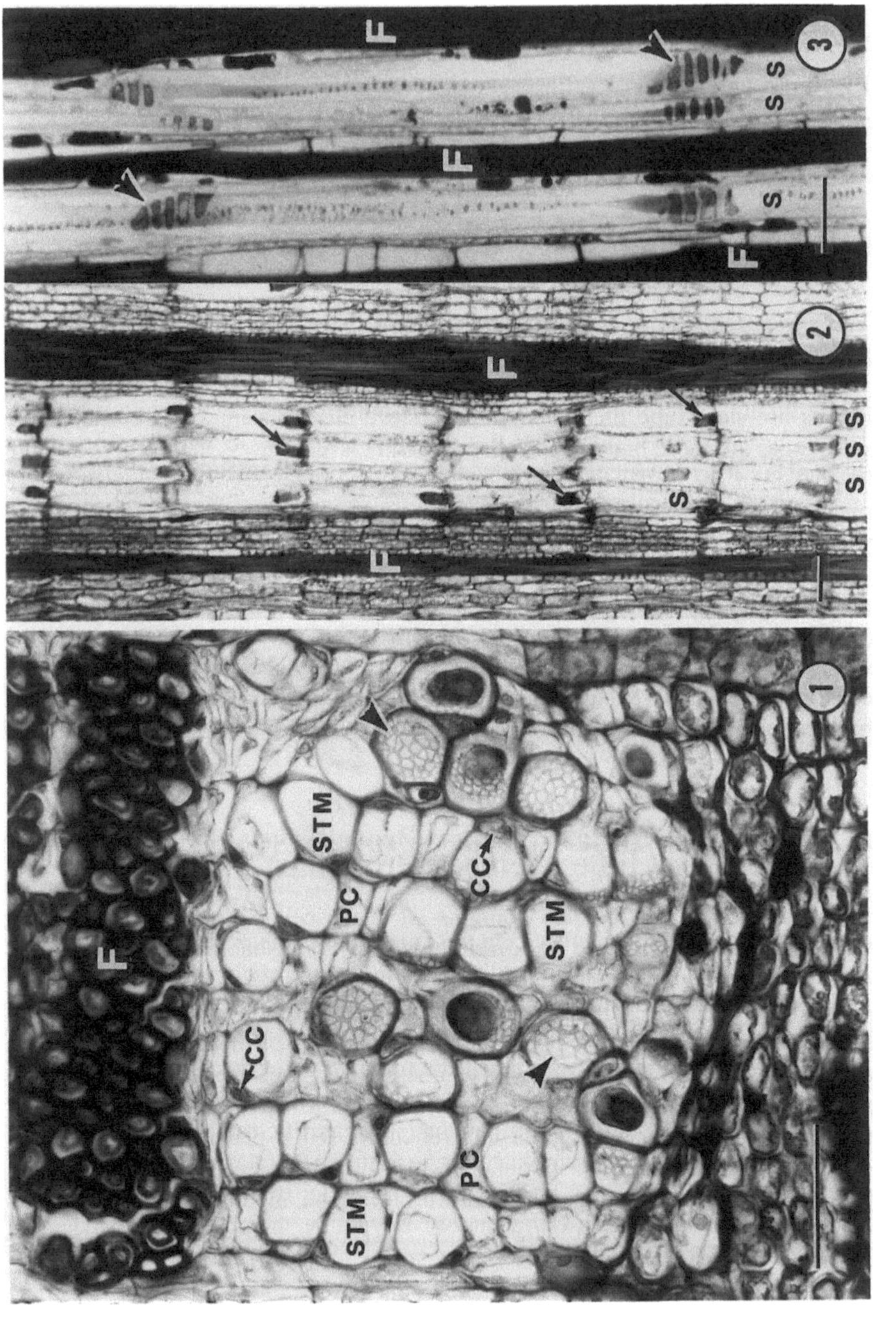
F
S
S
S
F
F
3
F
S
S
S
S
F
2
F
STM
PC
CC
STM
CC
PC
STM
1

6.2.1 Nucleus

Degeneration of the nucleus during the final stages of maturation has long been recognized as one of the major events of sieve-element ontogeny (Esau 1969). In dicotyledonous sieve-tube members, nuclear degeneration typically is by chromatolysis, a process involving the gradual loss of stainable contents (chromatin and nucleoli) and eventual rupture of the nuclear envelope (Esau and Gill 1971, 1972; Esau 1978b; Hoefert 1980; Thorsch and Esau 1981b; Behnke and Kiritsis 1983; Behnke 1986). Pycnotic degeneration, during which the chromatin forms a very dense mass prior to rupture of the nuclear envelope, has been reported, however, for the protophloem sieve-tube members of *Gossypium* (Thorsch and Esau 1981c). Although the nucleus usually disappears completely, either during or shortly after formation of the sieve-plate pores, remnants of nuclei may persist in some mature sieve-tube members (Evert et al. 1970; Deshpande 1975).

For several years it was believed that in some species of dicotyledons the nucleoli were released from the degenerating nuclei and remained intact within the mature sieve-tube member (Esau 1969). These structures, erroneously regarded as "extruded nucleoli," are now known to be proteinaceous cytoplasmic inclusions similar to P-protein (Ilker and Currier 1975; Oberhäuser and Kollmann 1977; Nehls et al. 1978) and not of nuclear origin (Deshpande and Evert 1970; Esau 1978a; Behnke and Kiritsis 1983).

The sieve-tube members of the Boraginaceae contain nuclear crystalloids that resemble the crystalline P-protein bodies that arise in the cytoplasm of fabaceous sieve-tube members (Esau and Magyarosy 1979a, b; Esau and Thorsch 1982). In *Amsinckia douglasiana* (Esau and Magyarosy 1979a, b), the nuclear crystalloids appear in the sieve-tube members before any other cytological features characterizing these cells. During nuclear degeneration, the crystalloids are released into the cell lumen where they may become fragmented or remain intact.

Figs. 6.1–6.3. Photomicrographs of secondary phloem of *Robinia pseudoacacia* (Figs. 6.1 and 6.2) and *Tilia americana* (Fig. 6.3). **Fig. 6.1.** Transverse section showing sieve-tube members *(STM)*, companion cells *(CC)*, and phloem parenchyma cells *(PC)*. Simple sieve plates of some sieve-tube members can be seen in face view *(arrowheads)*. The dark, thick-walled cells, above, are fibers *(F)*. × 384; *Bar line* = 50 μm. **Fig. 6.2.** Longitudinal (radial) section showing four sieve-tubes *(S)* which consist of sieve-tube members arranged end-on-end. The elongate bodies *(unlabeled arrows)* found in each of the sieve tubes (one per sieve-tube member) are regarded as a form of P-protein and appear as crystalline structures with the electron microscope (see Figs. 6.16 and 6.17). × 123; *Bar line* = 50 μm. **Fig. 6.3.** Longitudinal (radial) section showing three sieve tubes (S). The sieve-tube members in *Tilia* have compound sieve plates *(arrowheads)* which are borne on oblique end walls and are seen in face view in radial sections. × 196; Bar line = 50 μm. In the secondary phloem of *Robinia* and *Tilia*, groups of thick-walled fibers *(F)* alternate with groups of relatively thin-walled cells, including sieve-tube members, companion cells, and phloem parenchyma cells. (Figs. 6.1 and 6.2 from Evert 1984)

6.2.2 Endoplasmic Reticulum

In young sieve-tube members the endoplasmic reticulum is rough-surfaced and cisternal, the individual cisternae being randomly dispersed throughout the cytoplasm. At about the time the nucleus begins to degenerate, the ER cisternae begin to increase in number and to form stacks, some of which may become applied to the nuclear envelope (Hoefert 1980; Thorsch and Esau 1981b). In *Gossypium*, the ER of some stacks is combined with microtubules (MT) in an orderly arrangement, forming MT-ER complexes (Thorsch and Esau 1981b). In immature cells, the MT-ER complexes are associated exclusively with parts of the nuclear envelope. During disruption of the nuclear envelope the complexes remain associated with those parts and eventually are displaced toward the wall. In fully mature sieve-tube members, only a few remnants of the MT-ER complexes are to be found. Intercisternal MT have also been encountered in sieve-tube members of *Beta* and *Spinacia* at the beginning of degenerative changes leading to maturation of the cells (Esau and Hoefert 1980).

During the stacking of ER cisternae, the ribosomes disappear from the surfaces that face one another in a stack, although electron-dense material (intercisternal material) accumulates between contiguous cisternae (Esau and Hoefert 1980; Thorsch and Esau 1981a; Esau and Thorsch 1984). The intercisternal material sometimes appears as a single layer of alveoles in contact with the two membranes delimiting the intercisternal space. Ribosomes on the outer surfaces of the membrane stacks disappear concomitantly with the free ribosomes of the cytoplasm. Hence, in mature sieve-tube members, the ER typically is entirely smooth-surfaced.

The stacks of ER cisternae are most abundant at about the time the nucleus disappears and then occur near the cell wall, with the cisternae either parallel or perpendicular to the wall (Figs. 6.25-6.27). As the sieve-tube member becomes more mature, the ER may undergo further modification into convoluted, lattice-like, and tubular forms (Evert and Deshpande 1969; Oparka and Johnson 1978; Esau and Hoefert 1980; Thorsch and Esau 1981a). In most fully mature sieve-tube members, the ER is represented largely by a complex network - a parietal, anastomosing system - which lies next to the plasmalemma (Figs. 6.24-6.26; Esau 1978b; Hoefert 1980; Thorsch and Esau 1981a). The relationship between the parietal, anastomosing network and the other forms of ER is not clearly understood; however, the network appears during the later stages of differentiation and is the most consistent form of ER found in mature sieve-tube members. In freeze-fracture images of mature *Streptanthus* sieve-tube members obtained from callus cultures, the single, unstacked cisternae of the parietal network appear fenestrated and continuous with cisternae comprising the stacks (Sjolund and Shih 1983a). Only the mature protophloem sieve-tube members in root tips of *Phaseolus vulgaris* have been reported as lacking a parietal network of ER (Esau and Gill 1971).

Both plastids and mitochondria may become incorporated with the ER during differentiation (e.g., Thorsch and Esau 1981c). In mature sieve-tube members, the mitochondria may be partially surrounded by ER. The mitochondrial surfaces facing the cell lumen usually are free of ER in *Streptanthus* (e.g., Sjolund and Shih 1983a). ER cisternae are also associated with development of the sieve-plate pores (see Sect. 6.4).

Several roles have been suggested for the ER of the sieve-tube member. For example, the localization of acid phosphatase and nucleoside phosphatase activites (Evert 1977; Browning et al. 1980; Cronshaw 1980) on both the anastomosing system and stacked ER has led to the suggestion that the ER may be an important source of enzymes involved in autophagic phenomena (Oparka et al. 1981) and possibly a specific cytoplasmic differentiation related to the conducting function of the cell (Esau and Charvat 1975). Having calculated the ER-plasmalemma interface in mature sieve-tube members of *Streptanthus* to be almost 100%, Sjolund and Shih (1983a) suggested that the ER may play an important role in the phloem loading process, including the channeling of ATP from the mitochondria closely associated with the ER to proton-pumping ATPase in the plasmalemma. They noted that the fenestrae of the ER should enhance the passage of water and solutes from the plasmalemma to the interior of the cell. It has also been suggested that the parietal, anastomosing ER may serve in the distribution of substances entering the sieve-tube member from the companion cell along all wall surfaces, and thereby provide a greatly increased surface area for entry into the lumen of the sieve-tube member (Evert 1984). Wooding (1967) has suggested that stacking of the ER may represent a sequestering of the membrane system in an inactive form.

6.2.3 Plastids and Mitochondria

As mentioned previously, plastids and mitochondria are the only organelles to persist in the mature sieve-tube member. In very young sieve-tube members the two organelles are often difficult to distinguish from one another, for in such cells both organelles contain dense matrices and, in certain planes of section, the internal membranes of the plastids may appear numerous and resemble those of mitochondria. Early in development of the sieve-tube member, inclusions characteristic of the plastid type may appear and then they can serve as markers both to distinguish the plastids from the mitochondria and to distinguish the young sieve-tube member from other contiguous cell types of the phloem (Behnke 1981). As the plastid matures, its matrix becomes less dense. In mature sieve-tube members, the matrix is quite electron-transparent and often the internal membranes are sparse.

The plastids of dicotyledonous sieve-tube members accumulate two major kinds of ergastic substance, proteins and starch (Figs. 6.4–6.6). Accordingly, two major kinds of sieve-tube plastid can be recognized: S-type plastids, which contain only starch, and P-type plastids, which contain protein exclusively or in addition to starch. The P-type plastids can be divided into a number of subtypes

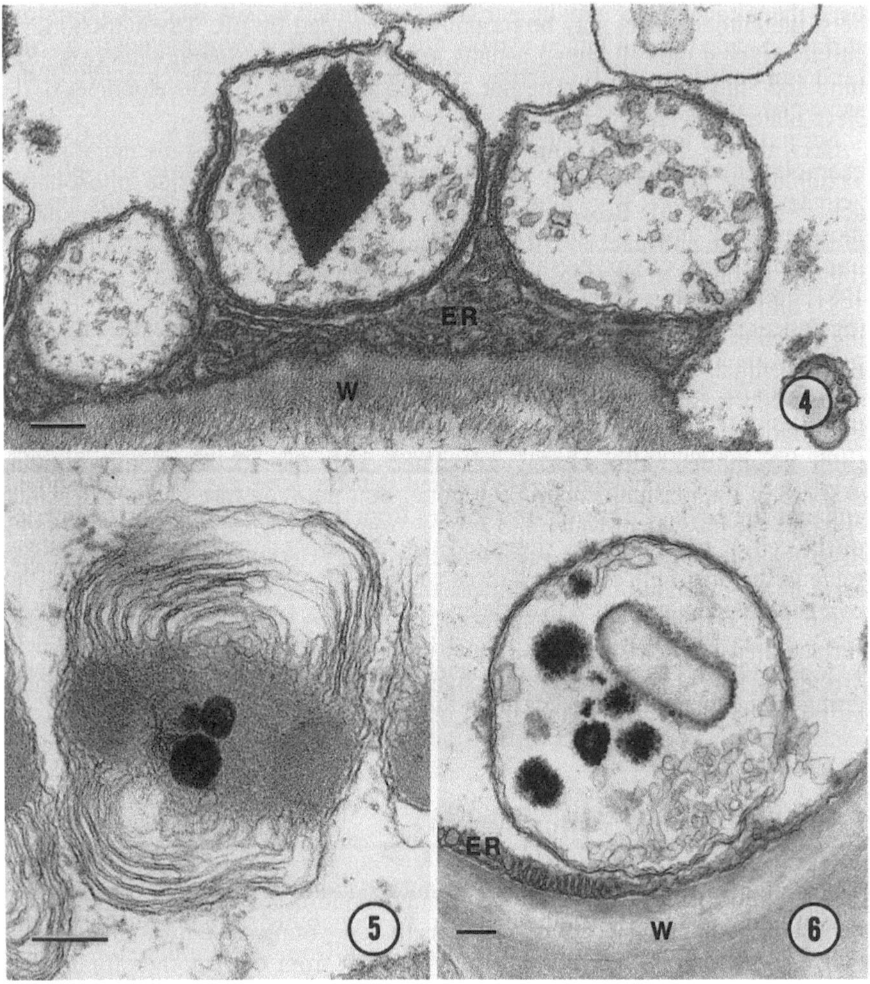

Figs. 6.4–6.6. Sieve-element plastids. **Fig. 6.4.** *Ulmus americana*. P-type plastid. × 40800; *Bar line* = 0.2 μm. **Fig. 6.5.** *Amaranthus retroflexus*. P-type plastid. × 52800; *Bar line* = 0.2 μm. **Fig. 6.6.** *Moricandia arvensis*. S-type plastid. × 26200; *Bar line* = 0.2 μm. *ER* endoplasmic reticulum; *W* wall. (Fig. 6.4 from Evert and Deshpande 1969; Fig. 6.5 from Fisher and Evert 1982)

and forms on the basis of their specific ultrastructural composition (Behnke 1981). Of all the ultrastructural features of sieve elements, their specific plastid types and forms provide the most useful characters for seed plant systematics and phylogeny. The development in wound sieve-tube members of specific plastid types from the amyloplasts of parenchyma cells has been cited as evidence that the structural composition of sieve-element plastids is "genetically fixed" (Behnke and Schulz 1983).

Unlike ordinary starch, sieve-tube starch stains brownish-red rather than blue-black with iodine (I_2KI). Using sequential enzymatic digestion and electron microscopy, Palevitz and Newcomb (1970) concluded that the sieve-tube starch in *Phaseolus vulgaris* is a highly branched molecule of the amylopectin type with numerous α (1→6) linkages at the branch points. The proteinaceous nature of the crystalline inclusions (Behnke 1975a) and of the filamentous material (Yapa and Spanner 1972) found in P-type plastids has been confirmed by use of proteolytic enzymes.

Although the release of starch grains or other inclusions from sieve-tube plastids has long been recognized by phloem workers as a response to injury and an indicator of disturbance in mature sieve-tube members (Esau 1969) and ultrastructural studies repeatedly have demonstrated that intact plastids survive in mature sieve-tube members (cf. Esau and Hoefert 1971; Behnke 1976), this conclusion has not been universally accepted. Some workers involved with the study of particle motion in sieve tubes of *Heracleum* included starch grains among "marker particles," which they envisaged attached to "wisps of plasmatic filaments" in what they believed to be undamaged sieve-tube members (Lee et al. 1971, Robidoux et al. 1973). Those involved with more recent studies of particle motion in *Heracleum* sieve tubes do acknowledge, however, that sieve tubes containing starch grains released from burst plastids are damaged at least "to some extent" (Barclay and Johnson 1982; Murphy 1986).

Among the surviving components of the sieve-tube member, the mitochondria undergo the least amount of structural change during differentiation. In some sieve-tube members the mitochondria may change from elongated to more nearly spherical forms and their matrix may become less dense. Nevertheless, in well-fixed sieve-tube members the mitochondria are normal in appearance, although they sometimes have somewhat dilated cristae.

Histochemical tests with diaminobenzidine (DAB) indicate that the mitochondria of mature sieve-tube members contain active cytochrome oxidases (Catesson and Liberman-Maxe 1974; Catesson 1980). In addition, nucleoside phosphatases (Gilder and Cronshaw 1973a, b) and acid phosphatase (Esau and Charvat 1975) have been localized in mitochondria of both mature and differentiating sieve-tube members.

6.2.4 Dictyosomes

Dictyosomes, or Golgi bodies, are abundant in young sieve-tube members, and they have clearly been implicated in cell wall formation. For example, in studies utilizing tritiated glucose (Pickett-Heaps 1967; Goff 1973; Arsanto and Coulon 1974, 1975), derivatives of the tritiated glucose are first found concentrated in the dictyosomes. The label then appears in smooth dictyosome vesicles, which subsequently fuse with the plasmalemma and discharge their contents into the wall. With Thiéry's method for the detection of polysaccharides, silver grains (indicating the presence of polysaccharides) are found in dictyosomes, dictysome vesicles, and the wall (Roland and Sandoz 1969; Catesson 1973; Arsanto and Coulon 1975).

Numerous smooth vesicles are produced by the dictyosomes during the period of cell wall formation. Coated vesicles bud off the smooth ones, and both kinds of vesicle apparently transport substances utilized in the synthesis of cell wall constituents (e. g., Hoefert 1979). With the cessation of cell wall formation, the dictyosomes become reduced in number and gradually disappear entirely.

6.2.5 Microtubules and Microfilaments

Microtubules (MT) have been reported next to the plasmalemma in differentiating sieve-tube members in several species of dicotyledons, but only one study - primarily on the protophloem sieve-tube members in the root of *Gossypium* (Thorsch and Esau 1982) - has been made relating changes in distribution, orientation, and abundance of MT to developmental changes in the differentiating cell.

In dividing protophloem elements of *Gossypium,* MT are found only as part of the phragmoplast. After division is completed, cortical MT, oriented at right angles to the long axis of the cell, begin to appear along the lateral wall. With the initiation of cell wall thickening, the number of MT increases sharply, reaches a plateau, and remains relatively constant throughout the period of wall thickening. During this period, the sieve-tube members contain normal nuclei, randomly distributed plastids and mitochondria, and scattered ER cisternae. With the cessation of cell wall thickening, the number of MT along the lateral walls declines. At the same time, nuclear chromatin aggregates, the plastids and mitochondria migrate to a parietal position in the cell, and ER cisternae accumulate into convoluted masses. During the final stages of maturation, the MT typically disappear entirely from the cell. MT also are found adjacent to developing sieve plates where they appear to be less numerous than those near the lateral walls during the plateau phase. The MT bordering the sieve plate lie parallel to the surface but otherwise are randomly oriented.

Microfilaments are common constituents of differentiating sieve-tube members (Parthasarathy and Pesacreta 1980). They usually are found in the peripheral regions of the cell and are almost always oriented parallel to the long axis of the cell. With few exceptions they do not persist beyond the stage of nuclear degeneration.

6.2.6 Plasmalemma and Tonoplast

In well-fixed sieve-tube members, the plasmalemma has a distinct three-ply appearance in thin sections. In differentiating sieve-tube members, it commonly has an undulating outline, in part a reflection of dictyosome vesicle activity associated with a delivery of substances for cell wall synthesis.

Presently, the osmotically generated pressure-flow mechanism is generally regarded as the only viable explanation for the long-distance transport of assimilates through the sieve tube (Evert 1982). With regard to that mechanism, the plasmalemma, which maintains its integrity and differentially permeable properties throughout the life of the sieve-tube member, is the most important protoplasmic component of the cell. Putative ATPase and other nucleoside phosphatase activity has been localized at the plasmalemma of sieve-tube members in a number of dicotyledons (Gilder and Cronshaw 1973a, b, 1974; Yapa and Spanner 1974; Catesson 1980). The presence of putative ATPase activity has been used as evidence in support of ATPase-mediated sucrose-proton cotransport in phloem loading (Giaquinta 1980). A high density of intramembrane particles (IMP) found in freeze-fracture replicas of the exterior half of the plasmalemma of *Streptanthus* callus sieve-tube members has been intrepreted as a morphological manifestation of the highly specialized transport activities carried out by such cells (Sjolund and Shih 1983b). The IMP potentially represent the sites of proton-pumping ATPase and of specific carriers for sucrose and amino acids.

The tonoplast, by contrast to the plasmalemma, breaks down during the final stages of maturation so that the delimitation between vacuole and cytoplasm ceases to exist. The absence of a tonoplast in mature sieve-tube members has been recorded in a wide variety of dicotyledons examined with both light and electron microscopes (Esau 1969). In some woody dicotyledons the tonoplast may persist until very late in maturation of the sieve-tube member (Evert and Deshpande 1969; Evert et al. 1969).

6.2.7 P-Protein

As mentioned previously, dicotyledonous sieve-tube members typically contain P-protein. With few exceptions [e.g., protophloem elements in roots of *Nicotiana tabacum* (Esau and Gill 1972); metaphloem elements in the aerial stem of the holoparasite *Epifagus virginiana* (Walsh and Popovich 1977)], P-protein has been encountered in the sieve-tube members of all dicotyledons examined thus far with the electron microscope, including those of such primitive Ranalean representatives as *Drimys* (Spanner and Moattari 1978; Behnke and Kiritsis 1983; Dute and Sharkey 1985; Thorsch and Esau 1985), *Trochodendron* (Jørgensen et al. 1975), *Austrobaileya* (Behnke 1986), and *Degeneria, Liriodendron,* and *Magnolia* (Friis and Dute 1983).

The presence of a distinct proteinaceous inclusion in the protoplast of dicotyledonous sieve-tube members was detected as early as 1854 by Hartig in sieve-tube members of *Cucurbita pepo.* At that time, Hartig called the entire contents of the sieve element Schleim, but gradually botanists began to distinguish between the cytoplasm and the proteinaceous inclusion, and the term Schleim or slime (in German and English, respectively) was adopted for the proteinaceous inclusion (Esau 1969). With the introduction of electron microscopy, slime was found to have several morphological forms (amorphous, filamentous

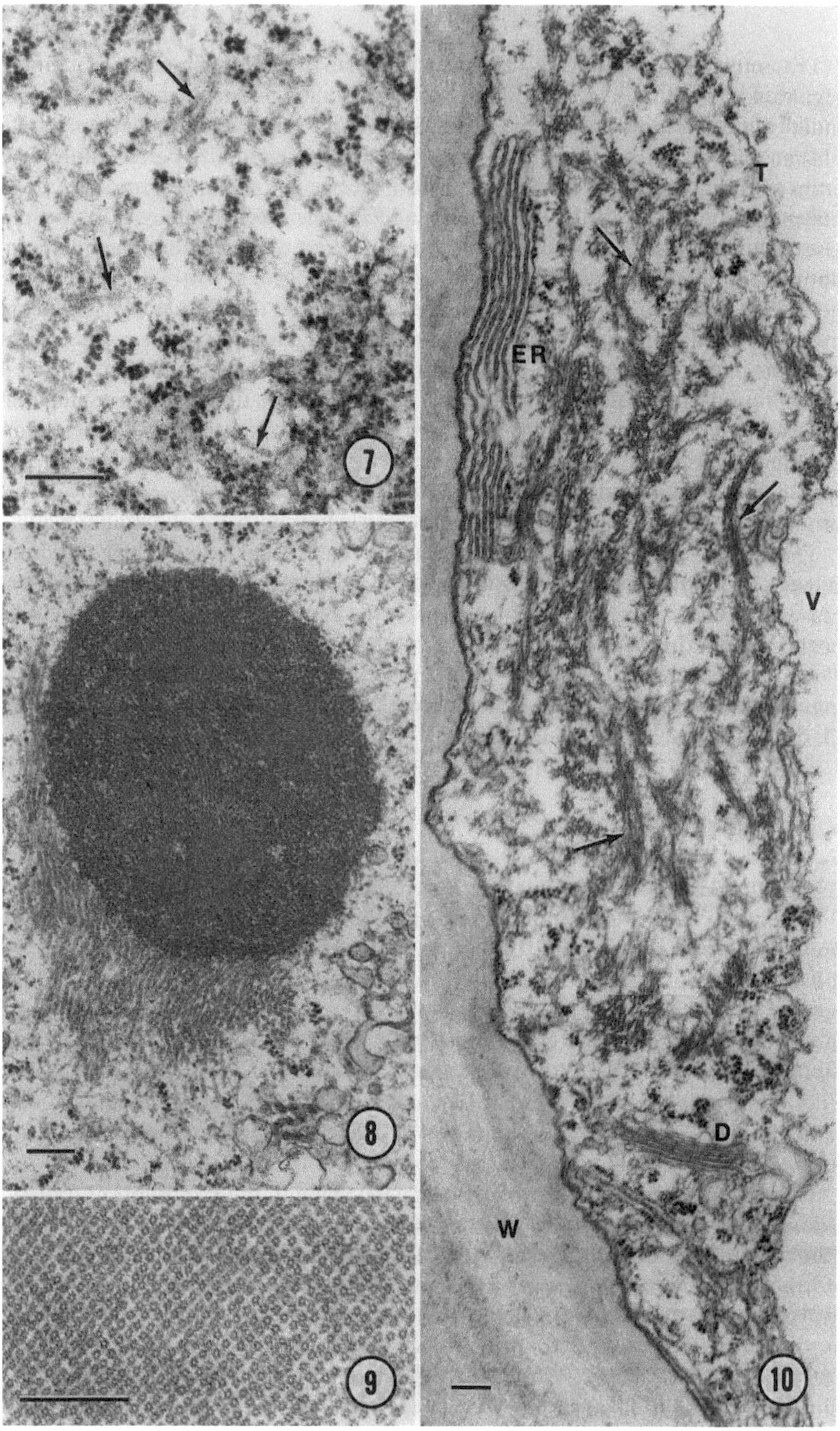

7
8
9
10
ER
T
V
D
W

or fibrillar, tubular, and crystalline) that vary from species to species and often within one species (Kollmann 1980), and this ultrastructurally distinguishable material was given its present name, P-protein (Cronshaw and Esau 1967).

Conformational changes from one morphological type of P-protein to another may take place within a differentiating sieve-tube member, and it has been suggested by several workers that the various forms of P-protein are constructed of basically similar or possibly identical subunits which undergo a self-assembly process (Parthasarathy and Mühlethaler 1969; Palevitz and Newcomb 1971; Cronshaw 1975a). The validity of this concept for P-protein interconvertibility has been questioned by Sabnis and co-workers, who have reported that the protein subunits in sieve-tube exudate clearly differ from species to species (see Chap. 13).

With the light microscope, P-protein is first discernible in the cytoplasm of young sieve-tube members as discrete bodies, of which there may be one or more per cell (Esau 1969). The P-protein bodies appear in the sieve-tube member protoplast after the precursor of the sieve-tube member has divided to give rise to the companion cell(s) and, as seen with the light microscope, their formation is the earliest indication of differentiation in the protoplast. Small at first, the P-protein bodies increase in size and eventually begin to fuse with one another. By this time chromatolysis of the nucleus has begun. When the P-protein is fully dispersed in the parietal layer of cytoplasm, the tonoplast breaks down and the sieve-tube member essentially is mature.

At the electron microscope level, precursors of the P-protein bodies may appear in the cytoplasm as small filaments in close spatial association with helical polysomes (Fig. 6.7; Behnke 1974). While the nucleus is still present, the protein filaments aggregate, forming small groups that enlarge to form prominent P-protein bodies (Fig. 6.8). Soon after P-protein synthesis is completed the ribosomes disappear. P-protein bodies often appear as aggregates of tubules (Fig. 6.9). In some species, the tubular aggregates are preceded by aggregates of fibrous material, from which it has been suggested the tubules may be assembled. Among the cytoplasmic components present at the time of P-protein synthesis, those most often implicated with that process, besides ribosomes, are the ER and spiny vesicles (Cronshaw 1975a; Arsanto 1982). It is pertinent to note that the implication of spiny vesicles in P-protein formation rests largely on

◁———

Figs. 6.7–6.10. P-protein. **Fig. 6.7.** *Cucurbita maxima*. Small filaments of P-protein *(arrows)* in cytoplasm of young sieve element. × 55250; *Bar line* = 0.2 µm. **Fig. 6.8.** *C. maxima*. Dense P-protein body composed of fibrillar components. Tubular components of P-protein, below, border the fibrillar components. × 35300; *Bar line* = 0.2 µm. **Fig. 6.9.** *C. maxima*. Part of a P-protein body consisting of tubular components. × 77750; *Bar line* = 0.2 µm. **Fig. 6.10.** *Ulmus americana*. Transverse section of portion of differentiating sieve element showing network composed of aggregates of striated, filamentous P-protein *(arrows)*. The free surface of a stack of endoplasmic reticulum cisternae *(ER)* is still partly studded with ribosomes. × 27050; *Bar line* = 0.2 µm. *D* dictyosome; *T* tonoplast; *V* vacuole; *W* wall (Fig. 6.10 from Evert and Deshpande 1969) their spatial association with P-protein in parenchyma cells associated with sieve-tube members (Esau 1971; Deshpande 1974b). In sieve-tube members such association has been demonstrated only once (in *Cucurbita maxima;* Cronshaw and Esau 1968).

As the P-protein bodies increase in size and begin to spread out in the cytoplasm, the filaments often assume a striated appearance (Fig. 6.10). This process begins at about the time degenerative changes have begun and the ER is being reorganized. It has been suggested that the striated appearance of the P-protein filaments results from stretching of the "super-double helix" proposed by some investigators as comprising the structure of P-protein tubules (D. M. Lawton and Johnson 1976; Arsanto 1982). According to some accounts, the individual filaments move apart and become dispersed in the surrounding cytoplasm, after which they become evenly distributed in the lumen of the sieve-tube member following breakdown of the tonoplast (Cronshaw 1975a). Other accounts indicate that dispersal of the P-protein bodies results in formation of a parietal network of very fine strands, which remains in the parietal position after the tonoplast disappears (Evert and Deshpande 1969; Evert et al. 1969).

The suggestion as to what constitutes the normal distribution of P-protein within the lumen and sieve-plate pores of mature sieve-tube members has been the subject of numerous investigations and long has represented the most controversial aspect of sieve-element structure (Esau 1969). The main problem is the extreme sensitivity of the sieve-tube member protoplast, with its high hydrostatic pressure, to manipulation and chemical fixation. When the sieve tube is severed at the time of sampling or the differentially permeable properties of its plasmalemma are destroyed by fixative, the contents of the sieve tube surge toward the site of pressure release, commonly resulting in the accumulation of P-protein as "slime plugs" on the sides of the sieve plates away from the sites of pressure release. Other protoplasmic components (mitochondria, plastids, and even ER) also may become displaced from their normal locations within the sieve tube during the surging phenomenon. Commonly, the sieve-tube plastids are ruptured and their contents released during surging. Many attempts have been made to prevent, or at least to minimize, surging and the consequent damage inflicted upon the sieve elements at the time of pressure release but with variable results (consult Spanner 1978a for references).

In well-preserved sieve-tube members showing little or no effects of surging (see Evert 1977 or 1982 for criteria used in judging quality of preservation), the P-protein either occurs as a loose network of filaments occupying the entire lumen of the cell (e.g., Cronshaw and Anderson 1971; Behnke and Kiritsis 1983; Behnke 1986), or it is parietal in position (Figs. 6.11 and 6.12). The first reports of a parietal distribution for P-protein came from Evert and co-workers after detailed light microscope (Evert et al. 1969) and electron microscope (Evert and Deshpande 1969) studies of sieve-tube member development in the secondary phloem of *Ulmus americana*. In another study (Evert et al. 1973), the P-protein in over 90% of the mature sieve-tube members in hypocotyls of

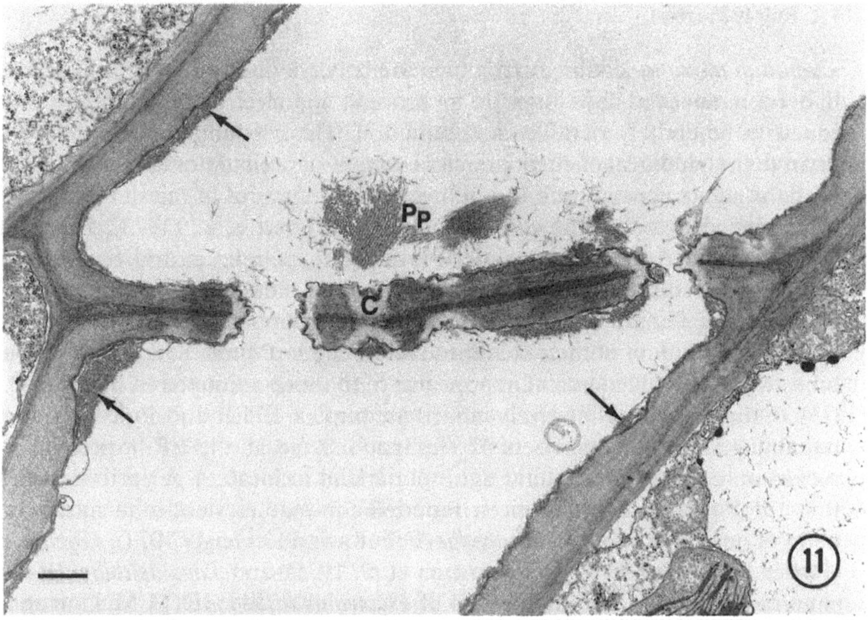

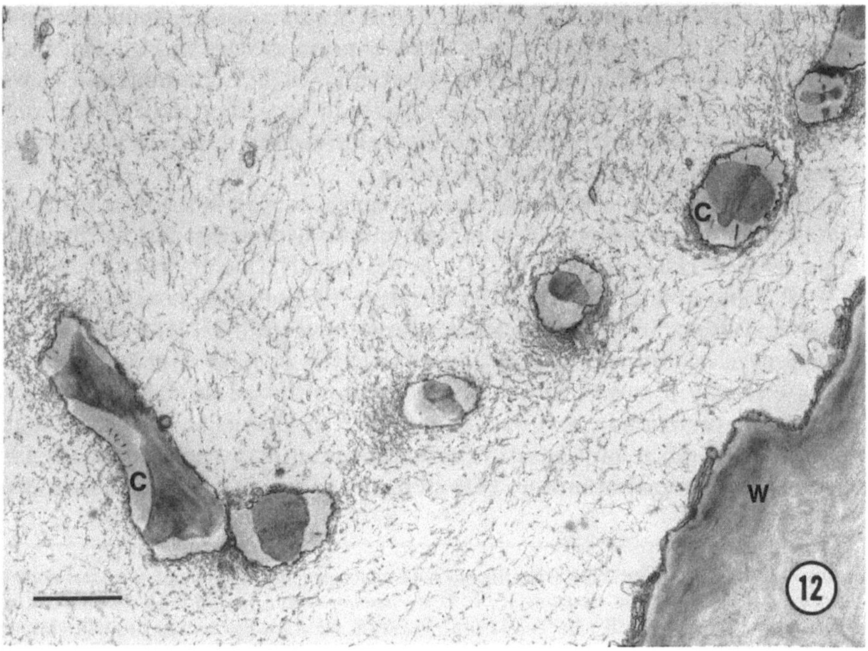

Figs. 6.11 and 6.12. Longitudinal sections of portions of mature sieve-tube members, showing distribution of P-protein and sieve plates with unoccluded pores. **Fig. 6.11.** *Cucurbita maxima.* Here the sieve-plate pores are largely devoid of P-protein, which is parietal in distribution within the sieve-tube members. × 14350; *Bar line* = 0.5 μm. **Fig. 6.12.** *Ulmus americana.* Here the P-protein is more or less evenly distributed throughout the lumina of the sieve-tube members and sieve-plate pores. × 24600; *Bar line* = 0.5 μm. *C* callose; *Pp* P-protein; *W* wall (Fig. 6.11 from Evert et al. 1973; Fig. 6.12 from Evert and Deshpande 1969)

Cucurbita maxima seedlings, from which the cotyledons and first foliage leaves had been removed 2 days prior to processing for electron microscopy, was found to be entirely parietal in distribution. The cotyledons were removed to deprive the seedlings of their principal source of assimilates and, hence, lower both the solute concentrations and hydrostatic pressures of the sieve tubes. In a study designed to test the results obtained by Evert et al. (1973), Deshpande (1984) examined the structure of sieve tubes from petioles and hypocotyls of *C. maxima* seedlings that had been kept in total darkness for 2 or 3 days. The most striking feature of the mature sieve-tube members of that study was the parietal distribution of their contents, including the P-protein. Some of the sieve tubes were virtually identical in appearance to those encountered by Evert et al. (1973), that is, with a relatively compact complex of ER and P-protein. In the remaining sieve-tube members of Deshpande's tissue, the ER/P-protein complexes were less compact, although still parietal in location. A parietal distribution for P-protein has also been reported for mature sieve-tube members in petioles or lamina veins of *Beta vulgaris* (Fellows and Geiger 1974), *Glycine max* (Fisher 1975), *Cucurbita pepo* (Turgeon et al. 1975), and *Populus deltoides* (Russin and Evert 1985), and in the stem of *Phaseolus multiflorus* (D. M. Lawton and Newman 1979). In addition, P-protein often occurs in a parietal position in senescing sieve-tube members whose sieve-plate pores have been occluded with definitive callose (Evert et al. 1969; Evert et al. 1973; D. M. Lawton and Newman 1979). Such elements might be expected to be least affected by the surging phenomenon that occurs when sieve tubes are severed, reflecting the normal distribution of P-protein within the lumen.

It is quite clear that P-protein does not normally exist in the form of transcellular strands or as a component of such strands as hypothesized by Thaine and his co-workers (Jarvis et al. 1973; De Maria and Thaine 1974; Thaine et al. 1975), Fensom and his co-workers (Fensom 1972; Robidoux et al. 1973; Fensom and Williams 1974), and others (Dempsey et al. 1975). The study of sieve-element development in *Ulmus americana* (Evert et al. 1969) referred to earlier was undertaken in part to find an explanation for the presence of strands in mature sieve-tube members, if such strands were determined not to be normal constituents of the protoplast. After examination of a very great many transverse and longitudinal sections of both fresh and variously treated tissues, it was

Figs. 6.13–6.18. Crystalline P-protein. **Fig. 6.13.** *Quercus alba*. Compound spherical body near sieve plate in mature sieve element. × 5850; *Bar line* = 1 μm. **Fig. 6.14.** *Q. alba*. Detail of spherical body. × 76050; *Bar line* = 0.2 μm. **Fig. 6.15.** *Rhus glabra*. Stellate body in mature sieve element. × 14050; *Bar line* = 0.5 μm. **Fig. 6.16.** *Robinia pseudoacacia*. Transverse view of spindle-shaped body in immature sieve element. × 12600; Bar line = 0.5 μm. **Fig. 6.17.** *R. pseudoacacia*. Longitudinal view of spindle-shaped body in mature sieve element. × 23800; *Bar line* = 0.5 μm. **Fig. 6.18.** *Tilia americana*. Portion of spherical inclusion. The peripheral region *(above)* is composed of rodlike components, while the more dense, central region *(below)* shows little or no substructure. × 50550; *Bar line* = 0.2 μm. The spherical bodies of *Quercus* and *Tilia* were once regarded as extruded nucleoli (Fig. 6.13 courtesy of S. E. Eichhorn. Figs. 6.14, 6.15 and 6.18 from Deshpande and Evert 1970)

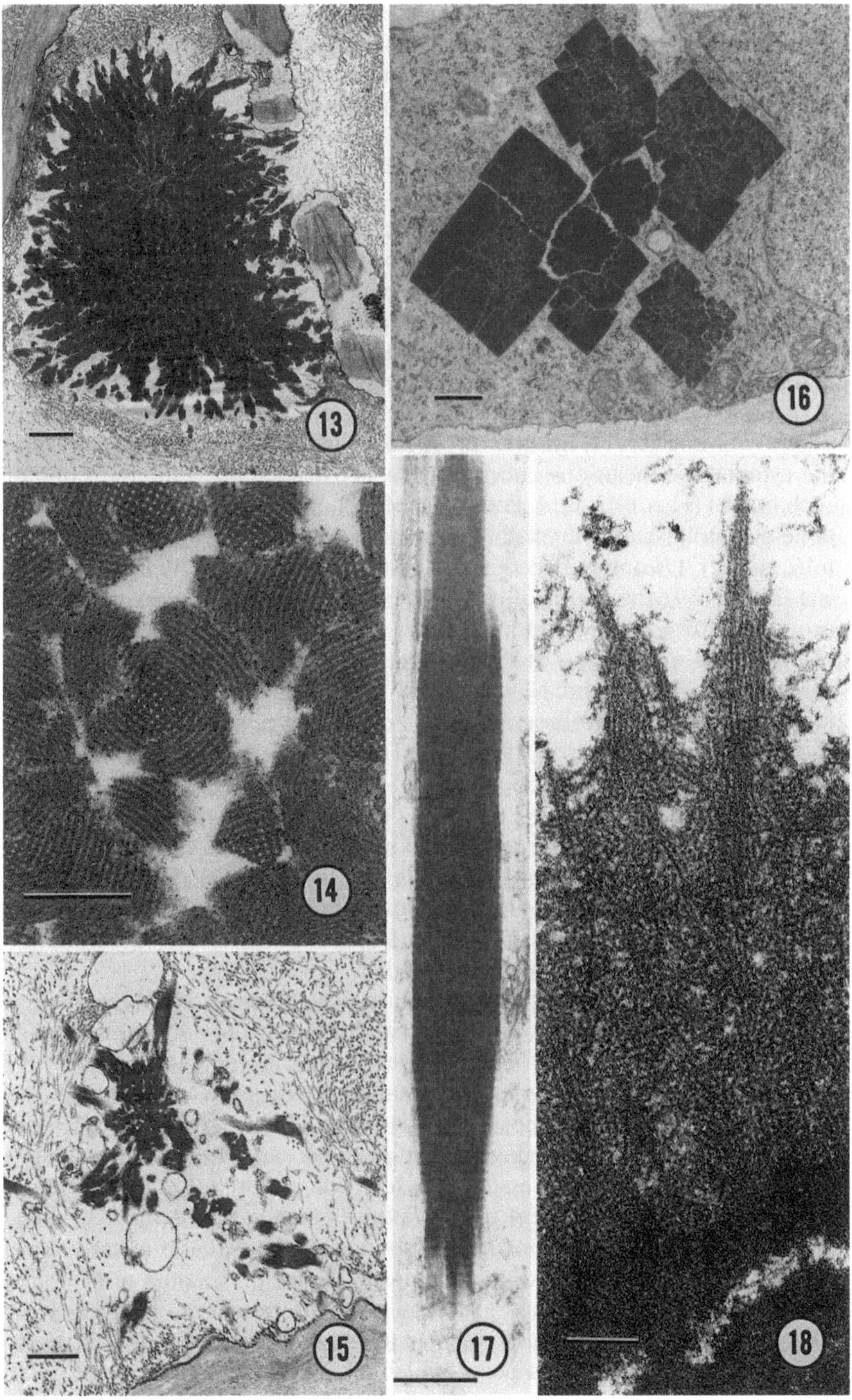
13
14
15
16
17
18

clear that the longitudinally oriented slime strands were artifacts arising through distortion of a network of very fine slime strands, which normally occupies a parietal position in mature sieve-tube members. Structures identified as slime strands during the *Ulmus* study were similar in appearance to structures earlier interpreted by Thaine and his associates (Thaine et al. 1967) to be either cytoplasmic strands or constituents of cytoplasmic strands. In *Ulmus*, some of these structures were slime sacs that extended the length of the sieve-tube members and had the same diameters as the pores through which they protruded. In a later study, Johnson et al. (1976) confirmed the conclusions of Evert et al. (1969) that the strands observed in sieve-tube members consisted of P-protein. They suggested that the strands may be formed by flow during translocation as well as by flow due to injury. No developmental evidence exists to support the normal presence of transcellular strands in mature sieve-tube members.

In some taxa (basically woody families), the P-protein bodies disperse only partially or not at all (Figs. 6.13-6.18; Behnke 1981). As mentioned previously, the cytoplasmic inclusions once regarded as extruded nucleoli are examples of such bodies (Figs. 6.13, 6.14, and 6.18). Most notable examples are the tailed or tailless spindle-shaped crystalline P-protein bodies of the Fabaceae (Figs. 6.2, 6.16, and 6.17). Long regarded as "persistent slime bodies" by light microscopists (Esau 1969), the results of early electron microscope investigations on fabaceous sieve-tube members indicated that these inclusions dispersed into a mass of fine fibrils or filaments in the mature cell (Laflèche 1966; Wergin and Newcomb 1970; Palevitz and Newcomb 1971). The results of more recent studies indicate, however, that these inclusions do in fact remain intact (Fisher 1975; Esau 1978a; D. M. Lawton 1978a, b). The crystalline P-protein bodies of the Fabaceae are a reliable taxonomic character of that family, having been found in all investigated species of the family. At present, four major kinds of crystalline P-protein bodies can be recognized in dicotyledonous sieve-tube members: spindle-shaped, compound spherical, stellate, and globular (Behnke 1981).

Biochemical data so far amassed clearly indicate that P-protein is neither actin- nor tubulin-like (see Chap. 13) and thus negate any force-generating role in assimilate transport proposed to it earlier.

What role does P-protein play in mature sieve-tube members? Together with callose, P-protein apparently serves to seal the sieve-plate pores of injured sieve tubes, preventing the loss of assimilates from the conduit (Eschrich 1975). Although callose may be deposited remarkably fast at the sieve plates in reaction to injury, its rate of deposition varies considerably from species to species (Evert and Derr 1964). Plugging of sieve-plate pores with P-protein is an almost instantaneous reaction to pressure release in active sieve tubes. P-protein plugs, or slime plugs, may represent the sieve tube's first line of defense against the loss of assimilates, with wound callose shoring up the defenses at variable rates. The smaller the size of the sieve-plate pores, the lesser the amount of P-protein needed to accomplish the task. The absence of P-protein in the sieve elements of gymnosperms and seedless vascular plants may be related to the relatively small size of their sieve-area pores, which often contain numerous membranes.

The absence of P-protein from the protophloem sieve tubes of some dicotyledons may also be related to the relatively small size of their sieve-plate pores. Besides P-protein serving a sealing function, the lectins of the P-protein may immobilize bacteria and fungi to the cross-linked filaments that seal wounded sieve tubes and thus maintain sterility (Read and Northcote 1983 a, b). Lectin bridges may also serve to anchor P-protein to the membranes of the sieve tube (Sabnis and Hart 1979; Kollmann 1980).

6.3 The Wall

The walls of dicotyledonous sieve-tube members commonly are described as primary and, although occasional reports of lignified walls appear in the literature, standard microchemical tests of the walls usually give positive reactions for only cellulose and pectin (Esau 1969). Although variable in thickness, the sieve-tube member walls frequently are thicker than those of contiguous parenchyma cells, a character that may facilitate recognition of the sieve-tube member.

In many species, the sieve-tube member wall consists of two morphologically distinct layers, a relatively thin outer layer and a more or less thick inner layer (Figs. 6.19 and 6.20). In fresh sections, the distinct inner layer exhibits a glistening or "pearly" appearance; hence, the use of nacreous or nacré in its description (Esau 1969). By comparison with the outer wall layer, the nacreous layer is less cellulosic and pectin-poor (Esau and Cheadle 1958; Botha and Evert 1981). The degree of thickening of the nacreous layer may be an indication of the stage of development of the cell. In the primary phloem, the nacreous wall commonly attains maximum thickness as the cell is approaching maturity and then disappears at about the time the sieve-tube member reaches maturity. In secondary phloem, the nacreous layer may or may not be reduced in thickness as the cell ages (Esau and Cheadle 1958; Gilliland et al. 1984). In some sieve-tube members, the nacreous layer becomes so thick as almost to occlude the lumen of the cell.

Utilizing mild extraction procedures for the removal of noncellulosic wall components, Deshpande (1976 b) was able to demonstrate that the nacreous thickenings in sieve-tube members of *Cucurbita* petioles are polylamellate; by comparison, the outermost parts of the walls are similar to the walls of ordinary parenchyma cells in having lamellae composed of thinly distributed microfibrils (Deshpande 1976 a). In addition, the microfibrils of the outer wall readily separated by certain treatments such as pectinase extraction, while the lamellae of the nacreous layer, with its densely packed microfibrils, did not. Catesson (1982) also found the inner parts of the sieve-tube member walls in the secondary phloem of *Populus* and *Acer* to have a polylamellate structure. Behnke (1971 a) suggested that the predominant parallel arrangement of the fibrils in nacreous walls of *Annona* and *Myristica* sieve tubes may be one condition for

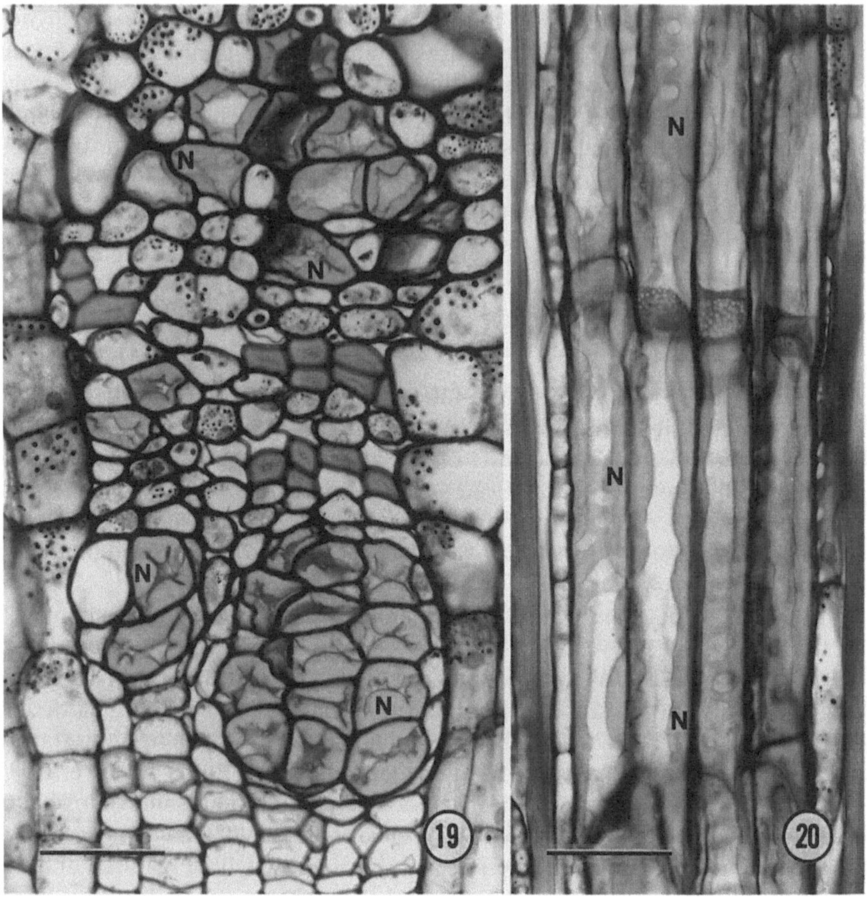

Figs. 6.19 and 6.20. Photomicrographs of secondary phloem of *Magnolia kobus*. Transverse (Fig. 6.19) and radial longitudinal (Fig. 6.20) sections showing mature sieve elements with nacreous thickenings *(N)*. × 350; *Bar line* = 50 μm

their pearly luster. He also noted that the fibrils at the boundary of the sieve-tube lumen appeared to be especially tightly packed together.

Use of the term nacreous has been extended to describe a distinctive inner wall layer discernible in sieve-tube members with the electron microscope (Esau 1969; Cronshaw 1975b). When sectioned obliquely, this layer - which may even be found in sieve-tube members with relatively thin walls - exhibits a distinct striate pattern. In addition, after fixation with glutaraldehyde-osmium tetroxide, it may be considerably more electron opaque than the rest of the wall. Lucas and Franceschi (1982), in an ultrastructural and histochemical study of the sieve-element walls in minor veins of *Beta vulgaris* and 13 other species, demonstrated the presence of a thin, dense, apparently pectin-rich layer of nonmicrofibrillar material adjacent to the plasmalemma. Although this pectin-rich layer appears to correspond to the layer exhibiting striations - that is, the nacreous layer in the extended use of the term - Lucas and Franceschi did not relate the two. They noted, however, that the pectin-rich layer does not represent nacreous wall as defined by Esau and Cheadle (1958) because it extends into the region of the sieve plate.

6.4 The Sieve Plate

As mentioned previously, sieve-tube members are characterized by the presence of sieve plates. Some sieve plates bear only a single sieve area (simple sieve plates; Figs. 6.1 and 6.2), while others bear two or more (compound sieve plates; Fig. 6.3). Among the dicotyledons, only the sieve elements of the primitive, woody species *Austrobaileya scandens* have been reported as lacking sieve plates (Srivastava 1970; Behnke 1986). However, the sieve elements of *A. scandens* share many of the characteristics of angiospermous sieve-tube members, most notably the presence of P-protein and association with companion cells. Hence, the sieve elements of *A. scandens* may be considered primitive sieve-tube members. (See end of Chap. 3 for discussion of terminology.)

In routine preparations of functional (conducting) phloem, the sieve-plate pores typically are lined with the wall constituent callose, a polysaccharide (β-1,3 glucan) that stains blue with resorcin blue and aniline blue and yields mainly glucose when hydrolyzed (Eschrich 1975). Once considered to be a normal constituent of sieve plates and lateral sieve areas of functional sieve-tube members, it is now assumed that most, if not all, of the callose associated with the pores of such elements is deposited there in response to mechanical injury or some other kind of stimulation (Esau 1969; Eschrich 1975). Not all of the callose associated with sieve pores, however, is "wound" callose. It has long been known that callose normally accumulates at the sieve plates and lateral sieve areas of senescing sieve elements. This "definitive" callose disappears some time after the sieve element dies. Callose may also accumulate at the sieve plates and lateral sieve areas of secondary phloem sieve elements that function for more than one growing season (Davis and Evert 1970). Typically, this "dormancy" callose is deposited in the fall and then removed in early spring during reactivation of the overwintering sieve elements. Callose also appears to play a role in the development of the sieve plate and lateral sieve areas.

Sieve-plate development has been thoroughly documented in ultrastructural studies, and a fairly uniform picture of its development has emerged (Deshpande 1974a, 1975; Esau and Thorsch 1984, 1985). In young sieve-tube members, the future sieve plates are penetrated by variable numbers of plasmodesmata, each of which is associated with an ER cisterna on both sides of the wall (Fig. 6.21). The pore sites first become distinguishable from the rest of the wall by the appearance of callose beneath the plasmalemma around each plasmo-

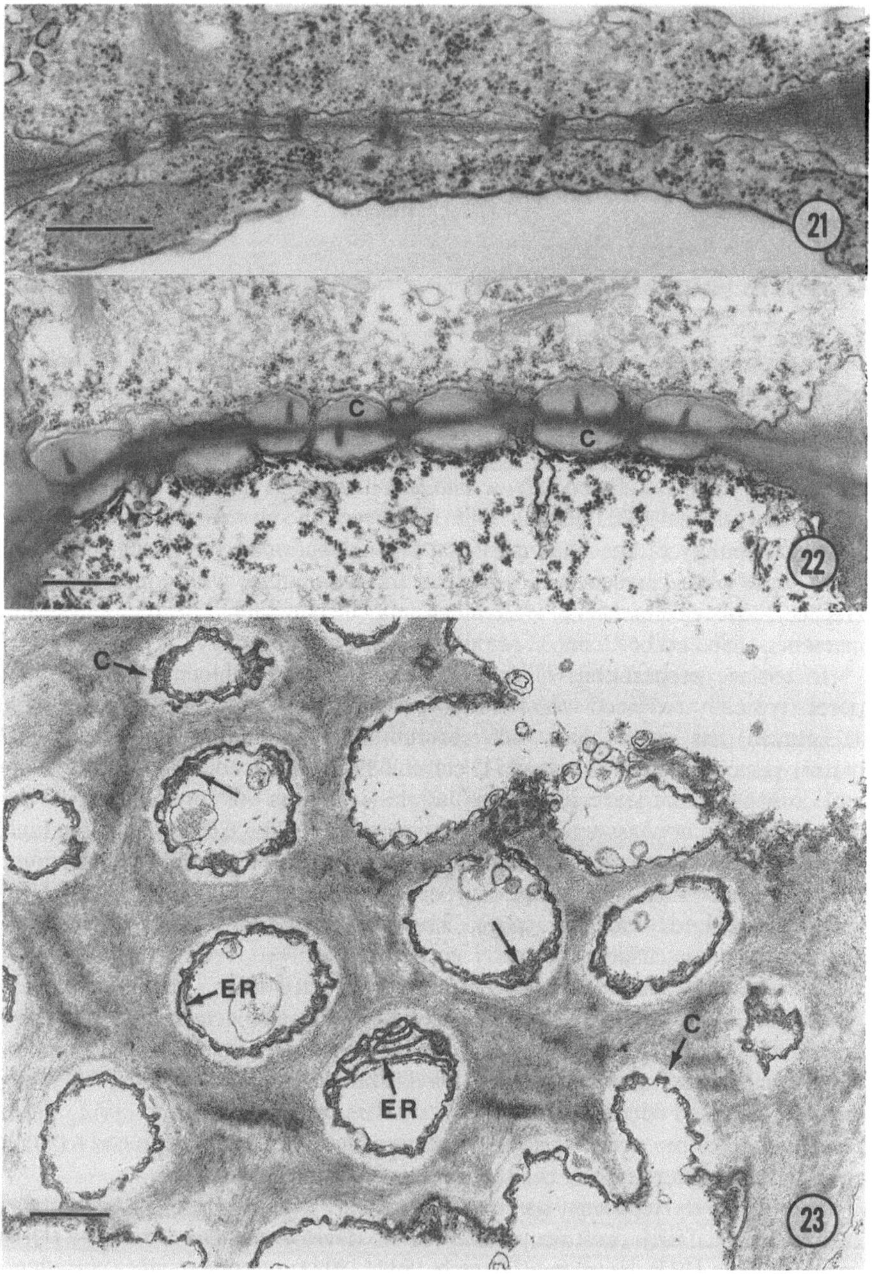

Figs. 6.21–6.23. Sieve plates. **Fig. 6.21.** *Beta vulgaris.* Differentiating sieve plate in wall between two sieve-tube members before deposition of callose platelets. The plasmodesmata mark the sites of the future sieve-plate pores. × 30400; *Bar line* = 0.5 μm. **Fig. 6.22.** *B. vulgaris.* Callose platelets *(C)* form collars around the ends of the plasmodesmata at the pore sites. The protoplast of the sieve-tube member below the differentiating sieve plate is more advanced

desma on both sides of the wall (Fig. 6.22). The localization of callose at the pore sites commonly is the first character distinguishing the plasmodesma-cell wall relation in a sieve plate from that in a parenchyma cell wall, although a barrel-shaped, electron dense "periplasmodesmatal" region may be apparent around the plasmodesmata of the future sieve plate before callose deposition is initiated (Deshpande 1975). The paired callose deposits, commonly called platelets, assume the form of collars or cones, which undergo rapid enlargement and initially may exceed the rest of the wall in their rate of thickening. Thickening of the wall between pore sites may overtake the callose platelets; then the pore sites appear as depressions in the plate. Once the pore sites are covered by callose, the cellulosic wall parts sandwiched between platelets remain thin. Although growth of the callosic and cellulosic parts of the sieve plate are separate processes, some substitution of callose for cellulose may take place during growth of the pore sites (Esau and Thorsch 1984, 1985).

Perforation begins at about the time of nuclear degeneration with the removal of wall material in the region of the middle lamella surrounding the plasmodesma. In some instances a median cavity is formed initially and then further simultaneous removal of the callose platelets and of the wall substance sandwiched between them results in formation of the pore (Evert et al. 1971; Deshpande 1974a, 1975). In others, lysis in the region of the middle lamella appears to result in a merging of the opposing callose platelets so that the young pore becomes uniformly lined with callose (Esau and Thorsch 1984, 1985). In either case, the ER cisternae remain closely appressed to the plasmalemma bordering the platelets throughout pore development. They are removed from the pore sites as the pores attain their full size.

The ER cisternae have been implicated in the deposition of callose at the pore sites. Both Catesson (1973) and Arsanto and Coulon (1975) have suggested that callose precursors may be transported in the ER and principally assembled in situ to the exterior of the plasmalemma. Esau (1969) also has suggested that the ER at the pore site may be involved with the laying down of the callose platelets and later with the release of enzymatic activity leading to the lysis of the callose and other wall constituents. Catesson (1973) noted that since acid phosphatase is present in the ER of young sieve-tube members, this enzyme may contribute to hydrolysis of the wall at the time of perforation. And what role does the callose platelet play? Possibly the platelets prevent or inhibit deposition of cellulose microfibrils at the sites of the future pores, although it has been suggested that the ER cisternae applied to the pore sites may play a similar role (Esau 1969). Esau and Thorsch (1985) have noted that since callose

in development than that of the sieve-tube member above the plate. × 19450; *Bar line* = 0.5 µm. **Fig. 6.23.** *Cucurbita maxima*. Face view of portion of mature sieve plate. The unoccluded pores are lined by narrow callose cylinders *(C)* and plasmalemma *(not labeled)*. Segments of endoplasmic reticulum *(ER)* and P-protein *(unlabeled arrows)* are also along margins of pores. × 22200; *Bar line* = 0.5 µm (Fig. 6.23 from Evert 1984)

is more easily digested enzymatically than cellulose, the deposition of callose may facilitate the eventual opening of the pores.

At maturity, the plasmalemma-lined sieve-plate pores in most well-preserved sieve-tube members contain little or no callose (Fig. 6.23). Some elements of ER may traverse the pores, lying to one side or closely appressed to the plasmalemma (Behnke 1971b; Evert et al. 1971). The distribution of P-protein within the pores of well-preserved sieve-tube members generally reflects the distribution of the P-protein within the lumina of those cells, i. e., either in a loose filamentous network or in a parietal position within the pore (Figs. 6.11, 6.12, 6.23). Hence, the pores are unoccluded by either callose or any cytoplasmic material (Evert et al. 1973; Giaquinta and Geiger 1973; Deshpande 1975, 1984; Fisher 1975; J. R. Lawton 1977; Walsh and Popovich 1977; Kallarackal and Milburn 1983; Sjolund et al. 1983). Some workers still contend that the pores of at least a fair proportion of sieve plates are normally occluded with P-protein (Spanner 1978a, b; see also discussion in Evert 1982).

6.5 The Lateral Sieve Areas

Sieve areas may occur on all lateral walls between sieve-tube members. In the walls between such cells, a sieve area on one side of the wall is opposed by a sieve area on the opposite side of the wall. The pores in the lateral walls are not always localized in distinct areas. For example, in *Ulmus americana* the pores of the radially oriented walls are localized in distinct areas with oval or round outlines, while those in the tangentially oriented walls are more or less diffuse in distribution (Evert et al. 1969).

Relatively little attention has been given to the structure and development of lateral sieve areas. In the secondary phloem of several dicotyledons, the pores of the lateral sieve areas were found to be basically similar to those of the sieve plates, differing from them primarily in size. In addition, development of the lateral sieve-area pores was essentially similar to that of the sieve-plate pores, involving callose platelets and formation of a median cavity as the first visible evidence that perforation of the pore site had begun (Evert et al. 1971). Many of the lateral sieve-area pores contained median cavities at maturity. In three of the seven woody dicotyledons examined *(Rhus, Robinia,* and *Tilia),* the median cavities of adjacent pores often merged during development and formed relatively large cavities, some of which were reminiscent of the large median cavities associated with the sieve areas of gymnospermous sieve cells. Such cavities were especially common in *Tilia,* where the pores had a tendency to aggregate in small groups within sieve areas. The merger of median cavities during formation of lateral sieve-area pores has also been reported for *Cucurbia maxima* but not as a common phenomenon (Deshpande 1975).

6.6 Parenchymatous Cells Associated with Sieve-Tube Members

The phloem of dicotyledons contains variable numbers of parenchymatous cells that differ from one another both structurally and functionally and in their degree of specialization in relation to the sieve-tube members (Esau 1969). Of the different kinds of parenchymatous cells found in the phloem, the companion cell is the one most intimately related to the sieve-tube member: not only is it derived from the same mother cell as its associated sieve-tube member, it remains alive only as long as its sieve-tube member does and typically has numerous cytoplasmic connections with the latter.

Unlike the sieve-tube member protoplast, which undergoes a selective autophagy and assumes a clear appearance during its ontogeny, the companion cell protoplast increases in density as it develops. This increase in density is due in part to an increase in the ribosome population and partly to an increase in density of the ground substance itself (Behnke 1975b; Esau 1978b). In addition to the high ribosome population, the companion cell contains numerous mitochondria, rough ER, plastids, and a prominent nucleus. Companion-cell plastids typically lack starch, although a few examples of starch-containing companion-cell plastids exist (e.g., in *Cucurbita,* Esau and Cronshaw 1968; *Amaranthus,* Fisher and Evert 1982; and *Solanum,* McCauley 1987). Some companion cells and phloem parenchyma cells also contain P-protein (Deshpande 1974b; Esau 1975, 1978a; Wergin et al. 1975; Couot-Gastelier 1982).

The connections between a sieve-tube member and a companion cell consist of a pore on the sieve-tube member side of the wall and much-branched plasmodesmata on the companion cell side (Figs. 6.24 and 6.25; Esau 1969; Behnke 1975b; Deshpande 1975; Esau and Thorsch 1985). During development of the connections, callose occurs on the sieve-tube member side of the wall at the site of the future pore. Pore development is initiated with the formation of a median cavity in the region of the middle lamella, and development of the branched plasmodesmata is associated with a build-up of cell wall on the companion cell side (Deshpande 1975; Esau and Thorsch 1985). Most workers assume that the parietal ER network of the mature sieve-tube member is connected with the ER of the companion cell via the desmotubules of the plasmodesmata in the companion cell wall (e.g., Kollmann 1973; Behnke 1975b). Recently, however, Esau and Thorsch (1985) have noted that the desmotubules of the plasmodesmatal branches appear to end blindly at the middle lamella and concluded that the cytoplasmic material of the branches has no connection with the living matter of the sieve-tube member except as represented by the plasmalemma.

Because of their numerous plasmodesmatal connections with the sieve-tube members and their general ultrastructural resemblance to secretory cells, it is believed that companion cells play a role in the delivery of assimilates to the sieve tube. In addition, it has been suggested that companion cells may maintain the enucleate sieve-tube members through the transfer of informational

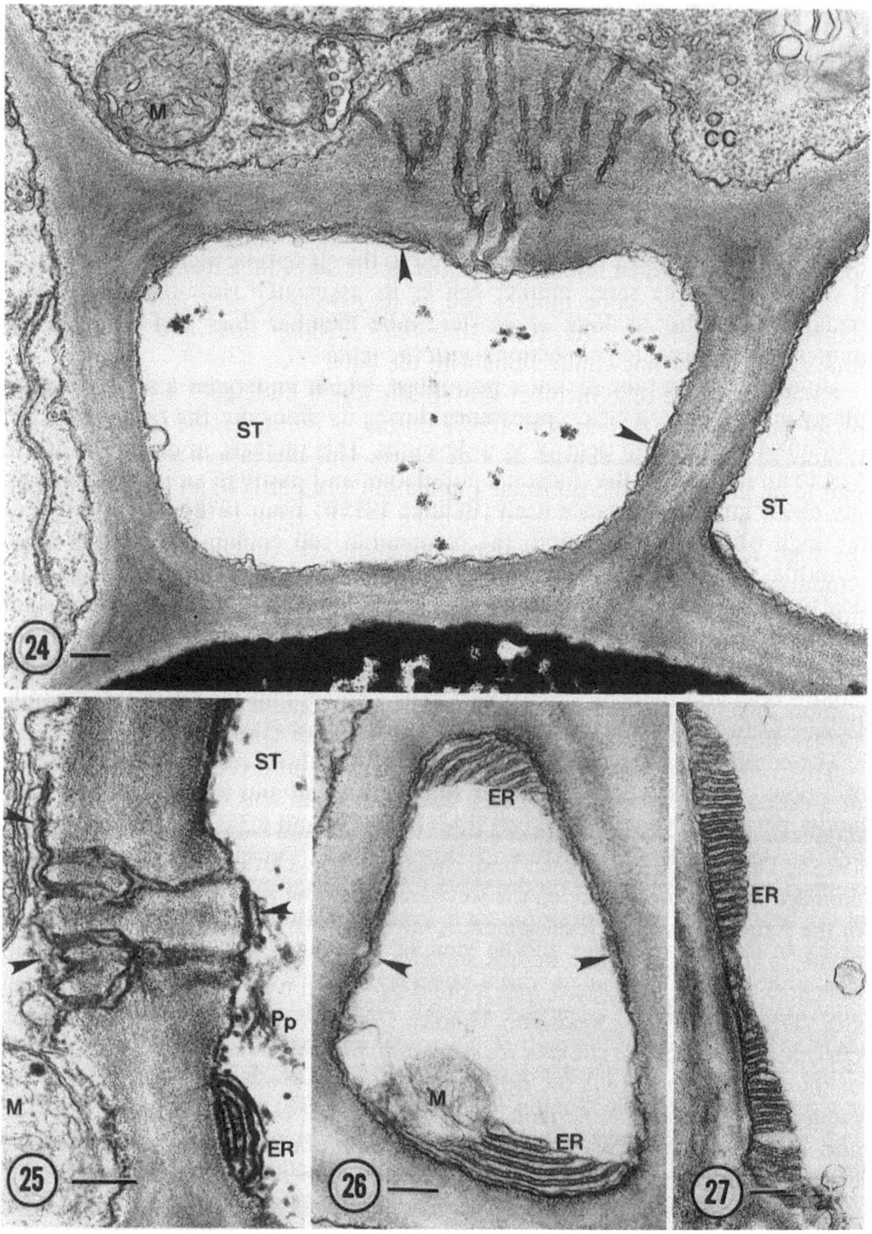

Figs. 6.24–6.27. Portions of sieve-tube members and companion cells from minor veins of *Populus deltoides* leaf. **Fig. 6.24.** Transverse section of sieve-tube member *(ST)* and part of associated companion cell *(CC)* with pore-plasmodesmata connections in thickened portion of their common wall. *Darts* point to discontinuous profiles of the parietal network of endoplasmic reticulum in the sieve-tube member. × 26800; *Bar line* = 0.2 μm. **Fig. 6.25.** Longitudinal view of plasmodesmata-pore connections between companion cell and sieve-tube member.

molecules (Gunning 1976) or other substances such as ATP (Lehmann 1979) from themselves to the sieve-tube members via the many connections in their common walls. The interdependence of these two cells is further supported by the fact that both cease to function and die at the same time. Companion cells frequently are absent in the protophloem in some shoots and roots (Esau 1969), a factor that may be associated with the short life span of many sieve-tube members in this earliest part of the primary phloem.

Companion cells and other parenchymatous elements of the phloem intergrade with one another in appearance, and the two are not always unequivocally distinguishable from one another even at the electron microscope level (Esau 1969). Some phloem parenchyma cells may arise from the same precursors as their contiguous sieve-tube members. The more closely related ontogenetically such parenchyma cells are to the sieve-tube members, the more closely they resemble companion cells both in appearance and in frequency of cytoplasmic connections with the sieve-tube members. The symplastic connection of parenchyma cells with sieve-tube members is brought about largely, however, through the companion cells. Parenchyma cells closely related to the sieve-tube members may die at the same time as their associated sieve-tube members.

Companion cells and phloem parenchyma cells are connected with one another and among themselves by plasmodesmata, as are the axial parenchyma cells and ray parenchyma cells in the secondary phloem of dicotyledons (Esau 1969). Often, e.g., in *Pyrus* (Evert 1960, 1963a), the companion cells of the secondary phloem abut the rays, presumably acting as intermediaries in the transfer of assimilates from the sieve-tube members to the ray parenchyma cells and vice versa.

Unlike most parts of the plant in which the parenchymatous cells are considerably smaller than their associated sieve-tube members, in the minor veins of dicotyledonous leaves the parenchymatous cells often are considerably larger than their associated sieve-tube members (Fig. 6.28). Many of the large parenchymatous cells have dense, organelle-rich protoplasts (Fisher and Evert 1982; Evert and Mierzwa 1986; Fisher 1986; McCauley 1987). Typically, these dense cells have numerous branched plasmodesmata connecting them with the sieve-tube members and are regarded as companion cells. The less dense parenchymatous cells are called phloem parenchyma cells. Fischer (1884) called the dense cells "Übergangszellen" or "intermediary cells" (Esau 1969) because he thought they served as intermediaries in the transfer of photosynthates between

◁

Darts point to discontinuous profile of endoplasmic reticulum in the sieve-tube member and to endoplasmic reticulum associated with plasmodesmata on companion-cell side of wall. × 45700; *Bar line* = 0.2 μm. **Fig. 6.26.** Transverse section of sieve-tube member. *Darts* point to discontinuous profiles of the parietal network of endoplasmic reticulum. The stack of endoplasmic reticulum *(ER)* cisternae above lie partly parallel and partly perpendicular to the wall, while that below lies entirely parallel to the wall. × 35850; *Bar line* = 0.2 μm. **Fig. 6.27.** Longitudinal section of portion of sieve-tube member with stacks of *ER* cisternae oriented perpendicular to the wall. × 24500; *Bar line* = 0.2 μm. *M* mitochondrion; *Pp* P-protein (Russin and Evert 1985)

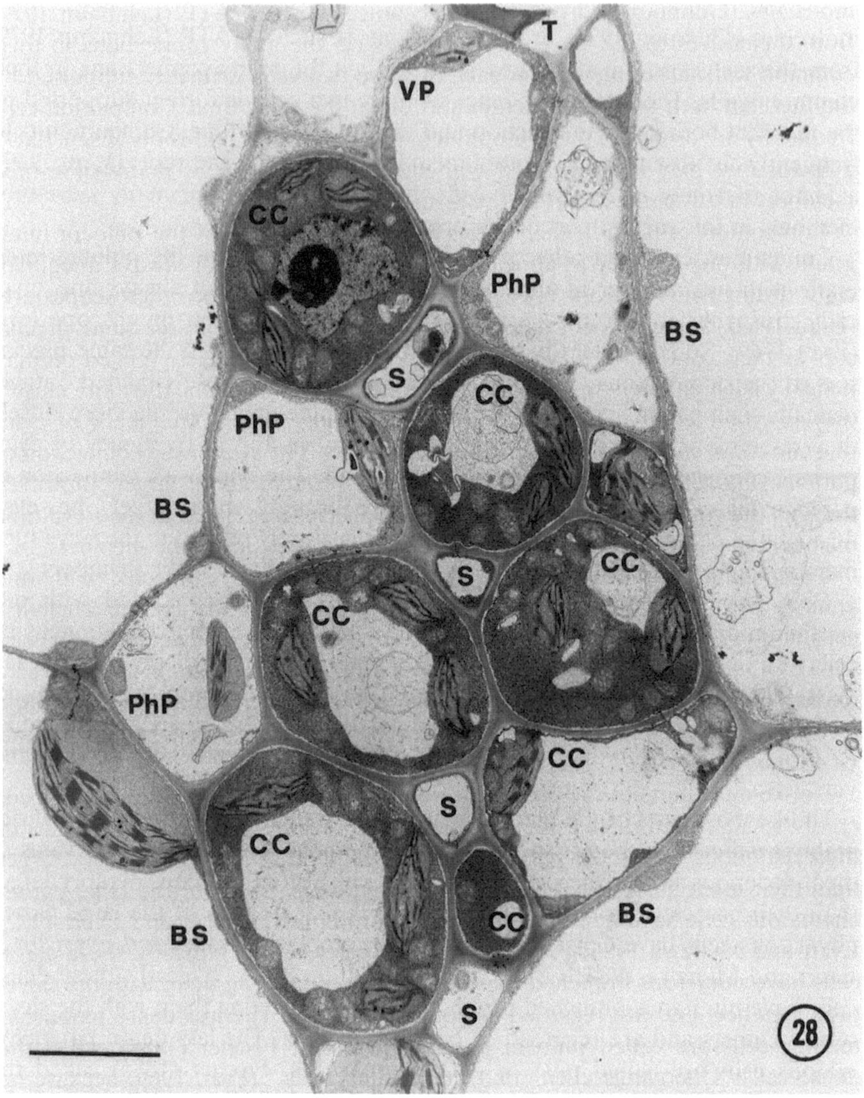

Fig. 6.28. Transverse section of portion of minor vein of *Beta vulgaris* leaf. The sieve elements *(S)* are narrower than the parenchymatous cells bordering them. The companion cells *(CC)* have very dense protoplasts. The less dense parenchymatous cells are phloem parenchyma *(PhP)* and vascular parenchyma *(VP)* cells. × 3500; *Bar line* = 5 μm. *BS* bundle-sheath cell; *T* part of tracheary element (Evert and Mierzwa 1986)

mesophyll cells and sieve tubes of the leaf. The less dense parenchyma cells may also be involved in the transfer or loading of photosynthates into the sieve tubes.

In the minor veins of many herbaceous dicotyledons, the parenchymatous cells of the phloem develop wall ingrowths that greatly increase the surface of the plasmalemma. Such cells are called transfer cells and are of two types, A and B (Pate and Gunning 1969, 1972). A-type transfer cells are companion cells with typically dense protoplasts and with wall ingrowths on all walls but generally less abundant on the common wall with the sieve tube. B-type transfer cells are phloem parenchyma cells with wall ingrowths best developed opposite the sieve tubes and companion cells.

Two main roles have been proposed for the transfer cells of minor veins: to collect and pass on photosynthates and to retrieve and recycle solutes that enter the leaf apoplast in the transpiration stream (Gunning et al. 1974). Pate and Gunning (1972) regard the wall-membrane apparatus of the transfer cell as an adaptation facilitating apoplast-symplast exchanges of solutes across the plasmalemma. Results of histochemical studies indicate that the wall ingrowths of minor vein transfer cells are rich in pectin (Botha and Evert 1981; Lucas and Franceschi 1982), suggesting that they are highly hydrated in vivo, a state that would facilitate solute transport in the apoplast. Developmentally, the appearance of wall ingrowths in the transfer cells of leaves coincides with the conversion of that portion of the leaf from an importer (sink) to exporter (source) of carbohydrates and with maturation of the contiguous sieve-tube members (Gunning et al. 1968; Bentwood and Cronshaw 1978).

Gamalei and Pakhomova (1983a, b) recognize three categories, or types, of minor veins in the leaves of dicotyledons, based largely on the distribution of plasmodesmata in the companion cell walls. In type I, many plasmodesmata (commonly in "fields") are located in thickened portions of the common walls between the companion cells and both the bundle sheath cells and sieve-tube members. Such companion cells (type I) are said to be of the open type. In type II minor veins, the companion cells (type II), which are called closed, have relatively few or virtually no plasmodesmata connecting them with the bundle sheath cells. In type III veins, the companion cells (type III) also are of the closed type but, unlike type II companion cells, have wall ingrowths, i.e., they are transfer cells. Both type II and type III companion cells have plasmodesmatal connections with their associated sieve-tube members. In some leaves, types I and II companion cells are found in the same minor vein. According to Gamalei (1985), leaves with open companion cells utilize a symplastic mechanism of phloem loading, while those with closed companion cells absorb sugars from the apoplast. Those with both open and closed companion cells would have a "mixed mechanism" of sieve-tube loading.

Typically, the walls of companion cells are neither sclerified nor lignified, and commonly the companion cells collapse when their associated sieve-tube members cease to function. A notable exception is found in *Tilia* (Evert 1963b) where some companion cells become sclerified and lignified before their protoplasts become disorganized.

6.7 Longevity of Sieve-Tube Members

The behavior of sieve-tube members and contiguous cells during the transition of the phloem from a functional to a nonfunctional condition has long been recognized (Esau 1969). Commonly, the earliest sign of the initiation of cessation of function is the appearance of definitive callose at the sieve plates and lateral sieve areas of the sieve-tube members. The callose may accumulate in massive amounts but eventually disappears entirely some time after the protoplasmic contents of the cell have degenerated. As mentioned previously, degeneration of the sieve-tube member is accompanied by degeneration of its companion cell(s) and sometimes other parenchymatous cells as well. The resultant loss of turgor pressure by the degenerating sieve-tube members and growth adjustments within the tissue may result in collapse and obliteration of the sieve tubes and closely associated parenchymatous cells. The sieve tubes may remain open, however, and become filled with air. Tylosoids (tylose-like protrusions from contiguous parenchyma cells) may invade the lumina of the dead sieve tubes. Interestingly, Lawton and Lawton (1971), in a study of the secondary phloem of six forest trees from Nigeria, found tylosoids in the secondary phloem of all trees that normally formed tyloses.

Undoubtedly, the shortest-lived sieve tubes in dicotyledons are those of the protophloem, which soon are replaced by those of the metaphloem. In plant parts with little or no secondary growth, most of the metaphloem sieve tubes probably remain functional for the life of the organ, a matter of months. The concept of short sieve-tube longevity in dicotyledons comes mainly from studies on the secondary phloem of woody species in which the sieve elements function for only one growing season or parts of two (Esau 1969).

In many temperate species of woody dicotyledons, the sieve tubes produced in the spring and early summer become nonfunctional and die in the fall, so that the phloem lacks living sieve tubes during winter (*Pyrus communis* and *P. malus*, Evert 1960, 1963c; *Robinia pseudoacacia*, Derr and Evert 1967; *Populus tremuloides*, Davis and Evert 1968; *Acer negundo*, Tucker and Evert 1969; *Parthenocissus*, Davis and Evert 1970; *Juglans nigra*, Schaad and Wilson 1970). In such species, the first functional sieve tubes in spring arise from undifferentiated cells that overwinter in the outer margin of the cambial zone. Similar growth patterns have been reported for a subtropical species from Taiwan (*Liquidambar formosana*, Lu and Chiang 1975) and three tropical species from India (*Polyalthia longifolia*, *Mimusops elengi*, and *Delonix regia*, Ghouse and Hashmi 1979, 1980a, b). In other woody species, all but the last-formed sieve tubes of the current season's growth increment become nonfunctional in fall. The former overwinter in a mature state and represent the first functional sieve tubes in spring in such species (e.g., *Salix fragilis* and *Acer pseudoplatanus*, J. R. Lawton 1976; *Fraxinus americana*, Zamski and Zimmermann 1979).

Exceptions to these patterns do exist, however. In *Tilia*, for example, some sieve tubes may function for five (*T. americana*, Evert 1962) or as many as ten

(*T. cordata*, Holdheide 1951) years. In *Vitis* (*V. vinifera*, Esau 1948; *V. riparia*, Davis and Evert 1970), the sieve tubes usually function two seasons, occasionally longer. In both *Tilia* and *Vitis* the overwintering sieve tubes deposit dormancy callose at the sieve plates and lateral sieve areas in the fall and remove it in the spring as they are reactivated. A similar pattern involving reactivation of two or more phloem increments has been reported for the tropical Indian tree *Grewia tiliaefolia* (Deshpande and Rajendrababu 1985). The longevity of the sieve tubes in some tropical trees may be much longer. In some species of Malaysian Dipterocarpaceae the sieve tubes may function for 11 or more years (Whitmore 1962). The age of the "conducting phloem" in these species was calculated from its presumed replacement time, however, and not from cytological evidence.

Acknowledgments. Thanks are expressed to Susan E. Eichhorn for her assistance with preparation of the manuscript, including the figures, and to Kathleen T. Delaney, typist. The preparation of this chapter was supported in part by NSF grant DCB87-01116.

References

Arsanto J-P (1982) Observations on P-protein in dicotyledons. Substructural and developmental features. Am J Bot 69: 1200-1212

Arsanto J-P, Coulon J (1974) Détections radio-autographique et cytochimique des sites d'élaboration ou de transit des précurseurs polysaccarides pariétaux dans les cellules criblées en cours de différenciation du métaphloème caulinaire de deux Cucurbitacées voisines (*Cucurbita pepo* L. et *Ecballium elaterium* R.). C R Acad Sci Paris Ser D 278: 2775-2778

Arsanto J-P, Coulon J (1975) Application des méthodes cytochimique et radioautographique de détection ultrastructurale des polysaccarides à l'étude de la différenciation des plateaux criblées du métaphloème caulinaire de deux Cucurbitacées voisines (*Ecballium elaterium* R. et *Cucurbita pepo* L.). C R Acad Sci Paris Ser D 280: 601-604

Barclay GF, Johnson RPC (1982) Analysis of particle motion in sieve tubes of *Heracleum*. Plant Cell Environ 5: 173-178

Behnke H-D (1971a) Über den Feinbau verdickter (nacré) Wände und der Plastiden in den Siebröhren von *Annona* und *Myristica*. Protoplasma 72: 69-78

Behnke H-D (1971b) The contents of the sieve-plate pores in *Aristolochia*. J Ultrastruct Res 36: 493-498

Behnke H-D (1974) Comparative ultrastructural investigations of angiosperm sieve elements: aspects of the origin and early development of P-protein. Z Pflanzenphysiol 74: 22-34

Behnke H-D (1975a) P-type sieve-element plastids: a correlative ultrastructural and ultrahistochemical study on the diversity and uniformity of a new reliable character in seed plant systematics. Protoplasma 83: 91-101

Behnke H-D (1975b) Companion cells and transfer cells. In: Aronoff S, Dainty J, Gorham PR, Srivastava LM, Swanson CA (eds) Phloem transport. Plenum, New York London, pp 153-175

Behnke H-D (1976) Ultrastructure of sieve-element plastids in Caryophyllales (Centrospermae), evidence for the delimitation and classification of the order. Plant Syst Evol 126: 31-54

Behnke H-D (1981) Sieve-element characters. Nord J Bot 1: 381-400

Behnke H-D (1986) Sieve element characters and the systematic position of *Austrobaileya*, Austrobaileyaceae - with comments to the distinction and definition of sieve cells and sieve-tube members. Plant Syst Evol 152: 101-121
Behnke H-D, Kiritsis U (1983) Ultrastructure and differentiation of sieve elements in primitive angiosperms. I. Winteraceae. Protoplasma 118: 148-156
Behnke H-D, Schulz A (1983) The development of specific sieve-element plastids in wound phloem of *Coleus blumei* (S-type) and *Pisum sativum* (P-type), regenerated from amyloplast-containing parenchyma cells. Protoplasma 114: 125-132
Bentwood BJ, Cronshaw J (1978) Cytochemical localization of adenosine triphosphatase in the phloem of *Pisum sativum* and its relation to the function of transfer cells. Planta 140: 111-120
Botha CEJ, Evert RF (1981) Studies on *Artemisia afra* Jacq.: the phloem in stem and leaf. Protoplasma 109: 217-231
Browning AJ, Hall JL, Baker DA (1980) Cytochemical localization of ATPase activity in phloem tissues of *Ricinus communis*. Protoplasma 104: 55-65
Catesson A-M (1973) Observations cytochimiques sur les tubes criblées de quelques angiospermes. J Microsc (Paris) 16: 95-104
Cateson A-M (1980) Localization of phloem oxidases. Ber Dtsch Bot Ges 93: 141-152
Catesson A-M (1982) Cell wall architecture in the secondary sieve tubes of *Acer* and *Populus*. Ann Bot (London) 49: 131-134
Catesson A-M, Liberman-Maxe M (1974) Les mitochondries des cellules criblées: réactions avec la 3,3'-diamino-benzidine. C R Acad Sci Paris Ser D 278: 2771-2773
Couot-Gastelier J (1982) Particularités fonctionnelles et infrastructurales du tissu phloémien du *Vicia faba* L. Beitr Biol Pflanzen 57: 257-268
Cronshaw J (1975a) P-proteins. In: Aronoff S, Dainty J, Gorham PR, Srivastava LM, Swanson CA (eds) Phloem transport. Plenum, New York London, pp 79-115
Cronshaw J (1975b) Sieve element walls. In: Aronoff S, Dainty J, Gorham PR, Srivastava LM, Swanson CA (eds) Phloem transport. Plenum, New York London, pp 129-147
Cronshaw J (1980) Histochemical localization of enzymes in the phloem. Ber Dtsch Bot Ges 93: 123-139
Cronshaw J, Anderson R (1971) Phloem differentiation in tobacco pith culture. J Ultrastruct Res 34: 244-259
Cronshaw J, Esau K (1967) Tubular and fibrillar components of mature and differentiating sieve elements. J Cell Biol 34: 801-816
Cronshaw J, Esau K (1968) P-protein in the phloem of *Cucurbita*. I. The development of P-protein bodies. J Cell Biol 38: 25-39
Davis JD, Evert RF (1968) Seasonal development of the secondary phloem in *Populus tremuloides*. Bot Gaz 129: 1-8
Davis JD, Evert RF (1970) Seasonal cycle of phloem development in woody vines. Bot Gaz 131: 128-138
De Maria ME, Thaine R (1974) Strands in sieve tubes in longitudinal cryostat sections of *Cucurbita pepo* stems. J Exp Bot 25: 871-885
Dempsey GP, Bullivant S, Bieleski RL (1975) The distribution of P-protein in mature sieve elements of celery. Planta 126: 45-59
Derr WF, Evert RF (1967) The cambium and seasonal development of the phloem in *Robinia pseudoacacia*. Am J Bot 54: 147-153
Deshpande BP (1974a) Development of the sieve plate in *Saxifraga sarmentosa* L. Ann Bot (London) 38: 151-158
Deshpande BP (1974b) On the occurrence of spiny vesicles in the phloem of *Salix*. Ann Bot (London) 38: 865-868
Deshpande BP (1975) Differentiation of the sieve plate of *Cucurbita*: a further view. Ann Bot (London) 39: 1015-1022
Deshpande BP (1976a) Observations on the fine structure of plant cell walls. II. The microfibrillar framework of the parenchymatous cell wall in *Cucurbita*. Ann Bot (London) 40: 439-442

Deshpande BP (1976b) Observations on the fine structure of plant cell walls. III. The sieve tube wall in *Cucurbita*. Ann Bot (London) 40: 443-446
Deshpande BP (1984) Distribution of P-protein in mature sieve elements of *Cucurbita maxima* seedlings subjected to prolonged darkness. Ann Bot (London) 53: 237-247
Deshpande BP, Evert RF (1970) A reevaluation of extruded nucleoli in sieve elements. J Ultrastruct Res 33: 483-494
Deshpande BP, Rajendrababu T (1985) Seasonal changes in the structure of the secondary phloem of *Grewia tiliaefolia*, a deciduous tree from India. Ann Bot (London) 56: 61-77
Dute RR, Sharkey CL (1985) Phloem of primitive angiosperms. III. Phloem of petioles of *Drimys granadensis* (Winteraceae). Proc Iowa Acad Sci 92: 104-110
Esau K (1948) Phloem structure in the grapevine, and its seasonal changes. Hilgardia 18: 217-296
Esau K (1969) The phloem. In: Zimmermann W, Ozenda P, Wulff HD (eds) Encyclopedia of plant anatomy, vol 5, pt 2. Borntraeger, Berlin Stuttgart, 505 pp
Esau K (1971) Development of P-protein in sieve elements of *Mimosa pudica*. Protoplasma 73: 225-238
Esau K (1975) The phloem of *Nelumbo nucifera* Gaertn. Ann Bot (London) 39: 901-913
Esau K (1978a) The protein inclusions in sieve elements of cotton (*Gossypium hirsutum* L.). J Ultrastruct Res 63: 224-235
Esau K (1978b) Developmental features of the primary phloem in *Phaseolus vulgaris* L. Ann Bot (London) 42: 1-13
Esau K, Charvat ID (1975) An ultrastructural study of acid phosphatase localization in cells of *Phaseolus vulgaris* phloem by the use of the azo dye method. Tissue Cell 7: 619-630
Esau K, Cheadle VI (1958) Wall thickening in sieve elements. Proc Natl Acad Sci USA 44: 546-553
Esau K, Cronshaw J (1968) Plastids and mitochondria in the phloem of *Curcurbita*. Can J Bot 46: 877-880
Esau K, Gill RH (1971) Aggregation of endoplasmic reticulum and its relation to the nucleus in a differentiating sieve element. J Ultrastruct Res 34: 144-158
Esau K, Gill RH (1972) Nucleus and endoplasmic reticulum in differentiating root protophloem of *Nicotiana tabacum*. J Ultrastruct Res 41: 160-175
Esau K, Hoefert LL (1971) Composition and fine structure of minor veins in *Tetragonia* leaf. Protoplasma 72: 237-253
Esau K, Hoefert LL (1980) Endoplasmic reticulum and its relation to microtubules in sieve elements of sugarbeet and spinach. J Ultrastruct Res 71: 249-257
Esau K, Magyarosy AC (1979a) A crystalline inclusion in sieve element nuclei of *Amsinckia*. I. The inclusion in differentiating cells. J Cell Sci 38: 1-10
Esau K, Magyarosy AC (1979b) A crystalline inclusion in sieve element nuclei of *Amsinckia*. II. The inclusion in maturing cells. J Cell Sci 38: 11-22
Esau K, Thorsch J (1982) Nuclear crystalloids in sieve elements of species of *Echium* (Boragniaceae). J Cell Sci 54: 149-160
Esau K, Thorsch J (1984) The sieve plate of *Echium* (Boraginaceae): developmental aspects and response of P-protein to protein digestion. J Ultrastruct Res 86: 31-45
Esau K, Thorsch J (1985) Sieve plate pores and plasmodesmata, the communication channels of the symplast: ultrastructural aspects and developmental relations. Am J Bot 72: 1641-1653
Eschrich W (1975) Sealing systems in phloem. In: Zimmermann MH, Milburn JA (eds) Transport in plants. I. Phloem transport. Encyclopedia of plant physiology, NS vol 1. Springer, Berlin Heidelberg New York, pp 39-56
Evert RF (1960) Phloem structure in *Pyrus communis* L. and its seasonal changes. Univ Cal Berkeley Publ Bot 32: 127-194
Evert RF (1962) Some aspects of phloem development in *Tilia americana*. Am J Bot 49: 659
Evert RF (1963a) Ontogeny and structure of the secondary phloem in *Pyrus malus*. Am J Bot 50: 8-37
Evert RF (1963b) Sclerified companion cells in *Tilia americana*. Bot Gaz 124: 262-264

Evert RF (1963c) The cambium and seasonal development of the phloem in *Pyrus malus*. Am J Bot 50: 149-159
Evert RF (1977) Phloem structure and histochemistry. Annu Rev Plant Physiol 28: 199-222
Evert RF (1982) Sieve-tube structure in relation to function. BioScience 32: 789-795
Evert RF (1984) Comparative structure of phloem. In: White RA, Dickison WC (eds) Contemporary problems in plant anatomy. Academic Press, Orlando, pp 145-234
Evert RF, Derr WF (1964) Callose substance in sieve elements. Am J Bot 51: 552-559
Evert RF, Deshpande BP (1969) Electron microscope investigation of sieve-element ontogeny and structure in *Ulmus americana*. Protoplasma 68: 403-432
Evert RF, Mierzwa RJ (1986) Pathway(s) of assimilate movement from mesophyll cells to sieve tubes in the *Beta vulgaris* leaf. In: Cronshaw J, Lucas WJ, Giaquinta RT (eds) Plant biology, vol 1. Phloem transport. Liss, New York, pp 419-432
Evert RF, Tucker CM, Davis JD, Deshpande BP (1969) Light microscope investigation of sieve-element ontogeny and structure in *Ulmus americana*. Am J Bot 56: 999-1017
Evert RF, Davis JD, Tucker CM, Alfieri FJ (1970) On the occurrence of nuclei in mature sieve elements. Planta 95: 281-296
Evert RF, Deshpande BP, Eichhorn SE (1971) Lateral sieve-area pores in woody dicotyledons. Can J Bot 49: 1509-1515
Evert RF, Eschrich W, Eichhorn SE (1973) P-protein distribution in mature sieve elements of *Cucurbita maxima*. Planta 109: 193-210
Fellows RJ, Geiger DR (1974) Structural and physiological changes in sugar beet leaves during sink to source conversion. Plant Physiol 54: 877-885
Fensom DS (1972) A theory of translocation in phloem of *Heracleum* by contractile protein microfibrillar material. Can J Bot 50: 479-497
Fensom DS, Williams EJ (1974) On Allen's suggestion for long-distance translocation in phloem of plants. Nature (London) 250; 490-492
Fischer A (1884) Untersuchungen über das Siebröhren-System der Cucurbitaceen. Borntraeger, Berlin
Fisher DB (1975) Structure of functional soybean sieve elements. Plant Physiol 56: 555-569
Fisher DG (1986) Ultrastructure, plasmodesmatal frequency, and solute concentration in green areas of variegated *Coleus blumei* Benth. leaves. Planta 169: 141-152
Fisher DG, Evert RF (1982) Studies on the leaf of *Amaranthus retroflexus* (Amaranthaceae): ultrastructure, plasmodesmatal frequency, and solute concentration in relation to phloem loading. Planta 155: 377-387
Friis J, Dute RR (1983) Phloem of primitive angiosperms. II. P-protein in selected species of the Ranalean complex. Proc Iowa Acad Sci 90: 78-84
Gamalei YV (1985) Characteristics of phloem loading in woody and herbaceous plants. Sov Plant Physiol 32: 656-665
Gamalei YV, Pakhomova MV (1983a) Minor veins of dicotyledonous leaves. I. Structure and typology. Bot Zh Leningrad 68: 287-301
Gamalei YV, Pakhomova MV (1983b) Minor veins of dicotyledonous leaves. II. Taxonomical distribution of the main types. Bot Zh Leningrad 68: 428-438
Ghouse AKM, Hashmi S (1979) Longevity of phloem in *Polyalthia longifolia* Benth. & Hook. Bull Torrey Bot Club 106: 182-184
Ghouse AKM, Hashmi S (1980a) Seasonal production of secondary phloem and its longevity in *Mimusops elengi* L. Flora 170: 175-179
Ghouse AKM, Hashmi S (1980b) Longevity of secondary phloem in *Delonix regia* Rafin. Proc Indian Acad Sci 89: 67-72
Giaquinta R (1980) Mechanism and control of phloem loading of sucrose. Ber Dtsch Bot Ges 93: 187-201
Giaquinta RT, Geiger DR (1973) Mechanism of inhibition of translocation by localized chilling. Plant Physiol 51: 372-377
Gilder J, Cronshaw J (1973a) The distribution of adenosine triphosphatase activity in differentiating and mature phloem cells of *Nicotiana tabacum* and its relationship to phloem transport. J Ultrastruct Res 44: 388-404

Gilder J, Cronshaw J (1973b) Adenosine triphosphatase in the phloem of *Cucurbita*. Planta 110: 189-204

Gilder J, Cronshaw J (1974) A biochemical and cytochemical study of adenosine triphosphatase activity in the phloem of *Nicotiana tabacum*. J Cell Biol 60: 221-235

Gilliland MG, van Staden J, Bruton AG (1984) Studies on the translocation system of guayule (*Parthenium argentatum* Gray). Protoplasma 122: 169-177

Goff CW (1973) Localization of nucleoside diphosphatase in the onion root tip. Protoplasma 78: 397-416

Gunning BES (1976) The role of plasmodesmata in short distance transport to and from the phloem. In: Gunning BES, Robards AW (eds) Intercellular communication in plants: studies on plasmodesmata. Springer, Berlin Heidelberg New York, pp 203-227

Gunning BES, Pate JS, Briarty LG (1968) Specialized "transfer cells" in minor veins of leaves and their possible significance in phloem translocation. J Cell Biol 37: C7-C12

Gunning BES, Pate JS, Minchin FR, Marks I (1974) Quantitative aspects of transfer cell structure in relation to vein loading in leaves and solute transport in legume nodules. Symp Soc Exp Biol 28: 87-126

Hartig T (1854) Über die Querscheidewände zwischen den einzelnen Gliedern der Siebröhren in *Cucurbita pepo*. Bot Z 12: 51-54

Hoefert LL (1979) Ultrastructure of devoloping sieve elements in *Thlaspi arvense* L. I. The immature state. Am J Bot 66: 925-932

Hoefert LL (1980) Ultrastructure of developing sieve elements in *Thlaspi arvense* L. II. Maturation. Am J Bot 67: 194-201

Holdheide W (1951) Anatomie mitteleuropäischer Gehölzrinden. In: Freund H (ed) Handbuch der Mikroskopie in der Technik , vol 5, pt 1. Umschau, Frankfurt am Main, pp 193-367

Ilker R, Currier HB (1975) Histochemical studies of an inclusion body and P-protein in phloem of *Xylosma congestum*. Protoplasma 85: 127-132

Jarvis P, Thaine R, Leonard JW (1973) Structures in sieve elements cut with a cryostat following different rates of freezing. J Exp Bot 24: 905-919

Johnson RPC, Freundlich A, Barclay GF (1976) Transcellular strands in sieve tubes; what are they? J Exp Bot 27: 1117-1136

Jørgensen LB, Møller JD, Wagner P (1975) Secondary phloem of *Trochodendron aralioides*. Bot Tidsskr 69: 217-238

Kallarackal J, Milburn JA (1983) Studies on the phloem sealing mechanism in *Ricinus* fruit stalks. Aust J Plant Physiol 10: 561-568

Kollmann R (1973) Cytologie des Phloems. In: Hirsch GC, Ruska H, Sitte P (eds) Grundlagen der Cytologie. Fischer, Jena, pp 479-505

Kollmann R (1980) Fine structural and biochemical characterization of phloem proteins. Can J Bot 58: 802-806

Laflèche D (1966) Ultrastructure et cytochimie des inclusions flagellées des cellules criblées de *Phaseolus vulgaris*. J Microsc (Paris) 5: 493-510

Lawton DM (1978a) P-protein crystals do not disperse in uninjured sieve elements in roots of runner bean *(Phaseolus multiflorus)* fixed with glutaraldehyde. Ann Bot (London) 42: 353-361

Lawton DM (1978b) Ultrastructural comparison of the tailed and tailless P-protein crystals respectively of runner bean *(Phaseolus multiflorus)* and garden pea *(Pisum sativum)* with tilting stage electron microscopy. Protoplasma 97: 1-11

Lawton DM, Johnson RPC (1976) A superhelical model for the ultrastructure of "P-protein tubules" in sieve elements of *Nymphoides peltata*. Cytobiologie 14: 1-17

Lawton DM, Newman YM (1979) Ultrastructure of phloem in young runner-bean stem: discovery, in old sieve elements on the brink of collapse, of parietal bundles of P-protein tubules linked to the plasmalemma. New Phytol 82: 213-222

Lawton JR (1976) Seasonal variation in the secondary phloem from the main trunks of willow and sycamore trees. New Phytol 77: 761-771

Lawton JR (1977) An investigation of the functional phloem in willow. New Phytol 78: 189-192

Lawton JR, Lawton JRS (1971) Seasonal variations in the secondary phloem of some forest trees from Nigeria. New Phytol 70: 187-196
Lee DR, Arnold DC, Fensom DS (1971) Some microscopical observations of functioning sieve tubes of *Heracleum* using Nomarski optics. J Exp Bot 22: 25-38
Lehmann J (1979) Nachweis von ATP und ATP-ase in den Siebröhren von *Cucurbita pepo*. Z Pflanzenphysiol 94: 331-338
Lu C-Y, Chiang S-HT (1975) Seasonal activity of the cambium in the young branch of *Liquidambar formosana* Hance. Taiwania 20: 32-47
Lucas WJ, Franceschi VR (1982) Organization of the sieve-element walls of leaf minor veins. J Ultrastruct Res 81: 209-221
McCauley MAM (1987) Structural studies on the leaf of potato (*Solanum tuberosum* L.). Ph D Thesis, Univ Wisc, Madison
Murphy R (1986) A reanalysis of particle motion in sieve tubes of *Heracleum*. Ann Bot (London) 57: 667-674
Nehls R, Schaffner G, Kollmann R (1978) Feinstruktur des Protein-Einschlusses in den Siebelementen von *Salix sachalinensis* Fr. Schmidt. Z Pflanzenphysiol 87: 113-127
Oberhäuser R, Kollmann R (1977) Cytochemische Charakterisierung des sogenannten "Freien Nucleolus" als Proteinkörper in den Siebelementen von *Passiflora coerulea*. Z Pflanzenphysiol 84: 61-75
Oparka KJ, Johnson RPC (1978) Endoplasmic reticulum and crystalline fibrils in the root protophloem of *Nymphoides peltata*. Planta 143: 21-27
Oparka KJ, Johnson RPC; Bowen JD (1981) Sites of acid phosphatase in the differentiating root protophloem of *Nymphoides peltata* (S. G. Gmel.) O. Kuntze. Plant Cell Environ 4: 27-35
Palevitz BA, Newcomb EH (1970) A study of sieve element starch using sequential enzymatic digestion and electron microscopy. J Cell Biol 45: 383-398
Palevitz BA, Newcomb EH (1971) The ultrastructure and development of tubular and crystalline P-protein in the sieve elements of certain papilionaceous legumes. Protoplasma 72: 399-426
Parthasarathy MV, Mühlethaler K (1969) Ultrastructure of protein tubules in differentiating sieve elements. Cytobiologie 7: 17-36
Parthasarathy MV, Pesacreta TC (1980) Microfilaments in plant vascular cells. Can J Bot 58: 807-815
Pate JS, Gunning BES (1969) Vascular transfer cells in angiosperm leaves. A taxonomic and morphological survey. Protoplasma 68: 135-156
Pate JS, Gunning BES (1972) Transfer cells. Annu Rev Plant Physiol 23: 173-196
Pickett-Heaps JD (1967) The use of radioautography for investigating wall secretion in plant cells. Protoplasma 64: 49-66
Read SM, Northcote DH (1983 a) Subunit structure and interactions of the phloem proteins of *Cucurbita maxima* (pumpkin). Eur J Biochem 134: 561-569
Read SM, Northcote DH (1983 b) Chemical and immunological similarities between the phloem proteins of three genera of the Cucurbitaceae. Planta 158: 119-127
Robidoux J, Sandborn EB, Fensom DS, Cameron ML (1973) Plasmatic filaments and particles in mature sieve elements of *Heracleum sphondylium* under the electron microscope. J Exp Bot 24: 349-359
Roland J-C, Sandoz D (1969) Détection cytochimique des sites de formation des polysaccharides pré-membranaires dans les cellules végétales. J Microsc (Paris) 8: 263-268
Russin WA, Evert RF (1985) Studies on the leaf of *Populus deltoides* (Salicaceae): ultrastructure, plasmodesmatal frequency, and solute concentrations. Am J Bot 72: 1232-1247
Sabnis DD, Hart JW (1979) Heterogeneitiy in phloem protein complements from different species. Consequences to hypotheses concerned with P-protein function. Planta 145: 459-466
Schaad NW, Wilson EE (1970) Structure and seasonal development of secondary phloem of *Juglans regia*. Can J Bot 48: 1049-1053
Sjolund RD, Shih CY (1983 a) Freeze-fracture analysis of phloem structure in plant tissue cultures. I. The sieve element reticulum. J Ultrastruct Res 82: 111-121

Sjolund RD, Shih CY (1983b) Freeze-fracture analysis of phloem structure in plant tissue cultures. II. The sieve element plasma membrane. J Ultrastruct Res 82: 189-197
Sjolund RD, Shih CY, Jensen KG (1983) Freeze-fracture analysis of phloem structure in plant tissue cultures. III. P-protein, sieve area pores, and wounding. J Ultrastruct Res 82: 198-211
Spanner DC (1978a) Sieve-plate pores, open or occluded? A critical review. Plant Cell Environ 1: 7-20
Spanner DC (1978b) The Münch hypothesis, freeze-substitution and the structure of sieve-plate pores. Ann Bot (London) 42: 485-488
Spanner DC, Moattari F (1978) The significance of P-protein and endoplasmic reticulum in sieve elements in light of evolutionary origins. Ann Bot (London) 42: 1469-1472
Srivastava LM (1970) The secondary phloem of *Austrobaileya scandens*. Can J Bot 48: 341-359
Thaine R, Probine MC, Dyer PY (1967) The existence of transcellular strands in mature sieve elements. J Exp Bot 18: 110-127
Thaine R, De Maria ME, Sarisalo HIM (1975) Evidence of transcellular strands in transverse cryostat sections of *Cucurbita pepo* sieve tubes. J Exp Bot 26: 91-101
Thorsch J, Esau K (1981a) Changes in the endoplasmic reticulum during differentiation of a sieve element in *Gossypium hirsutum*. J Ultrastruct Res 74: 183-194
Thorsch J, Esau K (1981b) Nuclear degeneration and the association of endoplasmic reticulum with the nuclear envelope and microtubules in maturing sieve elements of *Gossypium hirsutum*. J Ultrastruct Res 74: 195-204
Thorsch J, Esau K (1981c) Ultrastructural studies of protophloem sieve elements in *Gossypium hirsutum*. J Ultrastruct Res 75: 339-351
Thorsch J, Esau K (1982) Microtubules in differentiating sieve elements of *Gossypium hirsutum*. J Ultrastruct Res 78: 73-83
Thorsch J, Esau K (1985) An ultrastructural study of the phloem of *Drimys* (Winteraceae). IAWA Bull 6: 255-268
Tucker CM, Evert RF (1969) Seasonal development of the secondary phloem in *Acer negundo*. Am J Bot 56: 275-284
Turgeon R, Webb JA, Evert RF (1975) Ultrastructure of minor veins in *Cucurbita pepo* leaves. Protoplasma 83: 217-232
Walsh MA, Popovich TM (1977) Some ultrastructural aspects of metaphloem sieve elements in the aerial stem of the holoparasitic angiosperm *Epifagus virginiana* (Orobanchaceae). Am J Bot 64: 326-336
Wergin WP, Newcomb EH (1970) Formation and dispersal of crystalline P-protein in sieve elements of soybean (*Glycine max* L.). Protoplasma 71: 365-388
Wergin WP, Palevitz BA, Newcomb EH (1975) Structure and development of P-protein in phloem parenchyma and companion cells of legumes. Tissue Cell 7: 227-242
Whitmore TC (1962) Studies in systematic bark morphology. II. General features of bark construction in Dipterocarpaceae. New Phytol 61: 208-220
Wooding FBP (1967) Fine structure and development of phloem sieve tube content. Protoplasma 64: 315-324
Yapa PAJ, Spanner DC (1972) The effect of protease digestion (in situ) on the slime substance of mature sieve tubes. Planta 107: 89-96
Yapa PAJ, Spanner DC (1974) Localisation of adenosine triphosphatase activity in mature sieve elements of *Tetragonia*. Planta 117: 321-328
Zamski E, Zimmermann MH (1979) Sieve tube longevity in white ash *Fraxinus americana)* studies with a new histochemical test for the identification of sugar. Can J Bot 57: 650-656

7 Monocotyledons

ELEFTHERIOS P. ELEFTHERIOU

7.1 Introduction

The application of electron microscopy in biological research has yielded a wealth of information on phloem structure and function, with special emphasis being placed on the main conducting unit, the sieve element. One of the most controversial issues of phloem investigation has been the nature of the mature sieve-plate pores: open or occluded? A clear understanding of sieve-element structure at successive stages of development is thus of fundamental importance if one is to elucidate the mechanism of translocation in the sieve tubes. Considerable effort has been devoted during the last two decades to advancing our knowledge on the structure and function of the monocotyledons phloem. Although monocotyledons constitute a broad taxonomic group within the plant kingdom, including species wellknown through the centuries for their importance as food sources or as excellent pasture crops, they have been considerably less investigated in respect to phloem structure than the dicotyledons. From the evolutionary point of view, the monocotyledons are monophyletic on the basis of two character conditions: (1) they have a single cotyledon, whereas the dicotyledons have two; (2) their sieve tube plastids accumulate proteins in the form of triangular or cuneate bodies (Dahlgren et al. 1985). This chapter is therefore aimed at familiarizing the reader with the most salient structural features of sieve elements in monocotyledons and at comparing them with the corresponding structures of other groups, especially the dicotyledons.

7.2 Ontogeny

Most investigations on phloem of monocotyledons concern the sieve-element development, starting from where they can be distinguished from other meristematic cells. The first and most reliable identifying feature has so far been the appearance of cuneate electron-dense inclusions within their plastids (Behnke and Dörr 1967; Esau and Gill 1973; Parthasarathy 1974a; Walsh and Evert 1975; Kawahara et al. 1980; Singh 1980). Only a few studies deal with the very initial stages of sieve-element ontogeny. Psaras (1980) has traced the sequence of divisions giving rise to the vascular bundles of the leaves of *Zea mays* emerging from a single mother cell. In the leaf and stem vascular bundles of *Aegilops*

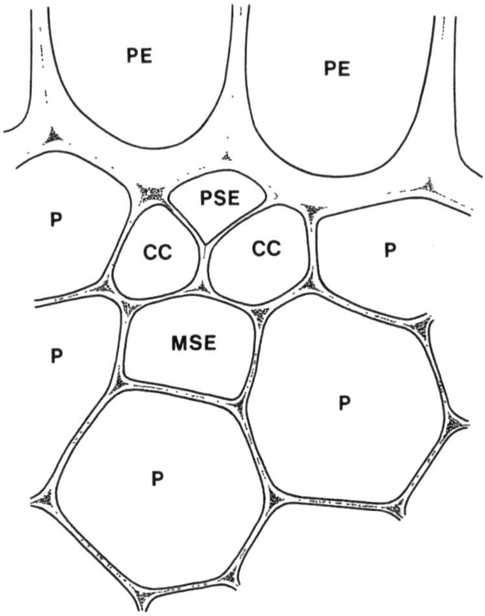

Fig. 7.1. General layout of a mature phloem pole of wheat seminal roots, and the surrounding cells, in transection. *PSE* protophloem sieve element; *CC* companion cells; *MSE* metaphloem sieve element; *PE* pericycle cells; *P* parenchyma cells of the vascular cylinder

comosa the metaphloem sieve elements originate from very long mother cells that divide longitudinally (Eleftheriou 1981).

The most detailed ontogenetic studies have been carried out in root protophloem of *Oryza* (Kawata et al. 1978), *Aegilops comosa* (Eleftheriou 1981; Eleftheriou and Tsekos 1982c) and *Triticum aestivum* (Eleftheriou 1985). The seminal roots of these grasses develop "phloem poles" occurring next to the pericycle and alternating with equal number of "xylem poles". Each phloem pole normally consists of four cells: one protophloem sieve element, two companion cells and one metaphloem sieve element (Fig. 7.1). A "protophloem pole" comprises the protophloem sieve element and the two companion cells flanking it on the internal side. Externally, the protophloem pole abuts the pericycle cells.

Each protophloem pole arises from a single protophloem mother cell (PMC). The PMC (Fig. 7.2a) undergoes two successive asymmetrical divisions: the first one produces two unequal cells (Fig. 7.2b–d): the smaller will later differentiate into a companion cell, while the larger divides once again asymmetrically, yielding another companion cell and a protophloem sieve element (Fig. 7.2e–i). Prior to initiating differentiation, the latter cell divides once more, now symmetrically (Fig. 7.2j–l), increasing the number of cells in vertical files (Eleftheriou 1985).

Both asymmetrical and symmetrical divisions are preceded by well-demarcated preprophase microtubule bands (PMBs) (Fig. 7.2b, e, j). The plane of a particular PMB coincides with those of the succeeding cell plate and of the daughter wall, which fuse with parent walls at sites previously occupied by the PMB. Furthermore, the planes of the PMBs (and therefore of divisions) change from one division to the next, so that the three successive divisions occur in

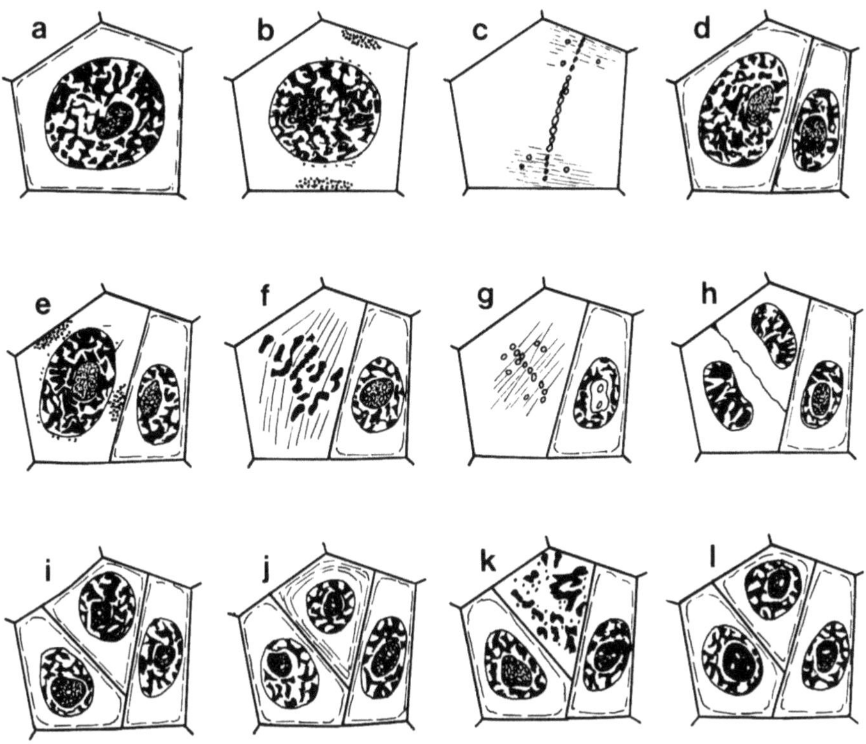

Fig. 7.2. Successive stages of protophloem ontogeny in seminal roots of wheat. The cells *a, d* and *i* are found at interphase, cells *b, e* and *j* at pre-prophase, cells *f* and *k* at metaphase, cells *c, g* and *h* at cytokinesis, while *l* is at the onset of differentiation. Note the change in distribution, organization and orientation of microtubules that reflects their dynamic role in the process (*dots* denote cross- and *lines* longitudinally sectioned microtubules) (Eleftheriou 1985, modified)

three spatial planes transversely to each other (cf. Eleftheriou 1985). The manner in which microtubules become highly involved in the ontogeny of root protophloem by continuously changing their distribution, organization and orientation is emphasized in Fig. 7.2.

7.3 The Protoplast

7.3.1 Nucleus

As in all angiosperms, sieve-tube members of monocotyledons become enucleate at maturity. The degeneration of the nucleus has been thoroughly documented with developmental studies at the ultrastructural level, and may involve

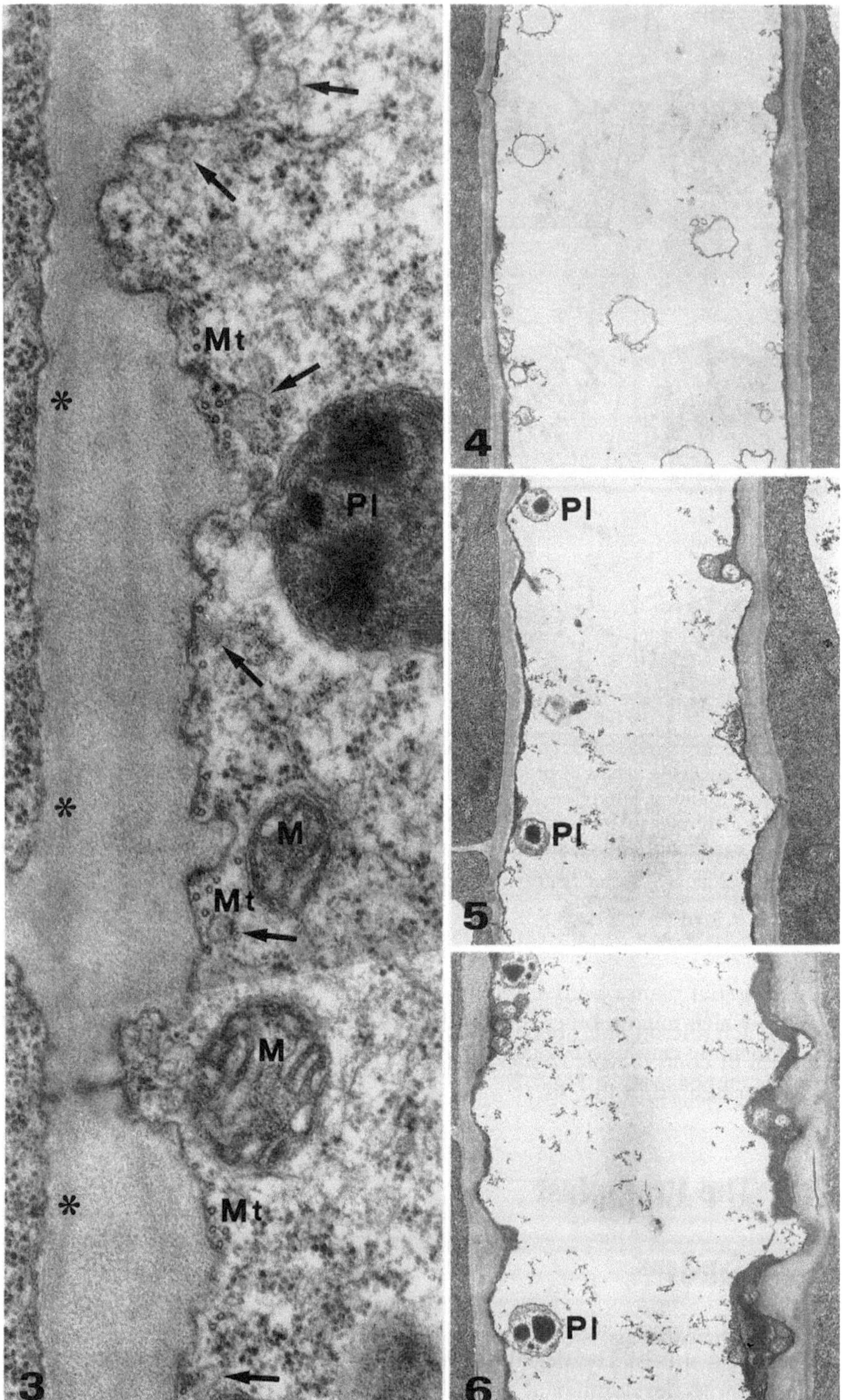

3
Mt
*
Pl
M
Mt
M
Mt
4
Pl
Pl
5
Pl
6

either the complete disappearance of the structure (Behnke 1969a, 1973; Parthasarathy 1974c; Melaragno and Walsh 1976; Singh 1980) or the retention of nuclear remnants for a long time after sieve elements attain a "clear" appearance (Ervin and Evert 1970; Esau and Gill 1973; Walsh and Evert 1975; Danilova and Telepova 1978; Eleftheriou and Tsekos 1982a, d; Eleftheriou 1986). Esau and Gill (1973), reporting on the presence of degenerated nuclei in mature protophloem sieve elements in roots of *Allium cepa,* concluded that "the loss of a functional nucleus in a differentiating sieve element may involve a complete breakdown of the organelle or a retention of the organelle in a necrotic state". To date there is no evidence against this conclusion.

The nucleus breaks down during the final stages of sieve-element differentiation undergoing a chromatolytic or a pycnotic type of degeneration. As is the case in dicotyledons (Hoefert 1980), chromatolysis seems to be the predominant way of nuclear desintegration in monocotyledons (Behnke 1969a; Singh and Srivastava 1972; Parthasarathy 1974b, 1980; Melaragno and Walsh 1976; Singh 1980; Danilova and Telepova 1980), although a pycnotic degeneration has been reported especially in differentiating protophloem sieve elements (Esau and Gill 1973; Danilova and Telepova 1978; Thorsch and Esau 1981c; Eleftheriou and Tsekos 1982a, d; Eleftheriou 1986), and occasionally in metaphloem ones (Walsh and Evert 1975). In *Aegilops comosa* (Eleftheriou 1981) and *Hordeum vulgare* (Danilova and Telepova 1980) both types of nuclear degeneration occur, the chromatolysis in the metaphloem sieve elements and pycnosis in the protophloem ones. Two types of nuclear degeneration are common in lower vascular plants (see Chap. 3).

The fine-structural details of nuclear degeneration remain largely unknown, presumably due to the fact that changes occur quite rapidly (Behnke 1969a; Esau and Gill 1973; Melaragno and Walsh 1976). The most complete view of the structural changes of the nucleus has probably been achieved in the root protophloem of *Triticum aestivum* (Eleftheriou 1986), characterized by chromatin condensation at the organelle periphery (Fig. 7.21), lobing and fragmentation (Fig. 7.22) and a close association of the nuclear envelope (NE) with the ER (Fig. 7.23). Occasionally, the ER-NE complexes assume a wavy conformation. At positions where chromatin masses aggregate the peri-nuclear or the ER intracisternal space becomes locally dilated (Figs. 7.22, 7.23). During the whole

Fig. 7.3. Part of the lateral wall of a differentiating sieve element of root protophloem of wheat, in longisection. Uneven wall development on the sieve-element side (*asterisks* mark the position of the middle lamella) creates broad ridges alternating with narrow depressions, the latter occupied by plasmodesmata. Microtubules *(Mt)* are seen next to thickenings only. Golgi vesicles *(arrows)* occur adjacent to the plasma membrane. *M* mitochondria; *Pl* plastid. × 46100 (Eleftheriou 1987a). **Figs. 7.4, 7.5, 7.6.** Portions of the first **(Fig. 7.6)**, second **(Fig. 7.5)**, and third **(Fig. 7.4)** protophloem sieve elements with "clear" lumen from the same sieve tube of a root of *Aegilops comosa* (longisection). A gradual smoothening from the highly irregular **(Fig. 7.6)** to the evenly thickened walls **(Fig. 7.4)** is obvious. *Pl* plastids. All × 11400 (Eleftheriou and Tsekos 1982b)

process of disorganization the nuclear envelope remains intact. In cells with "clear" lumen nuclear remnants occur as darkly staining masses frequently surrounded by membranous material apparently of ER origin (Fig. 7.24). Unusual threadlike inclusions exhibiting a unique fine structure consisting of short palisade-like units (Fig. 7.24) accompany the nuclear degeneration in wheat. Besides wheat (Eleftheriou 1986, 1987b), ER-NE complexes in monocotyledons

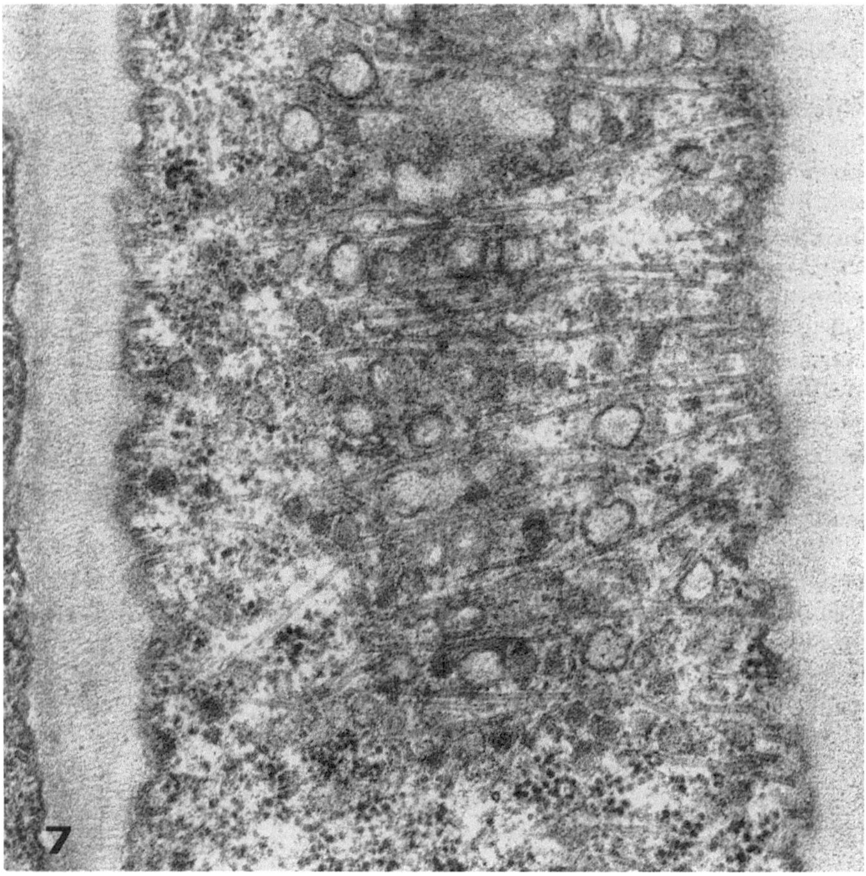

Fig. 7.7. A section tangential to the cell surface of a differentiating sieve element in root protophloem of wheat revealing a great number of cortical microtubules transversely aligned to the long axis (vertical direction). Numerous Golgi vesicles occur between the microtubules. × 52700 (Eleftheriou 1987a)

Fig. 7.8. Transection of sieve elements with nacreous walls *(NW)* in the leaf blade of the seagrass *Heterozostera tasmanica*. Note the irregular inner wall surface; × 7710 (Kuo 1983b).
Fig. 7.9. Thick-walled sieve elements *(asterisks)* in a large vascular bundle of the leaf blade of *Aegilops comosa* (transection); *V* xylem vessels; *P* phloem parenchyma cells; *MSE* thin-walled metaphloem sieve elements; *CC* companion cells. × 8380 (Eleftheriou 1981)

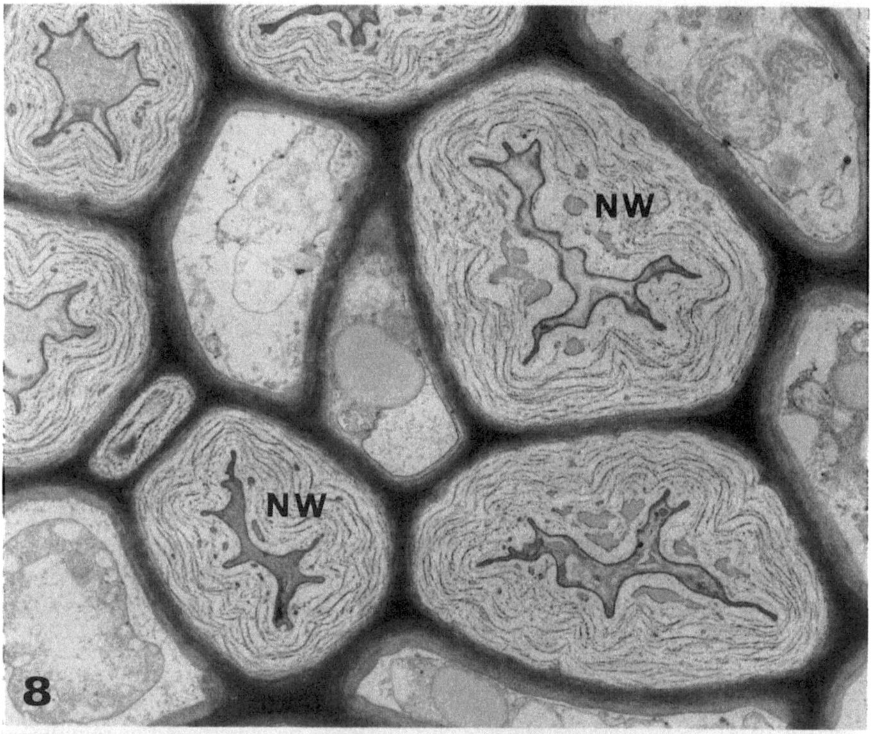

8

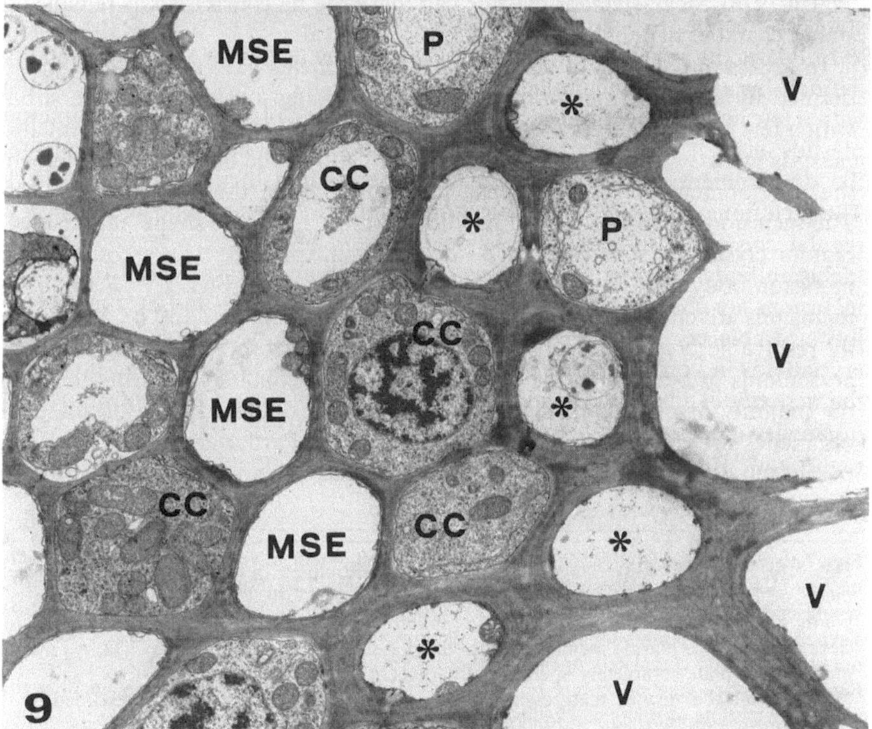

9

have also been noticed in palms (Parthasarathy 1974b), *Lemna minor* (Melaragno and Walsh 1976), *Hordeum vulgare* (Danilova and Telepova 1978, 1980) and *Aegilops comosa* (Eleftheriou 1981; Eleftheriou and Tsekos 1982d), but not in *Dioscorea* and *Smilax* (Behnke 1968, 1973b), *Allium cepa* (Esau and Gill 1973), *Zea mays* (Walsh and Evert 1975) and sugar-cane (Singh 1980). These examples support the conclusion that ER-NE complexes in monocotyledons are not very common nor as extensive as they are in dicotyledons (Esau and Gill 1971; Thorsch and Esau 1981b). But where they occur they involve one or a few ER cisternae in an intimate association with the nuclear envelope.

7.3.2 Plastids

Plastids are present over the entire lifetime of a sieve-tube member and may be found in sieve elements of some perennial monocotyledons even more than 6 years old (Parthasarathy 1974c, 1980). Being extremely sensitive to preparation and fixation procedures, they are frequently encountered with disrupted membranes and dispersed contents (Fig. 7.18). With careful tissue preparation, however, intact organelles have been demonstrated to be the normal condition of mature sieve elements (Fig. 7.17). All regions of a given plant contain the same type of sieve-element plastid (with an exception reported for *Zea mays*, Walsh 1974). Sieve-element plastids can therefore be considered as reliable and stable characters (Behnke 1975, 1981a).

In Poales the plastids of very young sieve elements have a small size and are defined by a double-membrane envelope enclosing a relatively dense stroma with a few internal membranes and ribosomes (Fig. 7.14). During differentiation plastids increase in size and undergo modifications, the most remarkable being the development of electron-dense inclusions within their stroma (Fig. 7.15). The inclusions occupy a central position, exhibit a cuneate shape and display a regular crystalline substructure. As development progresses, a second kind of inclusion body arises at the periphery of the organelle (Figs. 7.15, 7.16). These inclusions are characterized by smaller size, greater number and by a more easily resolvable crystalline substructure, thus they have been designated as thin crystalloids in contrast to the dense ones (Eleftheriou 1984). During this stage the majority of plastids display a circular or oval outline. Lobed or amoeboid organelles are, however, frequent (Behnke 1973; Parthasarathy 1974b; Melaragno and Walsh 1976; Singh 1980; Walsh and Melaragno 1981; Eleftheriou

Figs. 7.10–7.12. Successive stages of sieve plate development in root protophloem sieve elements of *Lemna minor* (Walsh and Melaragno 1976). **Fig. 7.10.** Plasmodesmata *(arrows)* of a young end wall developing into pores. × 36700. **Fig. 7.11.** Pores open to the more mature sieve element. The plasma membrane lines the pores *(arrows)*. × 74600. **Fig. 7.12.** A pore of a mature sieve plate completely open to both adjacent sieve elements. × 66000.

Fig. 7.13. Surface view of mature sieve plate of *Zea mays* showing open pores lined by the plasma membrane *(arrows)*. × 29400 (Walsh and Evert 1975)

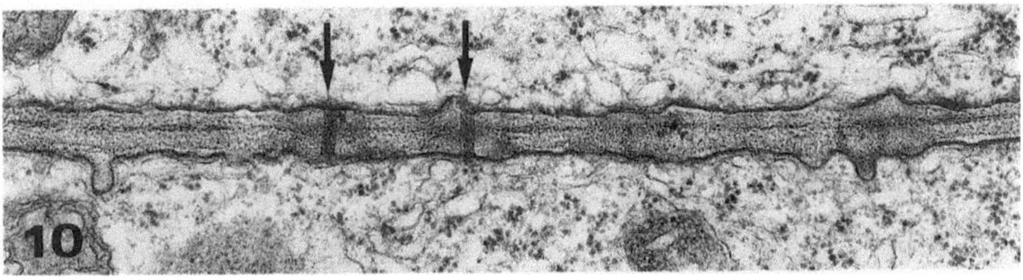
10

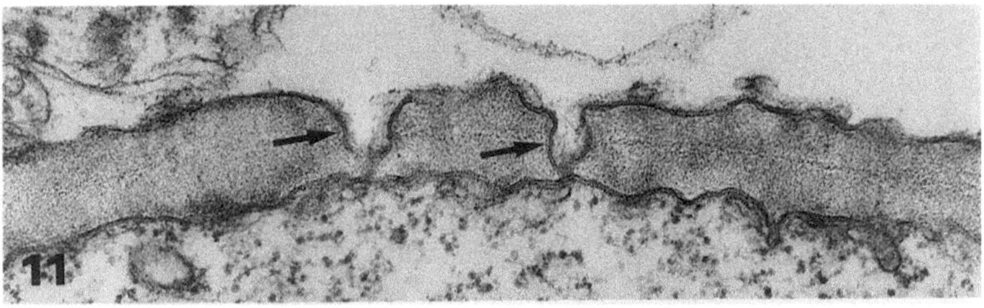
11

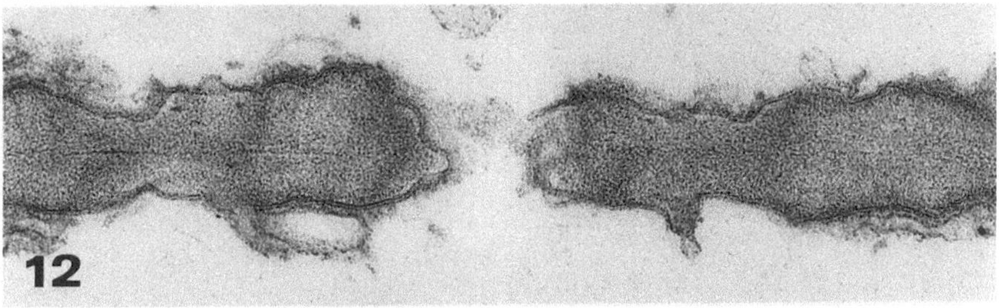
12

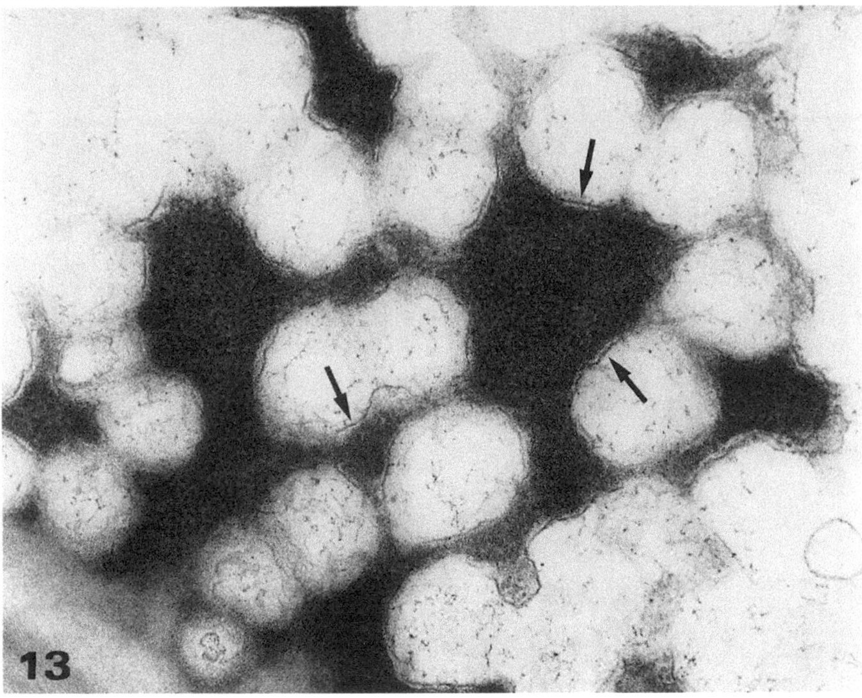
13

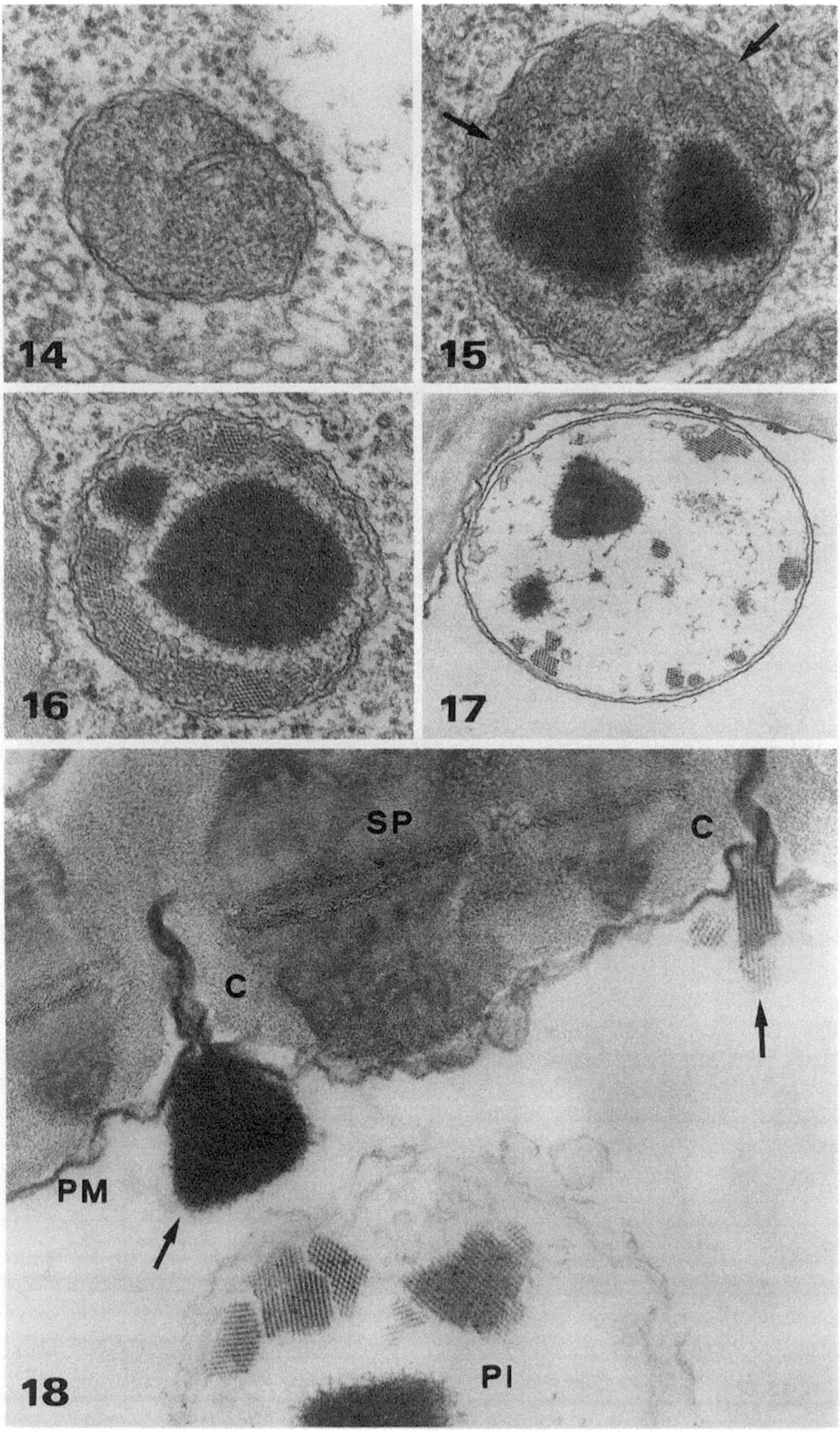
14
15
16
17
18
SP
C
C
PM
PI

1984); some may be dividing even while containing dense crystalloid inclusions (Singh 1980; Eleftheriou 1984).

The development of sieve-element plastids is accompanied by a progressive decrease and final loss of the electron density of the plastid stroma. The decrease commences around the dense crystalloids (Figs. 7.15, 7.16). The ultrastructural evidence gives the impression that the loss of stroma density results from the aggregation of the electron-dense material, which might be proteins, into the developing crystalloid (Eleftheriou 1984). The fully mature sieve elements bear intact plastids delimited by a double-membrane envelope, having circular or oval outline, and containing two kinds of crystalloid inclusions and some internal membranes in an electron-translucent stroma (Fig. 7.17).

Treatment of the ultrathin sections with proteolytic enzymes has demonstrated the inclusions to be proteinaceous in composition (Behnke 1975; Eleftheriou 1981, 1984).

Among the persistent components of the enucleate sieve elements only plastids, and to a minor extent P-protein, have thus far proved to be reliable ultrastructural features of taxonomic value for seed plant systematics (Behnke 1969b, 1972, 1981a, b).

More than 230 species from almost all families of monocotyledons have been screened for their sieve-element plastids; all have P-type plastids. While plastids only containing cuneate protein crystals (form PIIc) are dominant, those plastids containing starch (form PIIcs) and/or filaments (forms PIIcfs or PIIcf) in addition, have been reported in some families only (Behnke 1981a, b).

Behnke (1981a) has grouped the known forms of sieve-element plastids in a coordinate system according to their contents and evolutionary status. Figure 7.19 is part of this system and indicates the forms of monocotyledons sieve-element plastids as well as the basic route of evolution. The placement of forms cf/cfs in the innermost ring, close to the ancestral complex, conforms to the suggestion that these most complex forms are at the same time the most primitive ones. On the other hand, the PIIc' form, encountered primarily in the Poaceae, is regarded as the most derived.

But, what is the role of these proteinaceous inclusions? Structural evidence has been presented in support of the contention that they may be involved in a "sealing" mechanism of monocotyledons (Melaragno and Walsh 1976; Walsh

◁

Figs. 7.14–7.17. Successive stages of plastid development in the leaf metaphloem sieve elements of *Aegilops comosa*. **Fig. 7.14.** A simple-structure plastid of a very young sieve element. × 54000. **Fig. 7.15.** Development of dense crystalloid inclusions with detectable crystalline substructure. *Arrows* point to the very initial deposition of thin crystalloids. × 39600. **Fig. 7.16.** A plastid at a little later stage of development than that of Fig. 7.15, containing dense and thin crystalloids. × 38400. **Fig. 7.17.** A plastid in a fully mature sieve element with intact double-membrane envelope and two kinds of crystalloids. × 40800 (Eleftheriou 1981).

Fig. 7.18. Portion of a heavily disturbed leaf metaphloem sieve element near a sieve plate *(SP)* of *Triticum aestivum*. The envelope of the partly illustrated plastid *(Pl)* is ruptured, while released crystalloids of both kinds plug the sieve-plate pores *(arrows)*. Callose *(C)* has also been deposited within the pores. *PM* plasma membrane. × 70000 (Eleftheriou 1984)

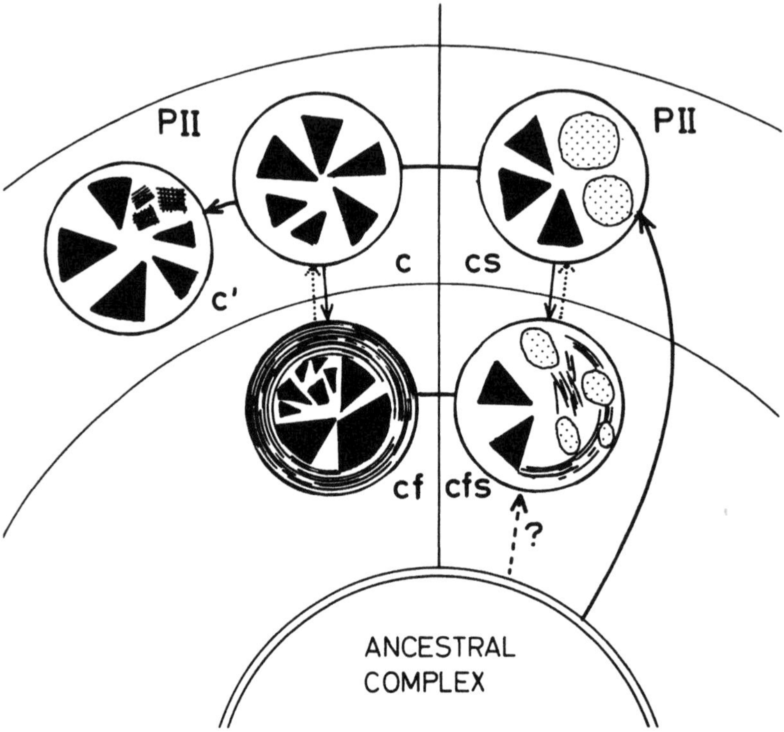

Fig. 7.19. Diagrammatic representation of the forms of sieve-element plastids of the monocotyledons and their presumed evolution. The basic route of evolution is directed from the *PIIcs* to *PIIc* and to *PIIc'* form, while a secondary route via the more primitive forms *PIIcf* or *PIIcfs* is also probable *(dotted arrows)* (Behnke 1981b)

and Melaragno 1981; Eleftheriou 1984). When sieve tubes are damaged, as, for instance, through browsing by herbivores, the existence and implementation of a rapid mechanism for blocking injured sieve tubes are vital for this food-conducting tissue. In dicotyledons this function may be served by P-protein (Eschrich 1975). In the monocotyledons, where P-protein is not so frequent (in Poaceae it is completely absent), an analogous role is presumably undertaken by the proteinaceous inclusions of sieve-element plastids. In heavily damaged sieve tubes the plastid envelope is disrupted and the released proteinaceous inclusions may become lodged in the sieve-plate pores, sealing them (Fig. 7.18). The quick sealing by plastid inclusions or by P-protein may be accomplished by heavy callose deposition.

7.3.3 Endoplasmic Reticulum and Mitochondria

In common with other angiosperms, the endoplasmic reticulum (ER) of differentiating sieve elements of monocotyledons undergoes morphological and apparently functional modifications, the most common of which involve the loss of ribosomes, the aggregation into stacks or the acquisition of a convoluted or a paracrystalline form (Behnke 1969a, 1973; Esau and Gill 1973; Walsh and Evert 1975; Melaragno and Walsh 1976). Occasional reports on the existence of ribosomes associated with ER membranes in sieve elements with a "clear" appearance (Esau and Gill 1973; Thorsch and Esau 1981c; Eleftheriou 1987b) should be regarded as late disappearing ribosomes. Stacking of ER may be against the nuclear envelope, free in the cytoplasm, or both (Eleftheriou 1986, 1987b). In mature sieve tubes of palms extensive tubular ER is found which is often dislocated against the sieve plates and also partly aggregated into complexes; paracrystalline structures and regular membrane stacks are likewise in contact with the tubular ER (Parthasarathy 1974b, c). In general, stacking of ER in monocotyledons occurs to a limited extent as compared to the massive and extensive ER aggregations of some dicotyledons (Esau and Gill 1971; Oparka and Johnson 1978; Thorsch and Esau 1981a, b; Fjell 1987a, b), or gymnosperms (Neuberger and Evert 1975). Structural evidence indicates that the ER of mature protophloem sieve elements of monocotyledons is reduced in quantity (Behnke 1973) and is considerably disorganized and degraded (Eleftheriou 1987b).

An association of ER with plastids was not observed in monocotyledons. In differentiating sieve elements ER is involved in sieve-plate pore development.

Mitochondria undergo less pronounced structural modifications during sieve-element differentiation and persist at maturity. In differentiating sieve elements of *Triticum aestivum* mitochondria are enveloped by single ER cisternae bearing ribosomes on their cytoplasmic face (Fig. 7.20), an association lost upon reaching maturity (Eleftheriou 1987b). In nearly mature sieve elements of palms the mitochondria become closely associated with the aggregated ER at the periphery of the cells (Parthasarathy 1974b), an association retained in mature sieve elements (Parthasarathy 1974c). In *Cocos nucifera* some mitochondria contain crystalline inclusions (Parthasarathy 1974b).

7.3.4 Microtubules and Dictyosomes

In the cortical cytoplasm of differentiating sieve elements of many monocotyledons microtubules have repeatedly been detected (Behnke 1969a, c, 1973; Esau and Gill 1973; Singh and Srivastava 1972; Parthasarathy 1974a, b, 1980; Walsh and Evert 1975; Melaragno and Walsh 1976; Psaras 1980; Singh 1980; Eleftheriou 1981, 1985, 1987a; Eleftheriou and Tsekos 1982b). The most detailed account in a monocotyledon is based on the root protophloem of wheat, where microtubules are involved both in the sieve-element ontogeny

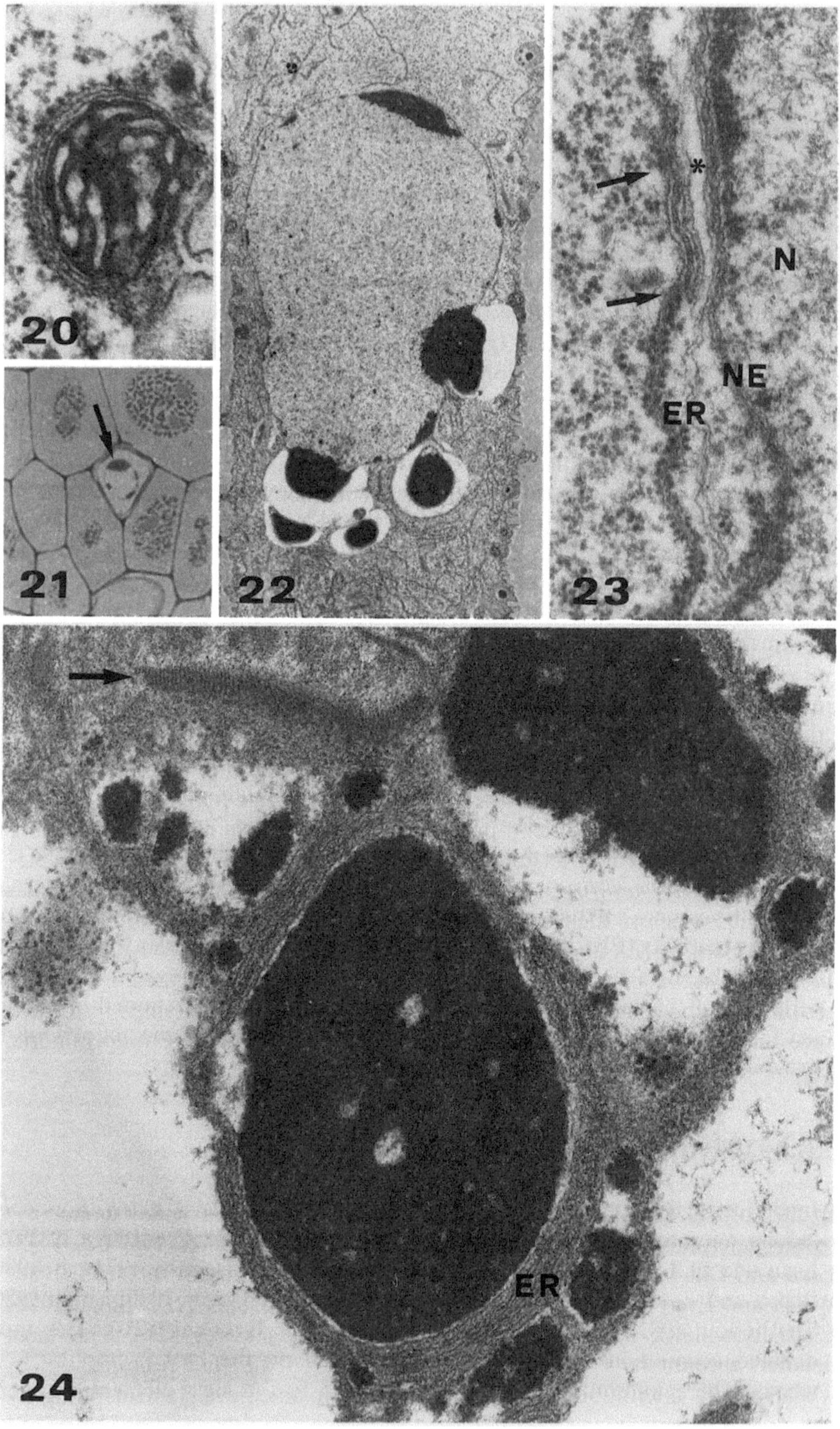
20
21
22
23
N
NE
ER
*
24
ER

(Fig. 7.2, and Eleftheriou 1985) and differentiation (Eleftheriou 1987a). The number of microtubules increases gradually in the very young sieve elements to reach a maximum just before initiation of wall thickening. During wall increment microtubules remain at high densities (more than 10 microtubules μm^{-1}), but their number declines abruptly when wall material deposition ceases in cells with a "clear" lumen. (For hints on the "maturity" condition of sieve elements, see discussion by Thorsch and Esau 1981c). Where wall increment is not uniform, microtubules overlie the thickenings only, being entirely absent from the nonthickened areas (Fig. 7.3). This relationship of microtubules to wall thickenings in sieve elements is similar to that seen in the differentiating tracheary elements. The orientation of microtubules is mirrored by that of the currently deposited cellulose microfibrils in the cell wall, both being perpendicular to the direction of cell expansion (Fig. 7.7). This accordance in orientation has also been pointed out in many monocotyledons, including *Zea mays* (Singh and Srivastava 1972; Psaras 1980), palms (Parthasarathy 1974a), sugarcane (Singh 1980) and *Aegilops comosa* (Eleftheriou and Tsekos 1982b). In conclusion, microtubules in sieve elements are involved in their ontogeny, cell wall pattern development, cellulose microfibril orientation and presumably in cell elongation.

The dictyosomes are fairly abundant in young sieve elements. In palms they are frequently in close association spatially with the nucleus (Parthasarathy 1974b). They become particularly active in producing numerous vesicles during the stage of intensive cell wall deposition. Two types of vesicles are produced by the dictyosomes, the smooth and the coated ones, both of which appear to fuse with the plasma membrane of differentiating sieve elements (Eleftheriou and Tsekos 1982b, d). The dictyosomes themselves do not have any association with the microtubules, but their vesicles, moving toward the plasma membrane, abound among the cortical microtubules (Figs. 7.3, 7.7). Although the spacing between adjacent microtubules is much smaller than the diameter of even the small-sized vesicles, the microtubules not only do not prevent but may even effect the movement of vesicles through their arrays (Eleftheriou 1987a). Microtubules may be involved in vesicle distribution along the protoplast surface, in that way presumably affecting the pattern of the wall thickening.

◁

Fig. 7.20. A mitochondrion of a differentiating protophloem sieve element of wheat enveloped by a single ER cisterna bearing ribosomes on its cytoplasmic face. × 51700 (Eleftheriou 1987b). **Figs. 7.21-7.24.** Aspects of nuclear degeneration in protophloem sieve elements of wheat *(Triticum aestivum)*. **Fig. 7.21.** A nucleus *(arrow)* with chromatin aggregated in dense masses at the periphery of the organelle (pycnotic degeneration). Transection, light micrograph. × 750. **Fig. 7.22.** Advanced chromatin condensation, particularly within the lobes, surrounded by electron translucent "caps" formed by dilation of the perinuclear space. × 4300. **Fig. 7.23.** ER-NE complex. The smooth ER cisterna appressed to the nuclear envelope *(NE)* is slightly dilated *(asterisk)*, while the ribosome-bearing ER cisterna loses the ribosomes in the region forming the stack *(between the arrows)*. *N* nucleus. × 45000. **Fig. 7.24.** Nuclear remnants (dense amorphous masses) surrounded by smooth membranes of ER in a sieve element with "clear" lumen. *Arrow* points to an inclusion consisting of short palisade-like units. × 44200 (Eleftheriou 1986)

The dictyosomes, like microtubules, decline in number and activity during the last stage of cytoplasm disorganization and they are absent in mature sieve elements.

7.3.5 P-Protein

P-protein has been encountered in the sieve elements of all dicotyledonous species investigated (see Chap. 5 and Cronshaw 1981), and in the majority of monocotyledons (see Behnke's 1981b Table for P-protein distribution among the monocotyledons). Typical filaments of P-protein have been observed in differentiating sieve elements of *Tradescantia albiflora* (Heyser 1971). In certain monocotyledons P-protein may be present in some sieve elements but absent in others within a given organ (Esau and Gill 1973; Parthasarathy 1974a). In others it may occur in such a small quantity that it is hardly identified. P-protein was not found at all in many palms (Parthasarathy 1974a, b), *Lemna minor* (Melaragno and Walsh 1976) and all members of the family Poaceae (Singh and Srivastava 1972; Evert et al. 1971; Kuo et al. 1972; Walsh and Evert 1975; Eleftheriou and Tsekos 1982a, b, d; Eleftheriou 1986, 1987a, b). In Poaceae the P-protein was presumably lost secondarily. The distribution of P-protein in perennial monocotyledons is sporadic (Parthasarathy 1980). Nondispersive (crystalline) P-protein bodies, like those encountered in dicotyledons in addition to the dispersive bodies, e. g., in *Gossypium hirsutum* (Esau 1978), have not been found in monocotyledons.

7.3.6 Vacuoles and Tonoplast

There is a general agreement among phloem workers that the tonoplast disorganizes and vacuoles disappear during the final stages of sieve-element differentiation. In some monocotyledons, however, the tonoplast or tonoplast remnants are reported to be preserved even at maturity (Kuo et al. 1972; Walsh 1980; Walsh and Evert 1975; Melaragno and Walsh 1976). According to Melaragno and Walsh (1976) and Walsh (1980) fixation artefacts may be responsible for the failure to preserve the tonoplast and vacuoles. When great care is taken to eliminate mechanical perturbation or to avoid severing of the sieve tubes, highly vacuolate, mature sieve elements can be observed. These sporadic reports are, however, not representative of the great majority of studies and may be based on abnormalities or instances in which the tonoplast was late to disappear from the cell (Evert 1977).

7.4 Cell Wall

Sieve elements of monocotyledons typically have nonlignified cellulosic walls. Differentiating protophloem sieve elements in roots of *Hordeum vulgare* (Danilova and Telepova 1978); *Oryza sativa* (Kawahara et al. 1980); *Aegilops comosa* (Eleftheriou and Tsekos 1982b) and *Triticum aestivum* (Eleftheriou 1987a) develop uneven wall thickenings consisting of broad ridges alternating with narrow depressions (Fig. 7.3). These peculiar wall thickenings are not permanent since they gradually disappear in the first sieve elements with "clear" lumen (Figs. 7.4–7.6) by a process involving stretching and smoothening. It was therefore suggested that the thickenings serve as stores of excess wall material that is used in subsequent growth enabling the enucleate sieve elements to keep pace with the active elongation of the neighboring cells (Danilova and Telepova 1978; Eleftheriou and Tsekos 1982b; Eleftheriou 1987a).

Nacreous thickenings are widespread, particularly in dicotyledons, but they have also been reported in some terrestrial (Singh and Srivastava 1972; Parthasarathy 1974a, c, 1980; Psaras 1980) and aquatic monocotyledons (Kuo 1983a, b). In palms the nacreous walls appear to be more prominent in protophloem than in metaphloem sieve elements. In seagrasses the nacreous-walled sieve tubes are restricted to certain genera and may serve as a taxonomic character (Kuo 1983b, and personal communication). Where they occur they are highly thickened with an irregular inner surface that constricts considerably the cell lumen (Fig. 7.8). Ultrastructurally, the nacreous thickenings contain many parallel microfibrils arranged approximately at right angles to the longitudinal axis of the cell. Chemically they consist of hemicelluloses or pectin and cellulose (Kuo 1983b).

7.5 Thick-Walled Sieve Elements

Besides the normal metaphloem sieve-elements with relatively thin walls a second type of sieve elements characterized by thick walls has been recognized in the longitudinal bundles of the leaves of some monocotyledonous species (Kuo and O'Brien 1974; Walsh 1974; Miyake and Maeda 1976; Evert 1980; Eleftheriou 1981). Walsh (1974) referred to those of *Zea mays* as late-formed metaphloem sieve elements, because they are formed later than the thin-walled ones. Thick-walled sieve elements of *Triticum aestivum* (Kuo and O'Brien 1974) and *Aegilops comosa* (Eleftheriou 1981) directly abut on the tracheary elements of the xylem, are separated from the thin-walled metaphloem sieve elements by parenchyma or companion cells (Fig. 7.9), and have been shown to be lignified. Lignified sieve elements may also be present in many other festucoid grasses but appear to be absent from the panicoid species (Kuo and O'Brien 1974). By contrast, in *Oryza sativa* they are not lignified and are separated from the tracheary elements by one or two intervening parenchyma cells (Miyake and

Maeda 1976). The counterpart cells of *Zea mays* are also nonlignified, they often abut on the xylem, lack companion cells but have numerous connections with vascular parenchyma cells (Evert 1980); their plastids contain unique crystalline structures which are different from the plastid inclusions of other monocotyledons (Walsh 1974).

Kuo and O'Brien (1974) have suggested that in *Triticum aestivum* the thick-walled sieve tubes may be specialized for long-distance transport or may serve as temporary storage reservoirs for sugars in excess of what can be transported by the normal sieve tubes. Experiments, however, of ^{14}C photosynthate transport in *Triticum* leaves (Cartwright et al. 1977) did not provide support to those suggestions. Evert et al. (1978) have suggested that the thick-walled sieve tubes in maize leaves may be involved in the retrieval of solutes entering the leaf apoplast in the transpiration stream.

7.6 Sieve Plates

Because of its importance for translocation sieve-plate structure and development in the angiosperms have been investigated extensively. Numerous studies conducted on dicotyledons, but also on monocotyledons (Behnke 1969c; Heyser 1971; Esau and Gill 1973; Parthasarathy 1974b; Walsh and Evert 1975; Walsh and Melaragno 1976) have corroborated a fairly uniform picture of development of sieve plates. The sieve-plate pores differentiate from single plasmodesmata which, during sieve-element differentiation, are associated with cisternae of ER, while callose platelets are deposited beneath the plasma membrane on either side. Widening of the developing pore is initiated in the region of the middle lamella and progresses toward the contiguous cells. Both the callose and the ER are removed as the pores attain their full size.

The generally accepted view of a universal callose participation in sieve-plate pore development has been questioned by Walsh and Melaragno (1976) in their study on root protophloem differentiation of *Lemna minor*. In this small vascular monocotyledon callose was not identified at any stage of sieve-plate pore development (Figs. 7.10–7.12), or during the development of lateral connections. Callose was, however, deposited when the sieve tubes were purposely injured. Walsh and Melaragno (1976) maintain that if the specimen is "small enough" and is fixed intact to eliminate traumatic effects, the developing pores do not exhibit callose formation. Mature sieve tubes of *Lemna minor* bear open sieve-plate pores, free of callose (Fig. 7.12), a feature shared by mature metaphloem sieve tubes in leaves of *Zea mays* (Fig. 7.13) (cf. Walsh and Evert 1975). Worley (1973), dealing with the movement of mycoplasma-like organisms, provided additional evidence in support of open sieve-plate pores in maize. Open pores have also been observed in sieve tubes of several other monocotyledons (Behnke 1969c; Evert et al. 1971; Kuo et al. 1972; Esau and Gill 1973; Eleftheriou 1981).

References

Behnke H-D (1968) Zum Aufbau gitterartiger Membranstrukturen im Siebelementplasma von *Dioscorea*. Protoplasma 66: 287-310

Behnke H-D (1969a) Aspekte der Siebröhren-Differenzierung bei Monocotylen. Protoplasma 68: 289-314

Behnke H-D (1969b) Die Siebröhren-Plastiden der Monocotyledonen. Vergleichende Untersuchungen über Feinbau und Verbreitung eines charakteristischen Plastidentyps. Planta 84: 174-184

Behnke H-D (1969c) Über den Feinbau und die Ausbreitung der Siebröhren-Plasmafilamente und über Bau und Differenzierung der Siebporen bei einigen Monocotylen und bei *Nuphar*. Protoplasma 68: 377-402

Behnke H-D (1972) Sieve-tube plastids in relation to angiosperm systematics. An attempt towards a classification by ultrastructural analysis. Bot Rev 38: 155-197

Behnke H-D (1973) Strukturänderungen des Endoplasmatischen Reticulums und Auftreten von Proteinfilamenten während der Siebröhrendifferenzierung bei *Smilax excelsa*. Protoplasma 77: 279-289

Behnke H-D (1975) P-type sieve-element plastids: a correlative ultrastructural and ultrahistochemical study on the diversity and uniformity of a new reliable character in seed plant systematics. Protoplasma 83: 91-101

Behnke H-D (1981a) Sieve-element characters. Nord J Bot 1: 381-400

Behnke H-D (1981b) Siebelement-Plastiden, Phloem-Protein und Evolution der Blütenpflanzen: II. Monokotyledonen. Ber Dtsch Bot Ges 94: 647-662

Behnke H-D, Dörr I (1967) Zur Herkunft und Struktur der Plasmafilamente in Assimilatleitbahnen. Planta 74: 18-44

Cartwright SC, Lush WM, Canny MJ (1977) A comparison of translocation of labelled assimilate by normal and lignified sieve elements in wheat leaves. Planta 134: 207-208

Cronshaw J (1981) Phloem structure and function. Annu Rev Plant Physiol 32: 465-484

Dahlgren RMT, Clifford HT, Yeo PF (1985) The families of monocotyledons. Structure, evolution and taxonomy. Springer, Berlin Heidelberg New York Tokyo

Danilova MF, Telepova MN (1978) Differentiation of protophloem sieve elements in seedling roots of *Hordeum vulgare*. Phytomorphology 28: 418-431

Danilova MF, Telepova MN (1980) Distinctive features of differentiation in proto- and metaphloem sieve elements of barley root. Phytomorphology 30: 380-387 [issued 1982]

Eleftheriou EP (1981) A light and electron microscopy study on phloem differentiation of the grass *Aegilops comosa* var. *thessalica*. Thesis, Univ Thessaloniki (in Greek with English summary)

Eleftheriou EP (1984) Sieve-element plastids of *Triticum* and *Aegilops* (Poaceae). Plant Syst Evol 145: 119-133

Eleftheriou EP (1985) Microtubules and root protophloem ontogeny in wheat. J Cell Sci 75: 165-179

Eleftheriou EP (1986) Ultrastructural studies on protophloem sieve elements in *Triticum aestivum* L. Nuclear degeneration. J Ultrastruct Mol Struct Res 95: 47-60

Eleftheriou EP (1987a) Microtubules and cell wall development in differentiating protophloem sieve elements of *Triticum aestivum* L. J Cell Sci 87: 595-607

Eleftheriou EP (1987b) Changes in the endoplasmic reticulum during sieve element differentiation in *Triticum aestivum*. Ann Bot (London) 60: 713-721

Eleftheriou EP, Tsekos I (1982a) The ultrastructure of protophloem sieve elements in leaves of *Aegilops comosa* var. *thessalica*. Ann Bot (London) 49: 557-567

Eleftheriou EP, Tsekos I (1982b) Developmental features of cell wall formation in sieve elements of the grass *Aegilops comosa* var. *thessalica*. Ann Bot (London) 50: 519-529

Eleftheriou EP, Tsekos I (1982c) Development of protophloem in roots of *Aegilops comosa* var. *thessalica*. I. Differential divisions and pre-prophase bands of microtubules. Protoplasma 113: 110-119

Eleftheriou EP, Tsekos I (1982d) Development of protophloem in roots of *Aegilops comosa* var. *thessalica*. II. Sieve-element differentiation. Protoplasma 113: 221-233
Ervin EL, Evert RF (1970) Observations on sieve elements in three perennial monocotyledons. Am J Bot 57: 218-224
Esau K (1978) The protein inclusions in sieve elements of cotton (*Gossypium hirsutum* L.). J Ultrastruct Res 63: 224- 235
Esau K, Gill RH (1971) Aggregation of endoplasmic reticulum and its relation to the nucleus in a differentiating sieve element. J Ultrastruct Res 34: 144-158
Esau K, Gill RH (1973) Correlations in differentiation of protophloem sieve elements of *Allium cepa* root. J Ultrastruct Res 44: 310-328
Eschrich W (1975) Sealing systems in phloem. In: Zimmermann MH, Milburn JA (eds) Transport in plants I. Phloem transport. Encyclopedia of plant physiology, NS. vol 1. Springer, Berlin Heidelberg New York, pp 39-56
Evert RF (1977) Phloem structure and histochemistry. Annu Rev Plant Physiol 28: 199-222
Evert RF (1980) Vascular anatomy of angiospermous leaves, with special consideration of the maize leaf. Ber Dtsch Bot Ges 93: 43-55
Evert RF, Eschrich W, Eichhorn SE (1971) Sieve-plate pores in leaf veins of *Hordeum vulgare*. Planta 100: 262-267
Evert RF, Eschrich W, Heyser W (1978) Leaf structure in relation to solute transport and phloem loading in *Zea mays* L. Planta 138: 279-294
Fjell I (1987a) Ultrastructural features of differentiating protophloem sieve elements in adventitious roots of *Salix viminalis*. Nord J Bot 7: 135-151
Fjell I (1987b) P-protein and inclusion bodies in root protophloem of *Salix viminalis*. Nord J Bot 7: 305-310
Heyser W (1971) Phloemdifferenzierung bei *Tradescantia albiflora*. Cytobiology 4: 186-197
Hoefert LL (1980) Ultrastructure of developing sieve elements in *Thlaspi arvense* L. II. Maturation. Am J Bot 67: 194-201
Kawahara H, Matsuda T, Chonan N (1980) Studies on morphogenesis in rice plant. XII. Ultrastructure of the phloem in the crown root. Jpn J Crop Sci 49: 330-339 (in Japanese with English summary)
Kawata S, Morita S, Yamazaki K (1978) On the differentiation of vessels and sieve tubes at the root tips of rice plants. Jpn J Crop Sci 47: 101-110 (in Japanese with English summary)
Kuo J (1983a) Notes on the biology of Australian seagrasses. Proc Linn Soc N S W 106: 225-245
Kuo J (1983b) The nacreous walls of sieve elements in seagrasses. Am J Bot 70: 159 164
Kuo J, O'Brien TP (1974) Lignified sieve elements in the wheat leaf. Planta 117: 349-353
Kuo J, O'Brien TP, Zee S-Y (1972) The transverse veins of the wheat leaf. Aust J Biol Sci 25: 721-737
Melaragno JE, Walsh MA (1976) Ultrastructural features of developing sieve elements in *Lemna minor* L. The protoplast. Am J Bot 63: 1145-1157
Miyake H, Maeda E (1976) Development of bundle sheath chloroplasts in rice seedlings. Can J Bot 54: 556-565
Neuberger DS, Evert RF (1975) Structure and development of sieve areas in the hypocotyl of *Pinus resinosa*. Protoplasma 84: 109-125
Oparka KJ, Johnson RPC (1978) Endoplasmic reticulum and crystalline fibrils in the root protophloem of *Nymphoides peltata*. Planta 143: 21-27
Parthasarathy MV (1974a) Ultrastructure of phloem in palms. I. Immature sieve elements and parenchymatic elements. Protoplasma 79: 59-91
Parthasarathy MV (1974b) Ultrastructure of phloem in palms. II. Structural changes, and fate of the organelles in differentiating sieve elements. Protoplasma 79: 93-125
Parthasarathy MV (1974c) Ultrastructure of phloem in palms. III. Mature phloem. Protoplasma 79: 265-315
Parthasarathy MV (1980) Mature phloem of perennial monocotyledons. Ber Dtsch Bot Ges 93: 57-70

Psaras GK (1980) On the development of the vascular bundles of *Zea mays*. Thesis, Univ Athens (in Greek with English summary)
Singh AP (1980) On the ultrastructure and differentiation of the phloem in sugarcane leaves. Cytologia 45: 1-31
Singh AP, Srivastava LM (1972) The fine structure of corn phloem. Can J Bot 50: 839-846
Thorsch J, Esau K (1981a) Changes in the endoplasmic reticulum during differentiation of a sieve element in *Gossypium hirsutum*. J Ultrastruct Res 74: 183-194
Thorsch J, Esau K (1981b) Nuclear degeneration and the association of endoplasmic reticulum with the nuclear envelope and microtubules in maturing sieve elements of *Gossypium hirsutum*. J Ultrastruct Res 74: 195-204
Thorsch J, Esau K (1981c) Ultrastructural studies of protophloem sieve elements in *Gossypium hirsutum*. J Ultrastruct Res 75: 339-351
Walsh MA (1974) Late-formed metaphloem sieve-elements in *Zea mays* L. Planta 121: 17-25
Walsh MA (1980) Preservation of the tonoplast in metaphloem sieve elements of embryonic roots of *Zea mays* L. Ann Bot (London) 46: 557-565
Walsh MA, Evert RF (1975) Ultrastructure of metaphloem sieve elements in *Zea mays*. Protoplasma 83: 365-388
Walsh MA, Melaragno JE (1976) Ultrastructural features of developing sieve elements in *Lemna minor* L. Sieve plate and lateral sieve areas. Am J Bot 63: 1174-1183
Walsh MA, Melaragno JE (1981) Structural evidence for plastid inclusions as a possible "sealing" mechanism in the phloem of monocotyledons. J Exp Bot 32: 311-320
Worley JF (1973) Evidence in support of "open" sieve tube pores. Protoplasma 76: 129-132

8 Sieve Elements in Internodal and Nodal Anastomoses of the Monocotyledon Liana *Dioscorea*

H.-DIETMAR BEHNKE

8.1 Introduction

The functional integrity of the phloem in vascular plants is dependent upon the establishment and maintenance of the structural continuity of its sieve elements throughout the entire plant body. The requirement for this integrity is nowhere better documented than in the formation of wound (repair) phloem (see Chaps. 10 and 11), but even the undamaged plant ensures that there are channels for a deviation of solutes in the case of an injury or if a demand emerges which is different from the main flow. Plants with cambial secondary growth may provide these deviations through extensive connections in their radial cell walls which contain lateral sieve areas or lateral sieve plates. Many herbaceous species develop anastomoses - cross-connections between primary phloem (and/or xylem) parts. Anastomoses may regularly connect all neighboring bundles; stem (cauline) bundles may be connected to leaf trace bundles as well as to medullary, cortical or other irregular bundles, thus providing linkages which are able to react to changes in physiological requirements with short-cuts, deviations, branches or combinations. In some plants the degree of development of these anastomoses may be so high that wound phloem will not be produced even if about half of all the vascular bundles are interrupted (Behnke and Sukkri 1971).

The most intensive connections between vascular tissues, however, occur in the nodes where the vascular systems of the leaf and the stem are joined together (see frontispiece, Fig. 24, by Hartig 1837). While the number of leaf vascular bundles entering a stem and the number of internodes passed before they contact a cauline bundle are taxon-specific, the vascular elements of the leaf trace are almost always directly connected to their stem counterparts. Largely due to the complexity of the nodal area, details of sieve-element connections are but rarely disclosed. One notable exception is in the nodal glomeruli of the Dioscoreaceae. In the annual vines of this monocotyledon family the sieve tubes of leaf traces are connected to those of cauline bundles via a series of consecutively smaller sieve elements, arranged in such a highly ordered array that in some species the phloem part of the nodal complex extends over an axial length of up to 4 mm. While this structure was mentioned already by the classical anatomists in the second half of the nineteenth century, Mason (1926) was the first to describe its anatomical composition and recognize its physiological consequence (for literature review see Behnke 1965a).

1

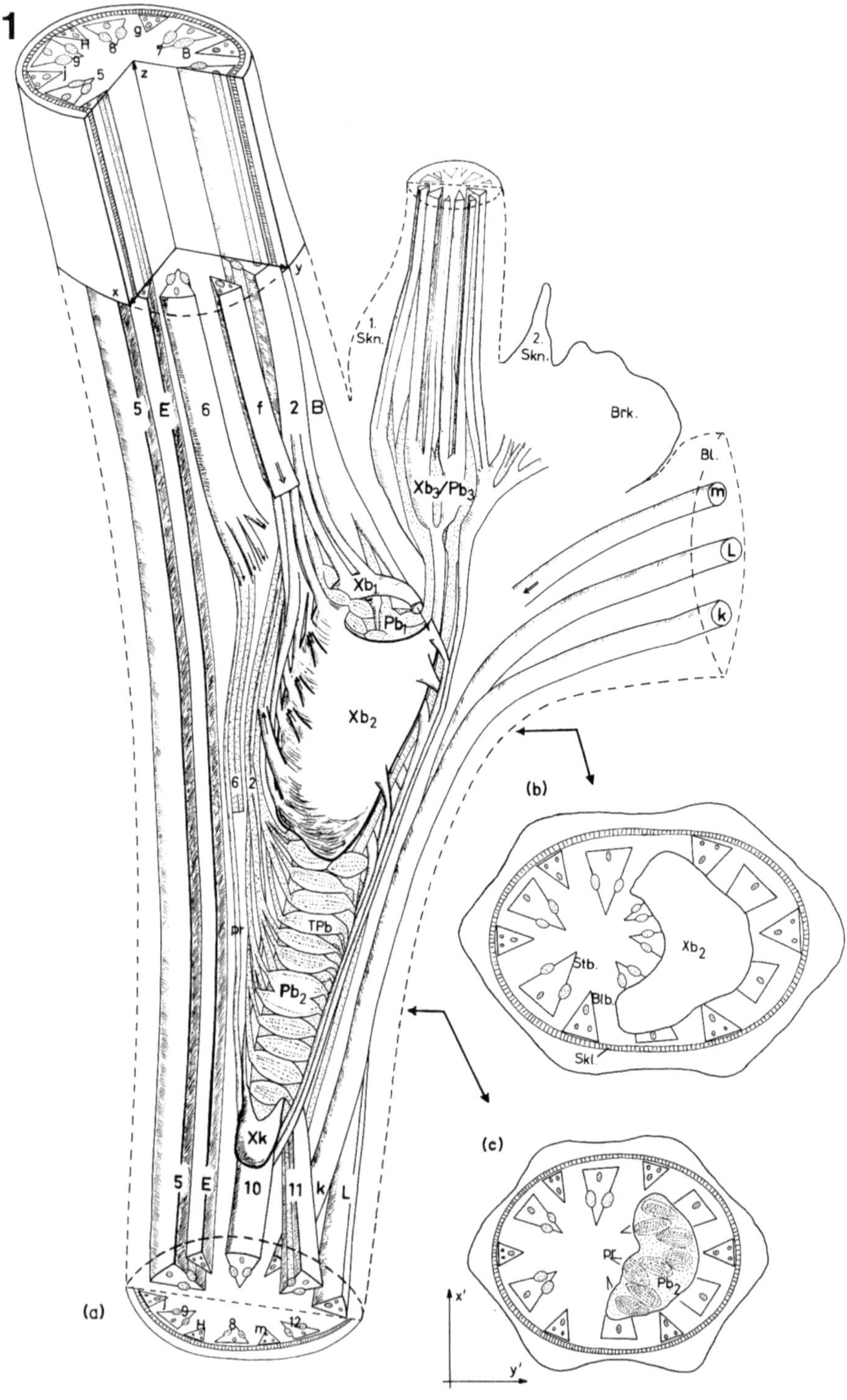
H
g
8
g
7
B
j
5
z
x
y
1.
Skn.
2.
Skn.
Brk.
5
E
6
f
2
B
Bl.
m
L
k
Xb3/Pb3
Xb1
Pb1
Xb2
6
2
(b)
pr
TPb
Pb2
Xb2
Stb.
Blb.
Skl.
Xk
(c)
5
E
10
11
k
L
(a)
9
H
8
m
12
Pb2
x'
y'

8.2 The Vascular Construction in the Aerial Stem of *Dioscorea*

The arrangement of the vascular bundles in two circles in the twining stems of the Dioscoreaceae is atypical for monocotyledons. The outer circle contains a number of so-called common (i. e., common to stem and leaf) vascular bundles situated opposite the ribs of the stem, the inner circle the same number of cauline vascular bundles positioned in the gaps between common bundles (Figs. 8.1, 8.2). Cauline and common vascular bundles are composed of an elliptical or V-shaped metaxylem and usually three phloem units. While the xylem develops centrifugally, starting in both cases from the protoxylem located at the far inner point of a bundle, the position of the phloem is distinct. In cauline bundles, two phloem units are found at the inside of the first large metaxylem vessels and one unit is situated at the external periphery and surrounded by small xylem elements. In common bundles, however, there is one phloem unit enclosed by the converging ends and two units near the outer periphery of the V-shaped metaxylem (see Fig. 8.2). This unusual distribution of phloem and xylem has been repeatedly used as an argument in favor of the presence of compound vascular bundles in the Dioscoreaceae (for literature see Behnke 1965a; Ayensu 1972; cf. also French and Tomlinson 1986).

The number of vascular bundles passing through the twining stem parts of the Dioscoreaceae varies within the different taxa of the family, but for an individual plant the number remains fixed throughout the entire axis. In the different specimens of *Dioscorea reticulata* studied, a total of 10 to 14 bundles make up the vasculature of the internode. At each node three leaf trace bundles, arriving from the petiole, become positioned between cauline bundles, replacing common bundles which had entered as leaf traces at the three upper nodes (Figs. 8.1, 8.4). The relatively small number of bundles found in *D. reticulata* facilitated the construction of diagrams designed to show the exact course and the sequence of replacement of the bundles, which is different for the median and the two lateral leaf traces (see Behnke 1965a, and Behnke and Sukkri 1971). Several other species studied showed a more complex pattern based on both a greater number of vascular bundles (e. g., five leaf trace bundles in *D. sativa,* see Behnke and Sukkri 1971) and another sequence of their incorporation into nodal complexes, being expressions of different phyllotaxies which in

◁

Fig. 8.1 a. Diagram showing the course of cauline (marked by *numbers*) and common (marked by *letters*) vascular bundles in the stem of *Dioscorea reticulata* and their connection to the vascular systems of a leaf *(Bl)*, a bulbil *(Brk)* and two axillary buds *(Skn)*. In the nodal area three cauline bundles *(2, 6, 7)* are combined with three common bundles *(B, f, g)* via a nodal complex composed of several xylem *(Xb, Xk)* and phloem parts *(Pb)* and connected to bundles of the axillary buds, while three leaf trace bundle *(k, L, m)* enter the stem to be integrated among the three bundles leaving the node *(10–12)*. **b** and **c** represent cross-sections through different areas of the node *(Blb)* leaf trace bundle, *(pr)* protoxylem, *(Stb)* cauline bundle (All from Behnke 1965a)

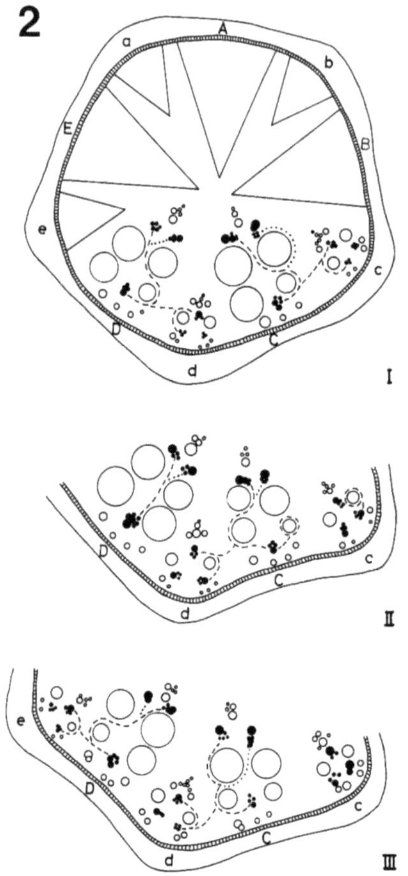
2
A
a
b
B
E
e
c
C
D
d
I
II
III

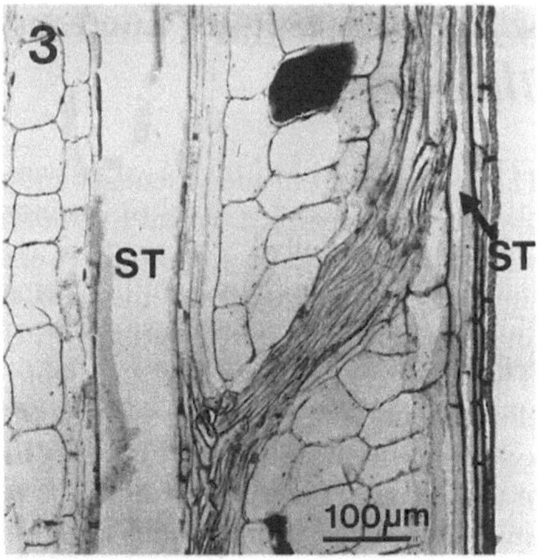
3
ST
ST
100µm

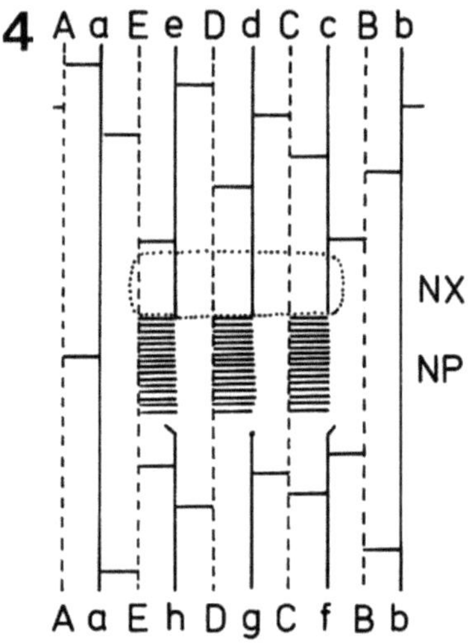
4
A a E e D d C c B b
NX
NP
A a E h D g C f B b

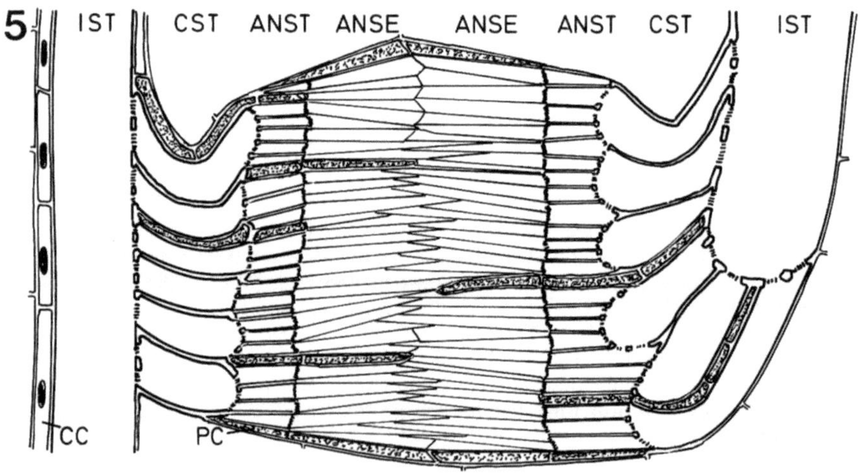
5
IST CST ANST ANSE ANSE ANST CST IST
CC
PC

some species approach a decussate pattern with two leaves inserted at each node (see also Ayensu 1972).

The integration of the leaf traces into the vasculature of the stem and their connection with the cauline bundles to form one complex system has been explained in detail only for a few *Dioscorea* species. Phloem (and xylem) anastomoses form the anatomical basis for all horizontal interconnections, in both the nodes and the internodes. While their composition is the same throughout the entire aerial stem (see Sect. 8.3), and can be regarded as a specific feature of this family, the frequency and arrangement of anastomoses differ greatly between the species and from nodal to internodal parts.

In the *internode* of *D. reticulata* an analysis of frequency (using about 3000 serial cross-sections through a total internodal length of about 28 cm) determined that there were ten anastomoses within each centimeter of internode length (Behnke and Sukkri 1971). The ten anastomoses which were distributed between the five common and the five cauline bundles in the specimen studied represent exactly one copy of each of the ten possible links between the ten adjacent bundles on every centimeter of internodal length. In addition to these "external" anastomoses twice as many "internal" anastomoses (Fig. 8.2) linking the different phloem units within a vascular bundle have been counted (Behnke and Sukkri 1971), i. e., the minimum necessary to connect three units. The presence of "internal" anastomoses in *Dioscorea* clearly supports the concept of compound vascular bundles.

In the *nodal area* common bundles (three or five, depending upon the number of leaf trace bundles entering at each node) are connected to cauline bundles. While the xylem units form a dense and rather obscure plexus (Fig. 8.1: Xb), the phloem connections are distinguished by a high frequency of distinct anastomoses which are highly concentrated in a horizontal succession (Fig. 8.1: Pb). The number of phloem anastomoses in the node (the phloem or bast "glomeruli" of Mason 1926, Brouwer 1953, Ayensu 1969; "Teilphloembecken" of Braun 1957, Behnke 1965a) and their arrangement differ considerably

---

Figs. 8.2–8.5. *Dioscorea reticulata.* **Fig. 8.2.** Three cross-sections through the internode of a stem showing the alternation of cauline bundles *(A to E)* with common bundles *(a to e)*. *I through III* display the course of anastomoses *(dashed lines)*, of which some parts *(dotted lines)* lie in another plane of sectioning. Phloem units marked by *black circles*, xylem by *open circles* (Behnke and Sukkri 1971). **Fig. 8.3.** Internodal phloem anastomosis between sieve-tube members *(ST)* of two different vascular bundles; ×100 (Behnke 1965a). **Fig. 8.4.** Simplified diagram showing the course of the ten vascular bundles seen in Fig. 8.2 and their interconnection by internodal and nodal anastomoses. While three common bundles *(c, d, e)* are connected in a nodal complex to cauline bundles and are subsequently replaced by new leaf-trace bundles *(f, g, h)*, the cauline bundles *C, D, E* pass through the entire stem part shown. *Horizontal lines* represent internodal and nodal anastomoses. The axial length of the stem shown is equivalent to about 2 cm. *NP* nodal phloem complex; *NX* nodal xylem complex. **Fig. 8.5.** Schematic drawing of an anastomosis extending between two internodal sieve-tube members *(IST)* and composed of connecting sieve-tubes members *(CST)*, anastomosal sieve-tube members *(ANST)*, anastomosal sieve elements *(ANSE)*, and phloem parenchyma cells *(PC)*. *CC* companion cells

between the species studied and most probably depend upon the number of sieve tubes to be connected between cauline and common bundles. Among those species which insert one leaf at each node, *D. reticulata* is advantageous for anatomical studies, since its nodes contain a comparatively high number of anastomoses. The approximately 20 anastomoses which are arranged in one plane within a 4-mm-long stack, as recorded by Behnke (1965 a), actually represent at least 60 or more anastomoses, if the third dimension and all connections between the three common and the respective cauline bundles are considered.

While there is no doubt that in the nodal area the vascular bundles of stem, leaf, axillary bud, and bulbil are united and reduced to the ordinary number found in the internode, the question of their mode of connection is not yet settled. At least two alternatives can be suggested:

1) In most previous investigations the assumption has been made that wherever two vascular bundles of different plant parts come together in the node they are combined and give rise to a new (cauline) bundle (cf. Fig. 8.1 and Behnke 1965 a), for which Braun (1957) and after him Ayensu (1969) even introduced the name "glomerulus bundle" (all cauline bundles would consistently have to be called glomerulus bundles). As regards the junction of the sieve tubes from different bundles, two slightly different schematic drawings were published: (a) Brouwer (1953) proposed that each of two arriving sieve tubes (e. g., from stem and axillary bud, respectively) forms a phloem glomerulus and the two glomeruli join and end in a single departing sieve tube (actually, his explanation starts at the base of the node and describes a division of the glomerulus); (b) Braun (1957) drew two arriving sieve tubes running parallel to each other and reduced them via a single glomerulus to one departing sieve tube.

2) Since so far neither Brouwer's (1953) nor Braun's (1957) concept could be verified (nor disproved) by microscopic studies, another alternative may be proposed. Considering the way that sieve tubes are connected in the internode by internal and external anastomoses (cf. Figs. 8.2–8.4), anastomoses could well be used to end the conducting elements of the leaf traces (i. e., common bundles) and to connect them laterally in the nodal area to the respective conduits of cauline bundles. This model implies that cauline bundles do not end after several internodes in a nodal complex but, instead, run through the entire axis

Fig. 8.6. *Dioscorea reticulata.* Longitudinal section through a nodal complex with xylem *(X)* and phloem *(P)* parts, the latter composed of many separate anastomoses. Within the phloem part *dark vertical lines* correspond to sieve areas of anastomosal sieve-tube members arranged in one plane, ×50. **Fig. 8.7.** *D. reticulata.* Longitudinal section through nodal phloem complex showing parts of anastomoses with irregularly shaped connecting sieve-tube members *(CST)*, anastomosal sieve-tube members *(1)* and anastomosal sieve elements *(2)*. *IST* internodal sieve-tube members, ×140. **Fig. 8.8.** *Dioscorea sp.* Part of nodal complex viewed with the fluorescence microscope after aniline-blue staining showing the extent of callose deposition in the different internodal and nodal sieve elements. The longitudinal walls of the innermost anastomosal sieve elements *(ANSE)* are evenly covered by callose, while within all other sieve elements separate sieve areas can be distinguished (see e. g., at *IST*). *X* autofluorescent walls of tracheids in xylem complex. ×175

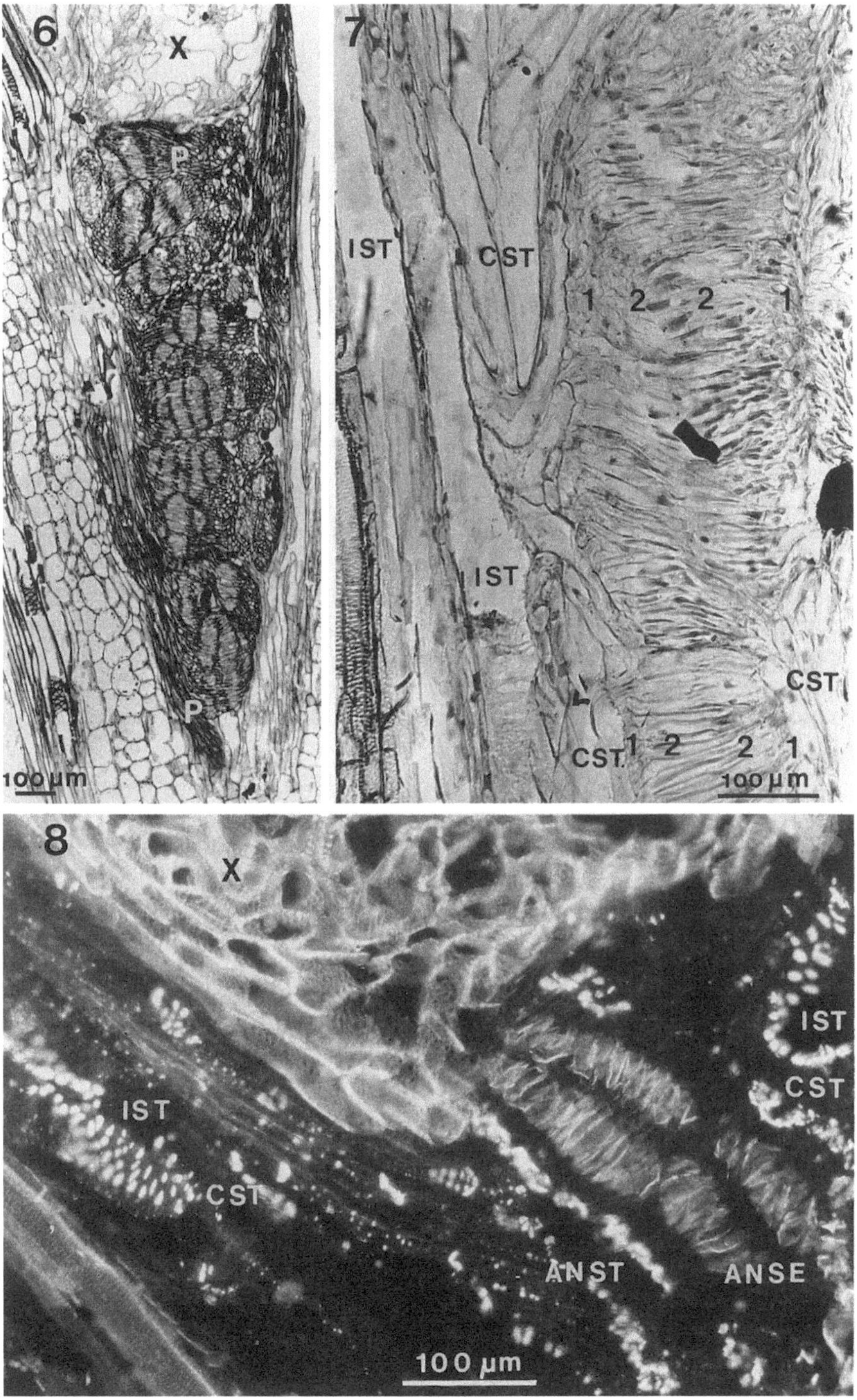
6
X
P
P
100 µm
7
IST
CST
1
2
2
1
IST
L
CST
1
2
2
1
CST
100 µm
8
X
IST
CST
IST
CST
ANST
ANSE
100 µm

of a stem (cf. Fig. 8.4). It is supported by our observations on serial sections of successive internodes, which showed that cauline bundles entering a node can also be identified by the same characters after passage through the node. In addition, micrographs published (see Fig. 8.7 and Behnke 1965a) confirm the splitting of sieve tubes into several categories of smaller sieve elements (see Sect. 8.3), but so far they have not shown the division of a glomerulus nor two internodal sieve tubes running into one glomerulus.

8.3 The Specific Composition of Phloem Anastomoses

In *Dioscorea* internodal and nodal anastomoses (the latter also named phloem glomeruli) are composed of symmetrically ordered layers of sieve elements interpolated between two large sieve tubes. Although the exact number of layers and their lengths depend upon the width of the gap to be bridged, it can be said that at least six layers are present and that the sizes of the sieve elements from the periphery to the center of any anastomosis are consecutively diminished, while their number increases. Thus, despite the large size of internodal sieve-tube members (diameters of 140 μm and lengths of several millimeters are not unusual, see Behnke 1965a; Ayensu 1972), their cross-sectional area even increases toward the plane of the innermost (only 5-10 μm wide and about 100 μm long) spindle-shaped anastomosal sieve elements. Brouwer (1953) estimated a five- to tenfold multiplication; Braun (1957) counted between 60 and 120 cells.

An anastomosis contains three different categories of sieve elements connected in series (cf. Fig. 8.5). Two internodal sieve-tube members (IST) are joined by symmetrical layers of five or more connecting sieve-tube members (CST), at least three times as many anastomosal sieve-tube members (ANST), and a great many anastomosal sieve elements (ANSE). Previous reports of a fourth category (see Braun 1957) situated in between the two innermost layers can be explained as overlaps between the spindle-shaped ends in the middle of an anastomosis (Behnke 1965a). All the different categories of sieve elements are connected by sieve pores (see following section for details).

Fig. 8.9. *Dioscorea* cf. *ethiopica*. Longitudinal section of connecting sieve-tube members *(CST)* and anastomosal sieve-tube members *(ANST)* connected via compound sieve plates. Typical monocotyledon sieve-element plastids reside in the parietal area of all sieve elements, ×3000.
Fig. 8.10. Schematic drawing showing the relative position of the sieve elements depicted in Figs. 8.9, 8.11, 8.12. Phloem parenchyma cells indicated by *hatching*

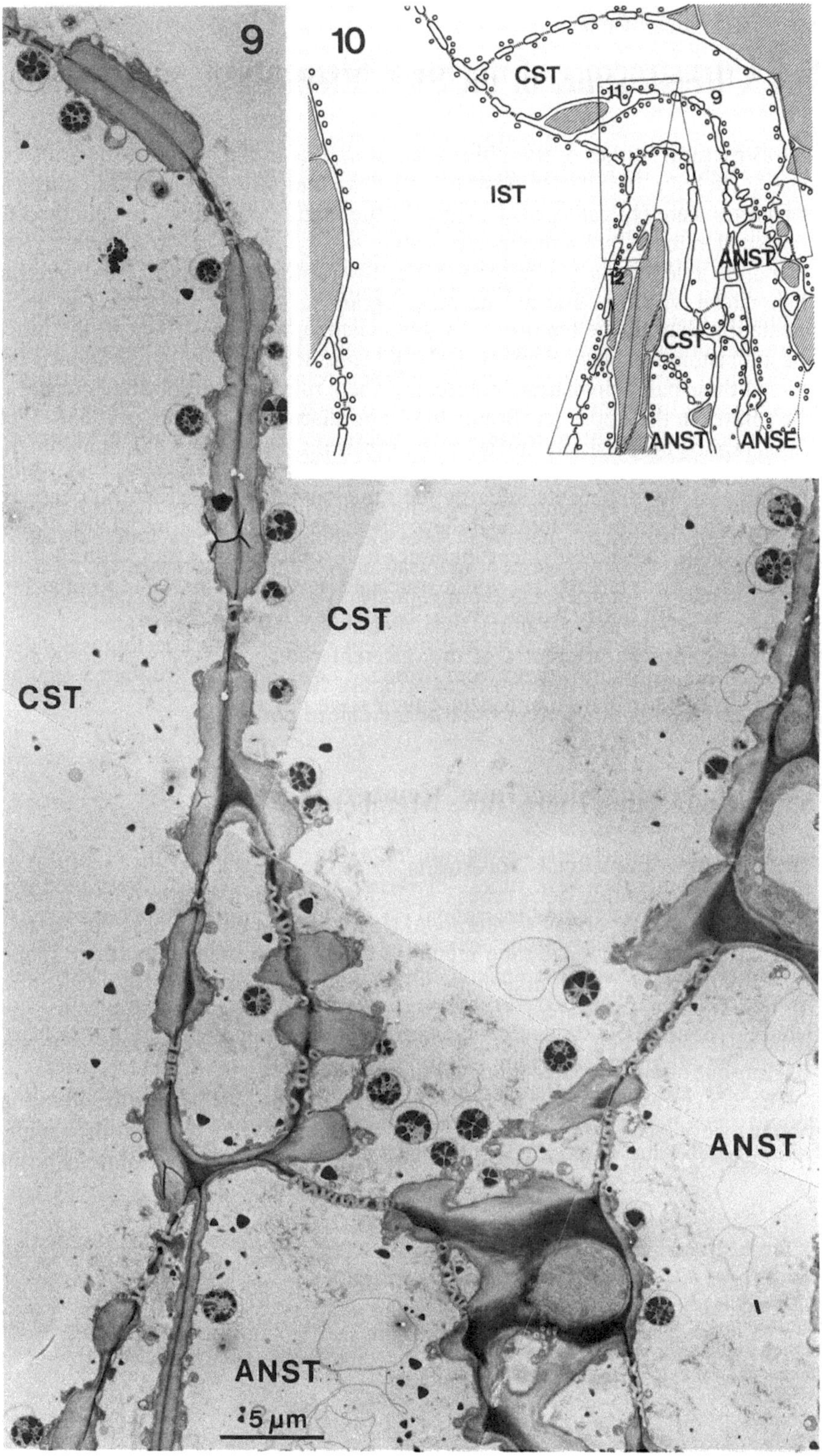
9
10
CST
11
9
IST
ANST
12
CST
ANST
ANSE
CST
CST
ANST
ANST
5 µm

8.4 Ultrastructure of the Sieve Elements of Anastomoses

Ultrastructural studies of the phloem anastomoses in the nodes of the Dioscoreaceae (Behnke 1965 b) demonstrated that the majority of cells within all three categories were typical sieve elements. In the mature state they are delimited by a plasmalemma, contain the specific P-type sieve-element plastids characteristic of monocotyledons, a few mitochondria, ER profiles, and some P-protein, but their nuclei, vacuoles and all the other organelles are degenerated. The documentation that all the features of a sieve element are found even in the innermost anastomosal sieve elements (ANSE) contradicts earlier studies which discussed their parenchymatous nature and their role as a cytoplasmic control of the solutes in the nodes (cf. Braun 1957, and discussion in Behnke 1965 b). Previous reports on persistent nuclei in the innermost cells of the phloem glomerulus (Braun 1957) were not supported. It has been shown, however, that while anastomosal sieve elements undergo the same characteristic structural changes as are seen in normal internodal sieve elements (cf. Behnke and Dörr 1967; Behnke 1969), the sieve-element maturation in some anastomoses within a single node may be retarded, i. e., young, nucleate sieve elements were found as far down as the 14th node (Walther 1986; Behnke and Walther 1987).

The specific characteristics of the different categories of anastomosal sieve elements that distinguish them from ordinary sieve elements are their irregular shape and the distribution and size of their sieve pores.

8.4.1 Connecting Sieve-Tube Members (CST)

The CST (the "funnel cells" of Mason 1926, the "funiculus cells" of Brouwer 1953, the "glomerulus sieve tubes" of Braun 1957 and Ayensu 1969) mediate between the axially oriented internodal (IST) and the more or less horizontally aligned anastomosal sieve-tube members (ANST). Consequently, their shapes are rather irregular and in general intermediate between the two categories (heavily bent or V-shaped forms are predominant; cf. Figs. 8.7, 8.10, and Behnke 1965 a). Their diameter is between 20 and 40 μm, their length is at least 100 μm but often is greater than 300 μm (see Fig. 8.7).

The CST are connected to both the IST and the ANST by (usually compound) sieve plates (Figs. 8.9-8.12, 8.15, 8.19, 8.20), with pore widths ranging from 0.4 to 0.8 μm, which are though often narrowed down by callose deposits.

Figs. 8.11 and 8.12. *Dioscorea* cf. *ethiopica*. Two parts of a connection between an internodal sieve-tube *(IST)* with several connecting sieve-tube members *(CST)* via compound sieve plates (for position of the figures within the anastomosis see Fig. 8.10). The sieve elements contain numerous plastids in their parietal parts and P-protein filaments distributed all over their lumina (*). *PC* phloem parenchyma cell, ×3000

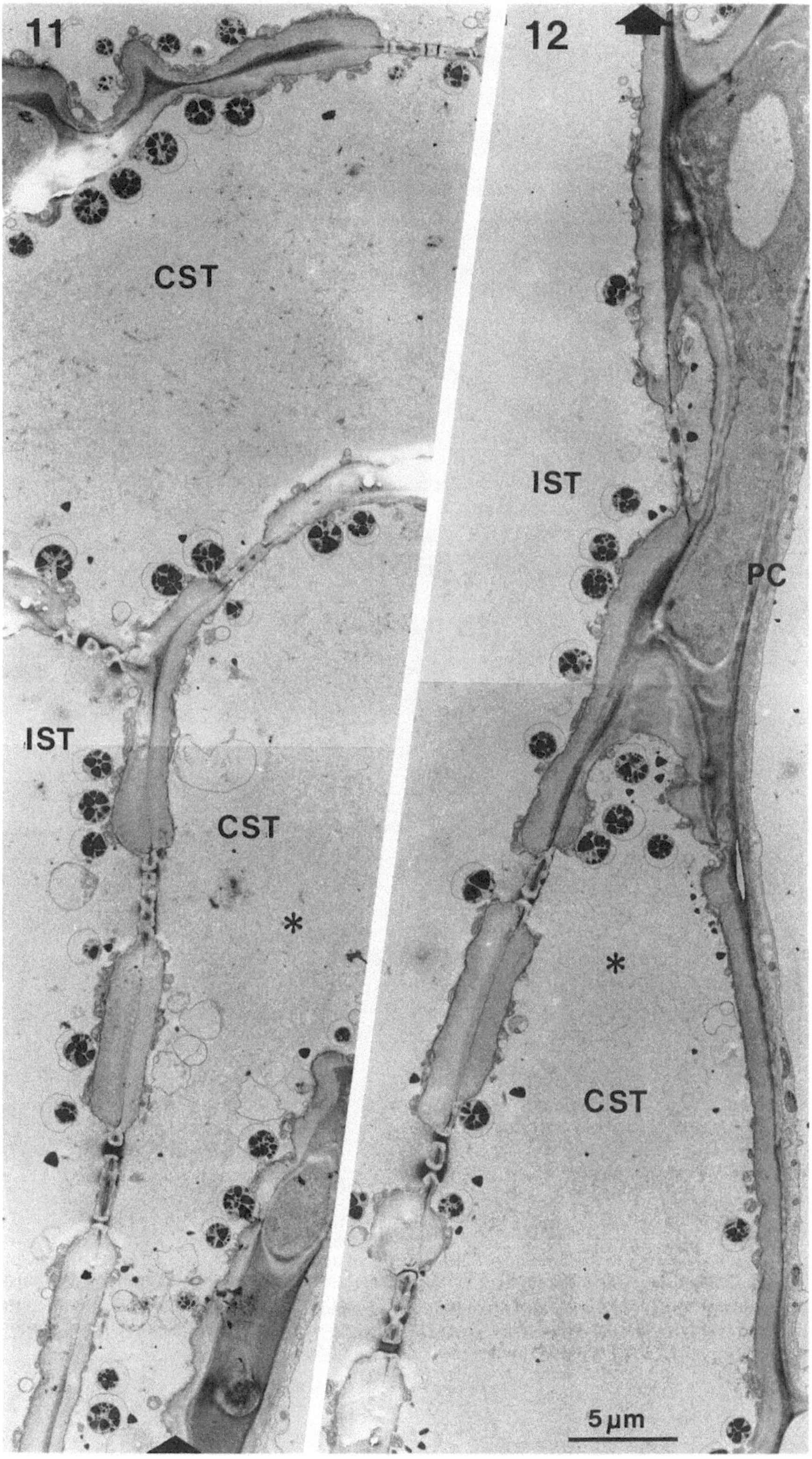

11
12
CST
IST
CST
*
IST
PC
*
CST
5 µm

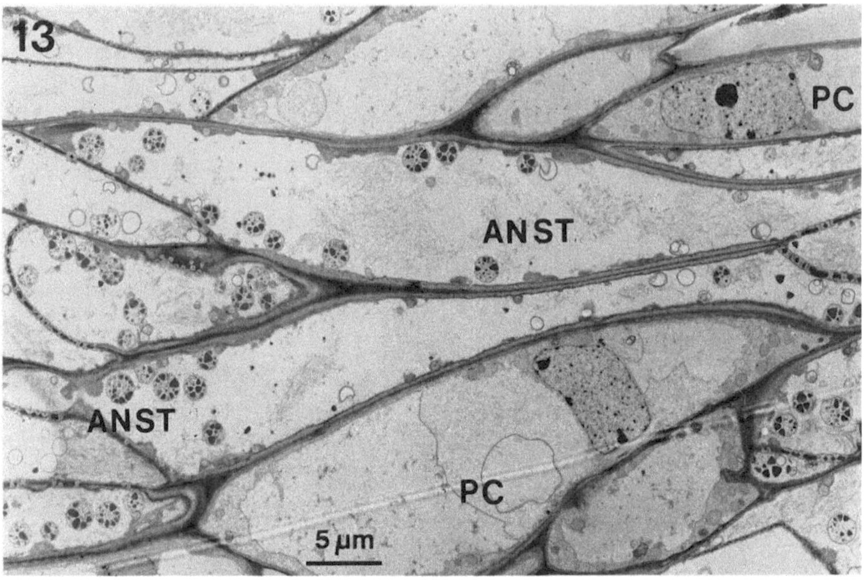

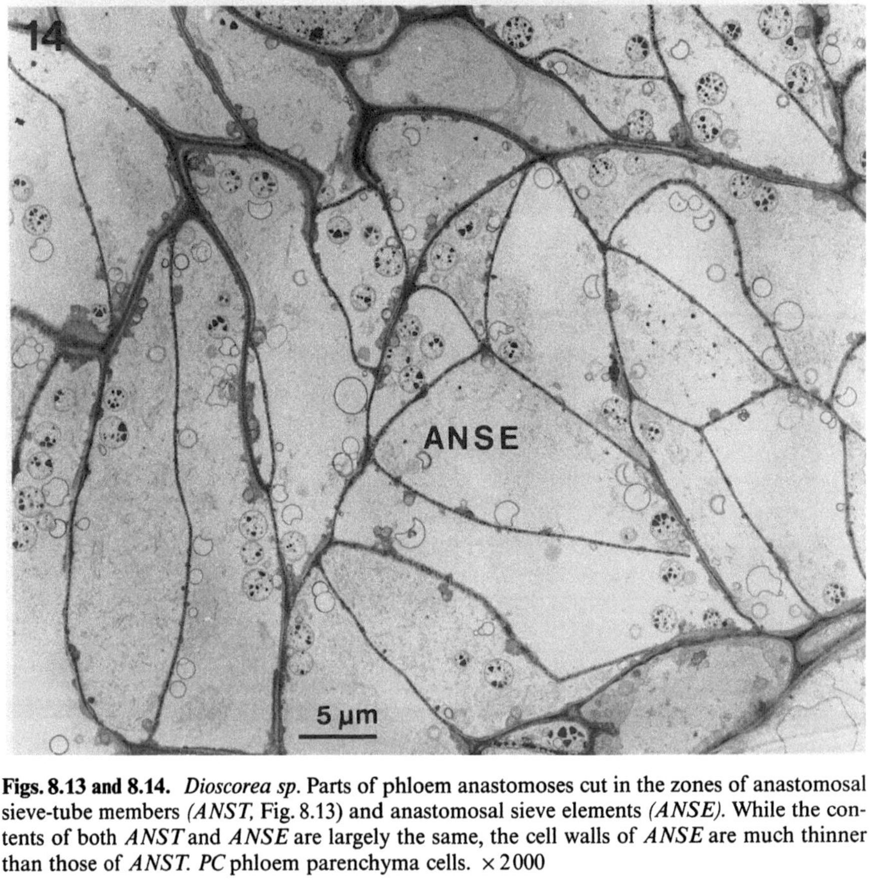

Figs. 8.13 and 8.14. *Dioscorea sp.* Parts of phloem anastomoses cut in the zones of anastomosal sieve-tube members *(ANST,* Fig. 8.13) and anastomosal sieve elements *(ANSE).* While the contents of both *ANST* and *ANSE* are largely the same, the cell walls of *ANSE* are much thinner than those of *ANST. PC* phloem parenchyma cells. ×2000

Their most typical contents are finely distributed filaments of P-protein (Fig. 8.12) and numerous plastids, which are found in the peripheral parts of the sieve elements and which contain up to ten or more of the monocotyledon-specific, cuneate protein crystals (Figs. 8.9, 8.11, 8.12). Filaments of P-protein are also seen filling sieve-plate pores to varying extents (Figs. 8.19, 8.20).

8.4.2 Anastomosal Sieve-Tube Members (ANST)

Depending upon the breadth of the gap between CST and the innermost anastomosal sieve elements (ANSE), one to several layers of ANST (the "bast tubules" of Mason 1926 and Brouwer 1953, the "phloem glomerulus cells 1" of Braun 1957 and Ayensu 1969, the "nodal sieve tubes" of Behnke 1965 a, b) are situated toward the center of the anastomoses. ANST are comparatively small sieve elements, characterized by rather constant sizes (diameters about 10 µm and lengths between 40 and 50 µm), and are connected to the CST via (often compound) sieve plates and to the ANSE by simple sieve plates. The sieve plates between the ANST and the ANSE are arranged in one plane within one anastomosis and even within the entire phloem part of a nodal complex. This makes this borderline especially prominent in light microscopic sections (Figs. 8.6–8.8). The contents of the mature ANST are similar to those of the other sieve elements of *Dioscorea:* P-type sieve-element plastids (Fig. 8.16), mitochondria, and filamentous P-protein (Fig. 8.13), the latter also found in sieve-plate pores. The size of the pores (often reduced by callose) decreases from about 0.5 µm at the CST side to between 0.2 and 0.4 µm at the ANSE side (Figs. 8.21, 8.22).

8.4.3 Anastomosal Sieve Elements (ANSE)

The innermost ANSE (the "glomerulus cells" or "phloem glomerulus cells 2" of previous authors, the "nodal sieve elements" of Behnke 1965 a, b) are the most peculiar cells of the entire anastomoses. Their diameter, 5–10 µm, is about the same as that of the ANST, but their length reaches 100 µm or more. Compared to the other sieve elements, the ANSE are characterized by thinner walls which are penetrated all over by plasmodesma-like small sieve pores (Figs. 8.23–8.26), except where they abut on the ANST (Figs. 8.16, 8.17, 8.21). The innermost end walls of the ANSE are tapered and the two symmetrically arranged layers of cells are, therefore, partly interdigitated. The sieve pores between the ANSE are of uniform diameters (about 60–80 nm) and are slightly narrowed down by callose (Figs. 8.23–8.26), which, on account of the dense distribution of the pores (Fig. 8.26), is responsible for the bright aniline-blue fluorescence of all lateral wall parts (Fig. 8.8). The content of the small pores has not been elucidated so far (Fig. 8.25), although some micrographs suggest that P-protein is associated with them. As is true for all other sieve elements, the ANSE are enucleate in the

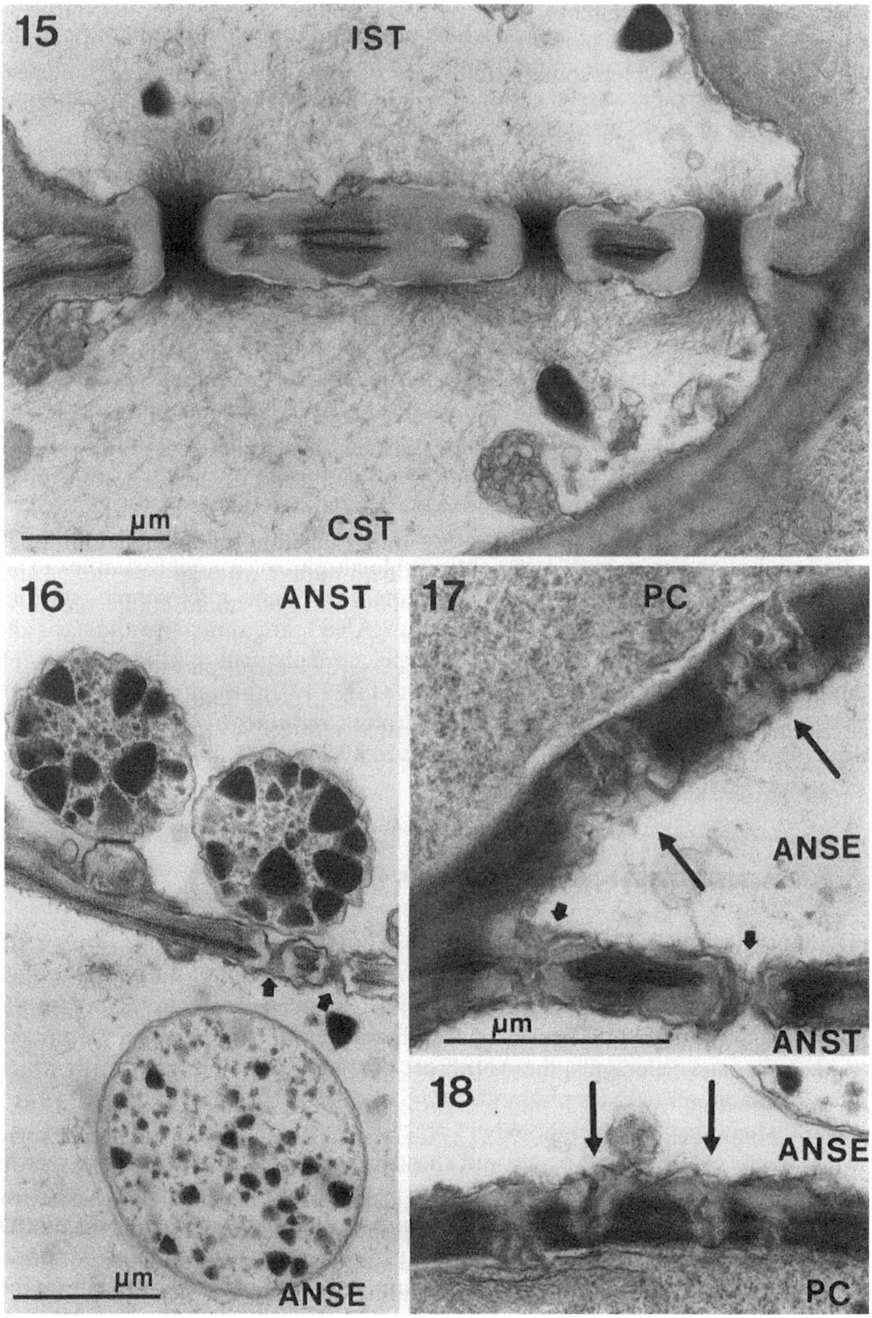

15
IST
µm
CST
16
ANST
µm
ANSE
17
PC
ANSE
µm
ANST
18
ANSE
PC

mature state and, in addition to a continuous plasmalemma (also lining all sieve pores; see Fig. 8.25), contain mitochondria, plastids, and some P-protein (Fig. 8.14; cf. Behnke 1965b). The plastids are of the specific P-type mentioned earlier, but, in contrast to the other sieve elements, contain only very small cuneate crystals (Fig. 8.16).

8.5 Parenchymatous Cells Associated with the Sieve Elements of Anastomoses

Internodal anastomoses are surrounded by parenchymatous cells of the stem parts traversed by them. Within a nodal complex phloem anastomoses (glomeruli) are separated from each other by narrow layers of parenchymatous cells (Behnke 1965a).

While internodal sieve-tube members are regularly associated with longitudinal series of small companion cells which are derived from the same precursor cell as the sieve elements, the origin of the parenchymatous cells which are found among the various anastomosal sieve elements has not been traced so far. Although some of these cells are linked to sieve elements by typical sieve pore/plasmodesmata connections (Figs. 8.17, 8.18) and exhibit other features of companion cells (e.g., a densely staining cytoplasm), they are called phloem-parenchyma cells because of their unresolved origin.

Phloem-parenchyma cells represent the only nucleate cells in a mature anastomosis, but their number is much less than that of all types of anastomosal sieve elements. While their overall distribution seems to be irregular, their frequency decreases toward the middle of an anastomosis (cf. Figs. 8.5 and 8.12–8.14). The extreme scarcity of phloem-parenchyma cells among the innermost ANSE may easily cause them to be overlooked and may lead to the conclusion that they are absent (Behnke 1965a). Recent reinvestigations, however, have demonstrated the presence of at least some parenchyma cells which are connected to the ANSE via a combination of branched plasmodesmata and sieve pores (Figs. 8.17, 8.18).

Figs. 8.15–8.18. *Dioscorea* cf. *ethiopica*. **Fig. 8.15.** Sieve area from a compound sieve plate between internodal sieve tube *(IST)* and connecting sieve-tube member *(CST)*. The three sieve pores cut are filled with P-protein. ×20000. **Fig. 8.16.** The sieve-element plastids of anastomosal sieve elements *(ANSE)* are distinct from those of all other sieve elements (e.g., *ANST*) by considerably smaller cuneate crystallois. *ANST* and *ANSE* are connected by small sieve pores *(arrows)*. ×20000. **Figs. 8.17, 8.18.** Sieve pore/plasmodesmata connections *(large arrows)* are found between anastomosal sieve elements *(ANSE)* and phloem parenchyma cells *(PC)*, while small sieve pores connect *ANSE* with *ANST (small arrows)*. ×30000

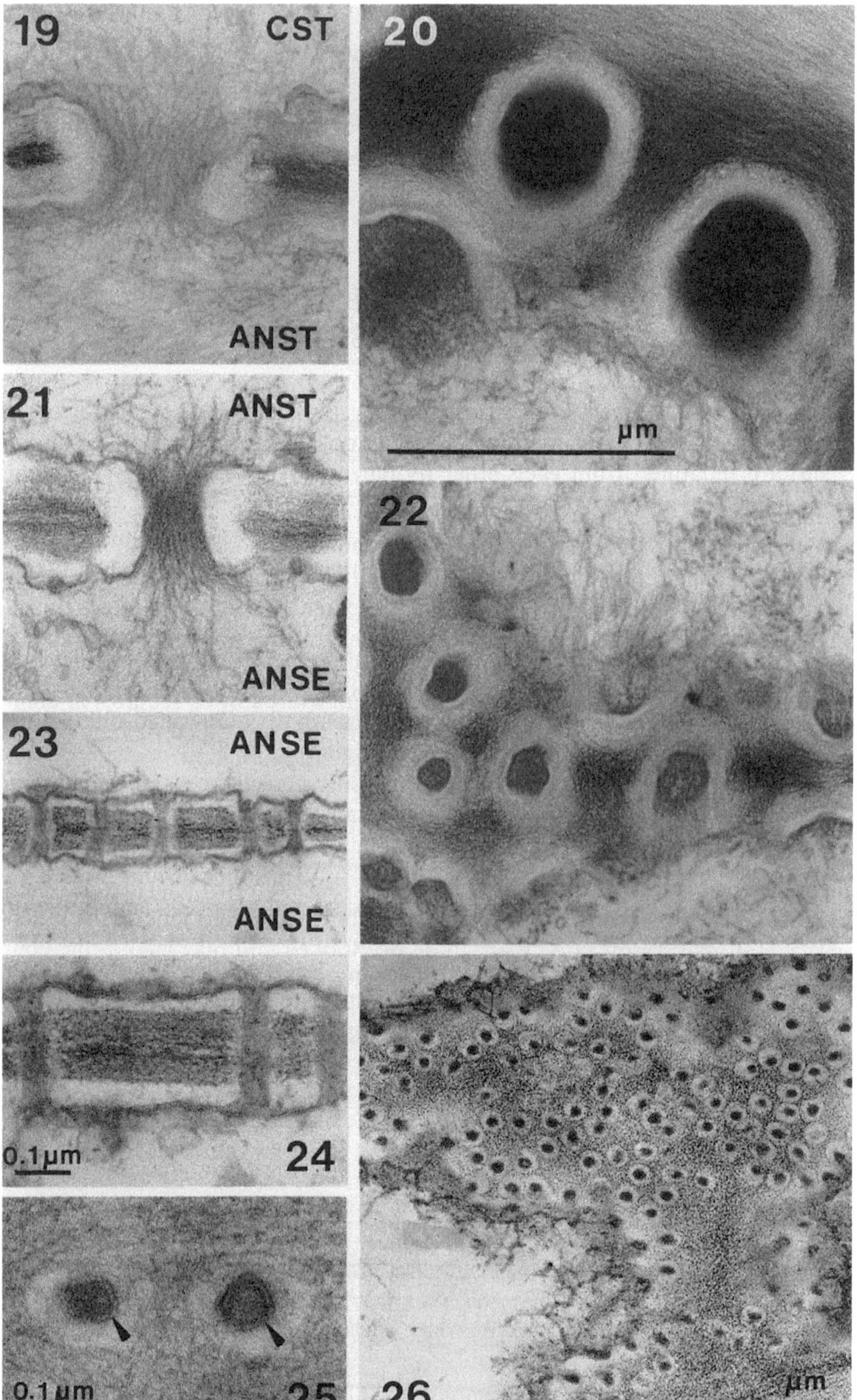
19
CST
ANST
20
μm
21
ANST
ANSE
22
23
ANSE
ANSE
0.1μm
24
0.1μm
25
26
μm

8.6 Some Physiological Implications of Nodal Anastomoses

Indisputably, nodal anastomoses are sites where assimilates can be distributed over more than one bundle and where the direction of flow can be changed. However, unlike those internodal anastomoses which function in cases where the flow of solutes is disturbed, requiring a quick and reliable deviation or short-cut, the nodal anastomoses of the kind present in the Dioscoreaceae have always been looked upon as obstacles to a rapid flow or as stations controlling the transport of assimilates (cf. Braun 1957; Braun and Sauter 1964). However, translocation rates of about 90 cm/h reported by Mason and Lewin (1926) and Lawton (1967) clearly indicate that the vigorously growing lianas of the Dioscoreaceae are not adversely affected by the presence of dozens of nodal complexes along the axial route of solutes [if the concept of previous authors be followed (Brouwer 1953, Braun 1957) in many species a cauline bundle may run into a nodal anastomosis at almost every third node; cf. Behnke 1965a]. Therefore, it may be argued that the total cross-sectional area of symplastic continuity is maintained or even increased to cope with a higher resistance to flow in the smaller pores by an increase of the number of cells from the periphery to the center of a nodal anastomosis (and the concomitant relative increase of the transsectional area; see Sect. 8.3) as well as by a denser distribution of the small pores over a greater part of the wall. However, neither exact measurements nor cross-sectional area calculations have been performed to date.

Acknowledgment. Supported in part by the Deutsche Forschungsgemeinschaft.

◁

Figs. 8.19–8.26. Sizes of sieve pores between the different sieve elements of an anastomosis, depicted from longitudinal (Figs. 8.19, 8.21, 8.23, 8.24) and cross-sections (Figs. 8.20, 8.22, 8.25, 8.26). The pore sizes gradually decrease from the *CST* to the *ANSE*. Sieve pores between two *ANSE* are of the size range of plasmodesmata (Figs. 8.23–8.26), they are lined by plasmalemma (*arrowheads* in Fig. 8.25) but their content is not resolved. All sieve pores, including the densely arranged pores of the *ANSE*, are narrowed down by varying amounts of callose *(white collars)*. **Figs. 8.19–8.25.** *Dioscorea* cf. *ethiopica;* **Figs. 8.19–8.23,** ×40000; **Fig. 8.24,** ×75000; **Fig. 8.25,** ×150000. **Fig. 8.26.** *Dioscorea macroura,* ×40000

References

Ayensu ES (1969) Aspects of the complex nodal anatomy of the Dioscoreaceae. J Arn Arbor 50: 124-132

Ayensu ES (1972) Dioscoreales. In: Metcalfe CR (ed) Anatomy of the monocotyledons, vol 6. Univ Press, Oxford

Behnke H-D (1965a) Über das Phloem der Dioscoreaceen unter besonderer Berücksichtigung ihrer Phloembecken. I. Lichtoptische Untersuchungen zur Struktur der Phloembecken und ihrer Einordnung in das Sproßleitsystem. Z Pflanzenphysiol 53: 97-125

Behnke H-D (1965b) Über das Phloem der Dioscoreaceen unter besonderer Berücksichtigung ihrer Phloembecken. II. Elektronenoptische Untersuchungen zur Feinstruktur des Phloembeckens. Z Pflanzenphysiol 53: 214-244

Behnke H-D (1969) Aspekte der Siebröhren-Differenzierung bei Monokotyledonen. Protoplasma 68: 289-319

Behnke H-D, Dörr I (1967) Zur Herkunft und Struktur der Plasmafilamente in Assimilatleitbahnen. Planta 74: 18-44

Behnke H-D, Sukkri B (1971) Anastomoses in the internode of *Dioscorea:* their frequency, structure, and function. Z Pflanzenphysiol 66: 82-92

Behnke H-D, Walther A (1987) Sieve elements of the stem nodes of *Dioscorea.* In: Abstr 14th Int Botanical Congr Berlin, p 228

Braun HJ (1957) Die Leitbündelbecken in den Nodien der Dioscoreaceae, mit besonderer Berücksichtigung eines neuen Typs assimilateleitender Zellen. Ber Dtsch Bot Ges 70: 305-322

Braun HJ, Sauter JJ (1964) Phosphatase-Lokalisation in Phloembeckenzellen und Siebröhren der Dioscoreaceae und ihre mögliche Bedeutung für den aktiven Assimilattransport. Planta 60: 543-557

Brouwer R (1953) The arrangement of the vascular bundles in the nodes of the Dioscoreaceae. Acta Bot Neerl 2: 66-73

French JC, Tomlinson PB (1986) Compound vascular bundles in monocotyledonous stems: construction and significance. Kew Bull 41: 561-574

Lawton JRS (1967) Translocation in the phloem of *Dioscorea* spp. II. Distribution of translocates in the stem. Z Pflanzenphysiol 58: 8-16

Mason TG (1926) Preliminary note on the physiological aspects of certain undescribed structures in the phloem of the greater yam, *Dioscorea alata* Linn. Sci Proc R Dublin Soc 18: 195-198

Mason TG, Lewin CJ (1926) On the rate of carbohydrate transport in the greater yam, *Dioscorea alata* Linn. Sci Proc R Dublin Soc 18: 203-205

Walther A (1986) Struktur und Entwicklung der Phloembecken bei *Dioscorea reticulata* Gay und *Dioscorea ethiopica.* Diplomarbeit, Fak Biol, Univ Heidelberg

9 Sieve Elements in Plant Tissue Cultures: Development, Freeze-Fracture, and Isolation

RICHARD D. SJOLUND

9.1 Introduction

Plant tissue culture techniques are widely used to grow populations of higher plant cells in vitro. The cultures may be grown as a callus tissue on media hardened with agar or as suspension cultures in liquid media agitated on a shaker platform. Tissue culture techniques have become established as a means for rapidly propagating plants on a commercial scale, as a source of secondary plant products, and as an important link between the incorporation of new genetic information into plant cells and the regeneration of plants with altered characteristics. A major goal of plant tissue culture research is the regulation of development and differentiation in vitro. The incorporation of specific growth factors in the culture medium can be used to regulate the production of roots and shoots from cultures of many plant cells. Progress in this area is responsible for the commercial success of plant propagation through the use of tissue cultures.

In addition to the regulation of organogenesis, however, it has long been recognized that differentiation at the cellular level also occurs in populations of cells grown as tissue cultures. Gautheret (1959) observed that islands of xylem and phloem cells were frequently found in callus tissues from many plant species. Wetmore and Rier (1963) demonstrated that xylem and phloem could be induced to form in cultures of homogeneous parenchyma cells through the addition of auxin and a sugar. This pioneering work also indicated that changing the concentration of auxin and sugar given to pieces of callus tissue resulted in changes in the proportion of xylem to phloem formed from the parenchyma cells: higher sugar levels favored phloem development over that of xylem. Jeffs and Northcote (1966) also indicated that high levels of sucrose resulted in the differentiation of more phloem, but other studies have failed to confirm these results (Bornman et al. 1977; Aloni 1987). Despite these and other studies of phloem formation in tissue cultures, the specific induction of phloem differentiation from undifferentiated callus parenchyma cells in vitro is not yet possible. The roles of plant hormones, sugars and other factors in triggering the differentiation of phloem in culture remain under investigation. The development of tracheary elements in vitro, on the other hand, has become established as an important system for the study of plant cell differentiation (Kohlenbach and Schmidt 1975; Fukuda and Komamine 1980, 1985; Falconer and Seagull 1988). The successful development of experimental systems for the study of xylem differentiation has resulted, in part, from the ease with which even early stages of xylogenesis can be recognized in the light microscope, including the observation of living cells undergoing xylogenesis. Phloem cells, on the other hand, are more difficult to see in the light microscope, especially during early stages of development and in the living state.

9.2 Phloem Function in Vitro

Compared to the vast literature on phloem in intact plants, relatively few studies have been done on phloem development or structure in plant tissue cultures. One apparent reason for the lack of interest in phloem formed in vitro is that callus sieve tubes are generally regarded to be nonfunctional. Callus sieve elements are discontinuous; they form relatively short sieve tubes, and they are scattered throughout the callus, usually as islands of vascular tissue surrounded by callus parenchyma cells. The nature of this callus phloem organization precludes the possibility of long distance transport of assimilates in vitro. Since the role of phloem sieve tubes in vivo is the translocation of photoassimilates over long distances, the demonstration (Hanson and Edelman 1970) that callus phloem of carrot did not conduct ^{14}C-labeled photoassimilates across a 2-cm piece of callus tissue has been cited (Spanner 1978) as evidence that callus sieve tubes do not function in vitro and are, therefore, relatively unsuitable for the study of phloem structure. Even Wetmore and Rier (1963) concluded that "The induction and differentiation in callus of xylem and phloem tissues has no relation to conduction."

It is important to note, however, that although long distance transport by callus phloem is precluded by the lack of sieve-tube continuity, it is entirely possible that sieve elements formed in vitro may function normally with respect to phloem loading (Sjolund and Cho 1986) and the development of high hydrostatic pressures (Sjolund et al. 1983). From a functional standpoint, the short callus sieve tubes, bathed in nutrients from the medium, are in a source region but are not connected to a sink area. Since phloem loading is a membrane transport event, and since plant cell cultures have been widely used for membrane transport studies (Komor et al. 1981), the study of phloem loading in vitro (Stanzel et al. 1988) and its relationship to sieve element membranes (Sjolund and Shih 1983 b) is justified despite the lack of translocation over long distances by callus phloem. Studies of membrane function, phloem-specific protein synthesis and the molecular biology of phloem differentiation using in vitro systems have been suggested as a means of avoiding some of the complexities associated with whole-plant organization (Daie 1989).

9.3 Phloem Development in Callus Tissues

Callus tissues from various plants differ greatly in morphology. Some callus grows as a compact, hard mass while others are loose, plate-like, or dissociated (Butenko 1964). My laboratory has used callus cultures of the California shield leaf, *Streptanthus tortuosus* (Brassicaceae) to study phloem development and structure because this callus tissue grows as a loose aggregation of small, spherical nodules (Fig. 9.1). Within the small nodules, scattered islands of phloem (Fig. 9.2) are formed, surrounded by callus parenchyma cells. Small clusters of phloem cells usually develop first and are later surrounded by xylem elements.

The development of sieve elements in callus tissue resembles closely that described for the plants from which the cultures are derived. The structure of the phloem cells which develop in vitro, except for cell shape and the length of the sieve tubes formed, is remarkably similar to that of comparable cells formed in the intact plant (Wooding 1968). The development of sieve elements in callus cultures of *S. tortuosus* is similar to that seen in pith cultures of tobacco (Cronshaw and Anderson 1971) and it follows the pattern established for phloem development in the intact plants of many dicot species (Esau and Gill 1972; Esau 1978). During development the nucleus, dictyosomes and vacuole disappear (Fig. 9.3) and the cell walls of the sieve elements are greatly increased in thickness compared to those of the surrounding callus parenchyma cells (arrows, Fig. 9.3). Companion cells are also formed in vitro and are connected to the sieve elements by plasmodesmata. The companion cell walls are not as thick as those of the sieve elements (Fig. 9.3), a feature also noted in the minor veins of several intact plants (Lucas and Franceschi 1982).

9.4 P-Protein, Callus Phloem and Wounding

One area of controversy concerning the structure and function of sieve elements in which callus phloem has been used to advantage concerns the distribution of P-protein filaments and their possible role in the movement of substances through sieve elements. Many early electron microscopy studies of sieve-element structure revealed that sieve-plate pores were blocked by P-protein filaments. Since the processing of intact plants or their organs for electron microscopy usually requires that the plant material be cut into small pieces for fixation, many investigators (see Chap. 13 for discussion) view the formation of plugs of P-protein in the sieve-plate pores as being the result of pressure-surge release due to injury at the time of cutting. According to this view, the P-protein filaments are dispersed into a random network in situ, but are broken loose and swept into the pores, plugging them when a rapid release of the hydrostatic pressure in the sieve element occurs at the time of cutting. Spanner (1978), on the other hand, interpreted the presence of occluding P-protein filaments in sieve-plate pores as being the normal condition in functioning sieve tubes. Since

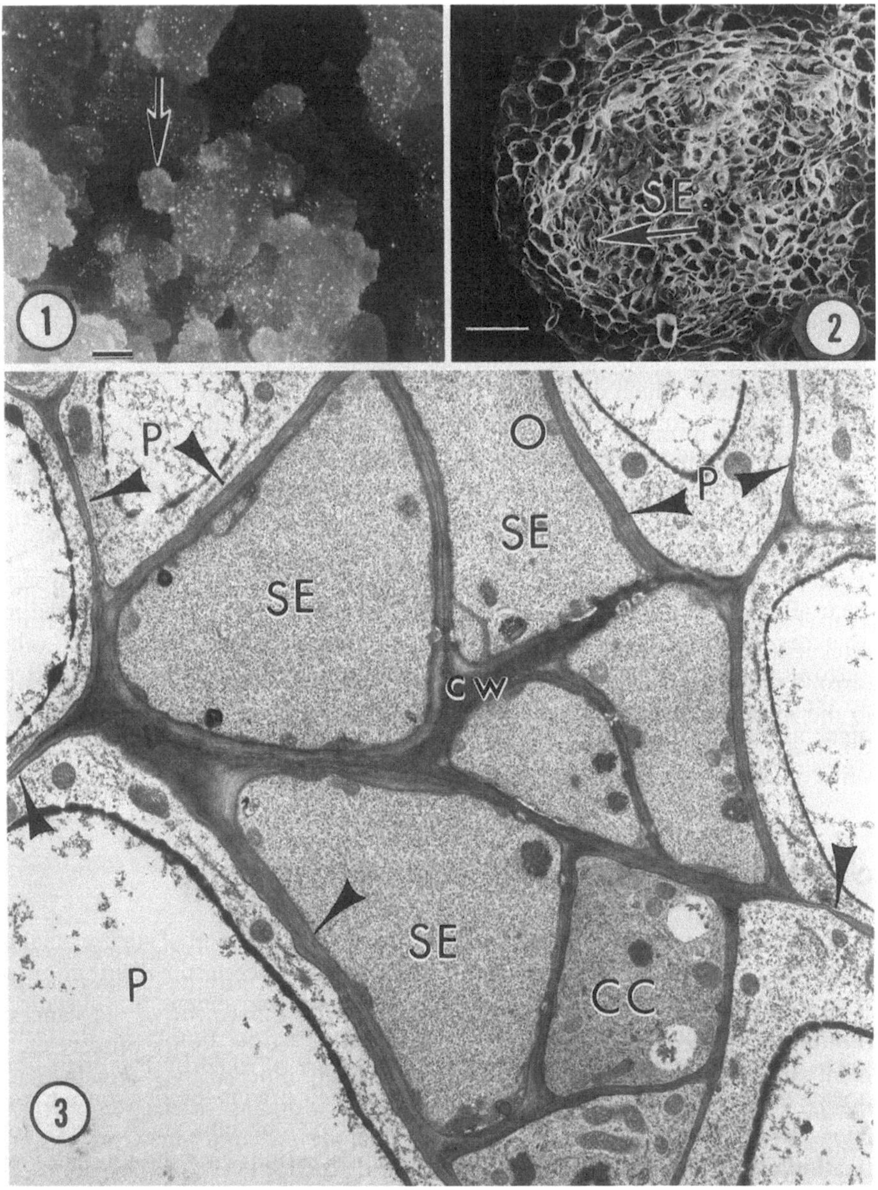

Fig. 9.1. Callus tissue of *Streptanthus tortuosus* grown on an agar medium. The *arrow* indicates a small nodule of callus tissue; *Bar line*=1.0 mm (Sjolund et al. 1983). **Fig. 9.2.** A scanning electron microscope view of a section through a small nodule of callus tissue similar to that shown in Fig. 9.1; *Bar line*=0.1 mm (Sjolund et al. 1983). **Fig. 9.3.** An island of phloem sieve elements *(SE)* and companion cells *(CC)* in a nodule of callus tissue. The cell walls *(cw)* of the sieve elements are thicker than are those of the surrounding callus parenchyma cells *(arrows)*. × 7900

the sieve tubes formed in vitro are much shorter than those found in intact plants, several investigators (Wooding 1968; Barnett 1978) have attempted to prepare callus phloem for examination in the electron microscope without causing pressure-surge disruptions of the P-protein. Although the sieve-plate pores of sieve elements formed in primary cultures of tobacco pith (Anderson and Cronshaw 1970; Cronshaw and Anderson 1971) were often plugged by P-protein filaments, many other sieve-plate pores of these phloem cells formed in vitro were seen to be open. In these cells, the P-protein was evenly distributed throughout the lumen of the sieve-tube member, suggesting that, in these short sieve tubes, no pressure-surge release had occurred to sweep the filaments into the pores to form a plug.

These studies were extended (Sjolund et al. 1983) to include the experimental formation of P-protein plugs in callus phloem of *S. tortuosus* by either fixing the small nodules of callus intact, or intentionally slicing the small pieces of callus (Fig. 9.2) before fixation, much as the organs of intact plants are cut before processing for electron microscopy. When the small nodules (Fig. 9.1) are processed intact, P-protein filaments are evenly distributed throughout the lumen of the callus sieve elements (Fig. 9.4) and the sieve-area pores are open (Figs. 9.4, 9.6). When the small nodules of callus were sliced into small pieces (Fig. 9.2) before fixation, however, dense plugs of P-protein were observed in the sieve-area pores (Sjolund et al. 1983). Thus, using these extremely small islands of phloem formed in vitro, images of either open or occluded pores could be obtained through the experimental manipulation of the fixation procedure. The observation that P-protein plugs could be induced by cutting prior to fixation also suggested that these callus sieve elements had developed high hydrostatic pressures, and that they, therefore, were active in phloem loading. Parenchyma cells adjacent to the cut sieve tubes did not show a pronounced disruption of their cellular contents when they were cut before fixation (Sjolund et al. 1983). The pressure-surge release was only noted in the sieve-tube members.

Since P-protein filaments in the callus sieve-tube members are evenly distributed throughout the cell lumen, and since this is interpreted to be the in-situ condition for sieve elements formed in vitro and in vivo, it is likely that the filaments are arranged in a three-dimensional network (Fig. 9.4) and that they are attached to the periphery of the sieve-tube member (Figs. 9.6, 9.12) so that they can be held in position during the normal flow of assimilates through the lumen of the cell. The P-protein network of fine filaments would not impede normal assimilate movement through the lumen of the cell, but it could be "torn loose" when a rapid flow results from the sudden release of pressure due to injury. This wound-sealing property is of obvious benefit to the plant.

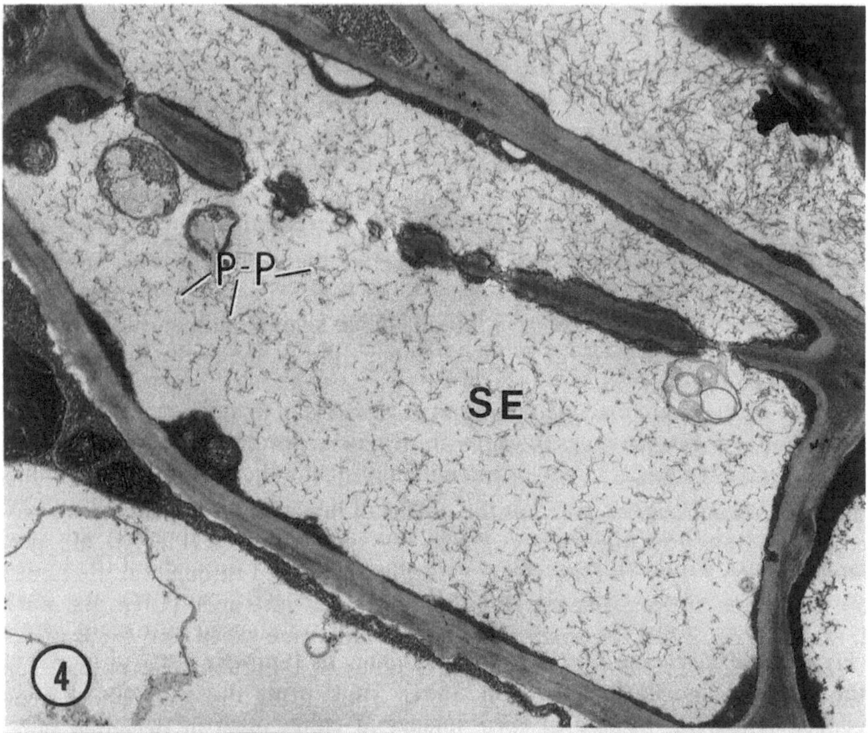
P-P
SE
4

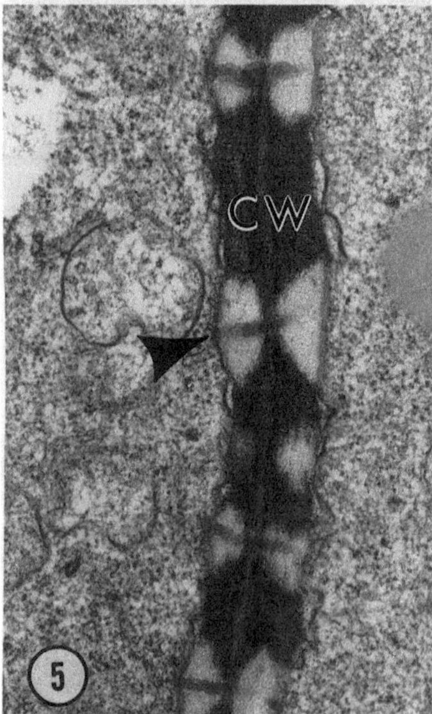
CW
5

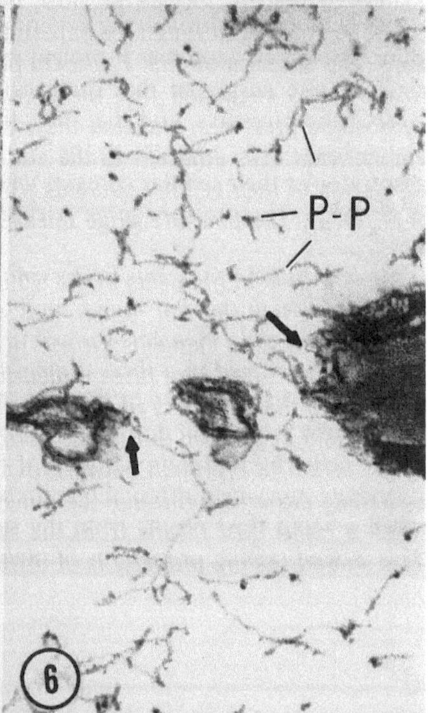
P-P
6

9.5 Freeze-Fracture Studies Using Callus Sieve Elements

The small size of the nodules of *S. tortuosus* callus has also facilitated the study of phloem through the use of freeze-fracture techniques (Sjolund and Shih 1980, 1983 a, b; Sjolund et al. 1983). The preservation of fine structure detail by freezing is dependent on the size of the tissue samples employed. Previous applications of freeze-fracture techniques to the study of phloem utilized petioles of *Nymphoides peltata* (Johnson 1973) and *Vinca rosea* (Braun 1977). Since the islands of callus phloem can be prepared for freeze-fracture without cutting, the distribution of P-protein filaments seen in sectioned material (Fig. 9.4) is also observed in callus sieve-tube members examined in freeze-fracture replicas (Fig. 9.7). The filaments of P-protein are observed to be evenly distributed through the lumen of the sieve element, and no plugs of P-protein are observed in the pores of sieve elements frozen without slicing (Fig. 9.9). When, however, the small nodules of tissue are sliced prior to freeze-fracture preparation, the P-protein filaments are observed to organize into large bundles of filaments; the even distribution normally seen in vitro is not found in freeze-fracture replicas following intentional damage and pressure-surge release (Sjolund et al. 1983).

9.6 Sieve-Area Pores

The formation of sieve-area pores in callus phloem (Hanson and Edelman 1970; Anderson and Cronshaw 1970; Sjolund and Shih 1983 b) duplicates the well-documented developmental process studied in intact plants (see Chap. 6). The sieve-area pores develop from plasmodesmata which are enlarged as the surrounding wall material is removed and replaced by callose (β 1,3 glucan). Callose deposition at the pore site occurs as a pair of platelets which, in the case of *S. tortuosus* callus phloem, are cone-like, forming an hourglass configuration (Fig. 9.5) around the plasmodesma. The deposition of callose raises the surface of the cell wall (arrow, Fig. 9.5). Freeze-fracture techniques have also been useful in the study of pore formation. When the inner half of the plasma mem-

◁

Fig. 9.4. Sieve elements *(SE)* from a small nodule of callus tissue fixed for electron microscopy without prior cutting or injury. The P-protein *(P-P)* is evenly dispersed throughout the lumen of the cells and the pores are open. × 12300 (Sjolund et al. 1983). **Fig. 9.5.** A developing sieve plate in callus phloem of *S. tortuosus*. The *arrow* points to a plasmodesma which is surrounded by callose platelets. Note that the cell wall is raised at the site of callose deposition. × 37200. **Fig. 9.6.** An open pore in a sieve element (Fig. 9.4) fixed without injury and seen at higher magnification. The P-protein *(P-P)* filaments are attached *(arrows)* to membranes at the periphery of the sieve element. × 77700

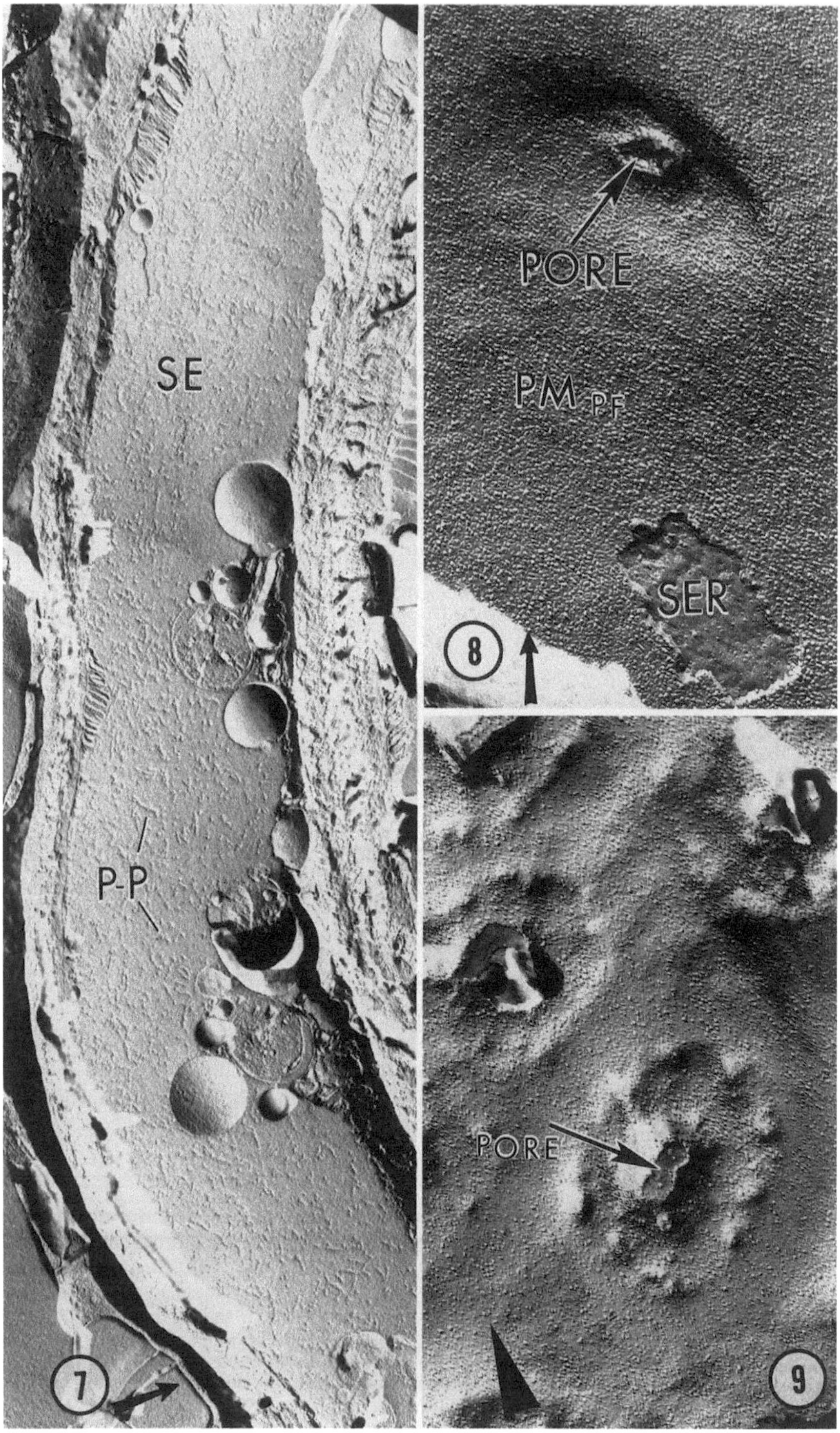
SE
P-P
7
PORE
PM PF
SER
8
PORE
9

brane (the protoplasmic fracture face, PF) at a developing pore site is viewed (Sjolund and Shih 1983 b) in a freeze-fracture replica, the plasma membrane (PF, Fig. 9.8) is seen to have a depression where it is pushed down by the callose platelet. The enlarging plasmodesma, which now represents a collar of plasma membrane extending upward from the P fracture face, is sheared off during fracture and can be seen as a torn edge (Fig. 9.8). When the complementary, exterior half (the E fracture face, EF) of the plasma membrane is viewed at a developing pore site, the plasma membrane of the plasmodesma is seen as a descending sheath which extends downward into the developing pore (Fig. 9.15). As the wall material is replaced by callose, the diameter of the plasma membrane-lined pore is increased, forming the sieve-area pores of the mature sieve-tube member (Fig. 9.9). Thus, during development, the entire sieve-tube is maintained as a single, membrane-bound compartment, while the individual tube members become connected by openings large enough to facilitate unimpeded assimilate flow. Since the translocation process is essentially based on osmotic pressure, the maintenance of plasma membrane integrity during development and at maturity is vital to the functioning of the phloem conduit. This membrane continuity is apparent in the freeze-fracture replicas which allow large areas of membranes to be visualized.

9.7 The Sieve-Element Reticulum (SER)

During the development of sieve-tube members a membrane system is formed that is unique to the sieve tube. Although the membranes resemble the endoplasmic reticulum (ER) membrane system in some ways, the origin of the SER membranes, their lack of ribosomes, and the appearance of the membranes in freeze-fracture replicas (Sjolund and Shih 1983 a) suggest that this phloem-specific membrane system deserves a special designation. Srivastava and O'Brien

---

Fig. 9.7. A freeze-fracture view of a callus sieve element *(SE)* processed without cutting or injury. The P-protein *(P-P)* filaments form a finely dispersed network in the lumen of the sieve element. The center of the cell is devoid of organelles; the mitochondria, plastids and SER membranes line the periphery of the cell. The *unlabeled arrow at the bottom* of all freeze-fracture images indicates the direction of shadowing. × 14700 (Sjolund et al. 1983). **Fig. 9.8.** A freeze-fracture view of a sieve-element pore. The plasma membrane *(PM)* of the sieve element is seen as a P fracture face *(PF)*, the inner half of the plasma membrane. A fracture in the plasma membrane reveals an *SER* membrane below the plasma membrane. The callose platelet at the site of the sieve area pore has made an impression in the plasma membrane, and the collar of membrane that extended into the pore is seen to be sheared off. This freeze-fracture view should be compared to the thin-sectioned view of a pore as seen at the *arrow* in Fig. 9.5. × 56000 (Sjolund and Shih 1983 b). **Fig. 9.9.** A portion of a callus sieve plate showing an open sieve-plate pore. Note that the pore is open and that only a few, evenly spaced filaments of P-protein are seen in the pore. × 80000 (Sjolund et al. 1983)

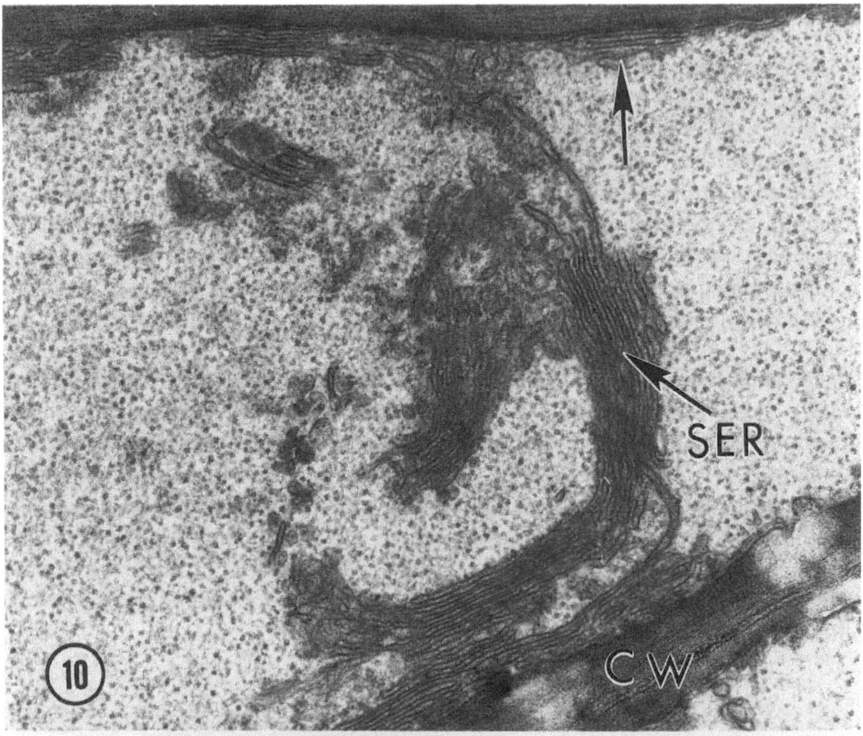
SER
CW
10

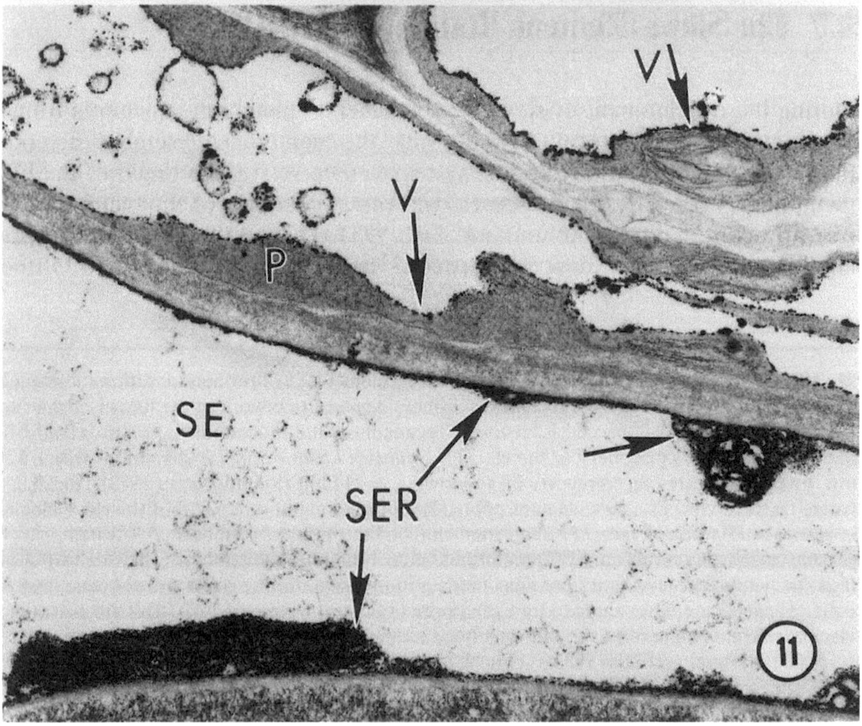
V
V
P
SE
SER
11

(1966) termed this membrane system the sieve element reticulum (SER), a term adopted by some authors (Sjolund and Shih 1983 a; Arsanto 1986) but rejected by others (Esau 1972). In the original designation of the SER, Srivastava and O'Brien (1966) described the membrane system as being distinct from that of the ER present at the beginning of phloem development. They suggested that the pre-existing ER of the cell was degraded and that the SER had a separate origin. Other authors (Thorsch and Esau 1981; Eleftheriou 1986) have interpreted this membrane system as being formed from the pre-existing endoplasmic reticulum of the developing sieve elements (ER) through the formation of membrane stacks, some of which become attached to the nuclear envelope when the nucleus is degenerating.

In the case of *S. tortuosus* callus phloem, the SER is also associated with the nuclear envelope (Fig. 9.10) during sieve element differentiation, but few stages have been found to support the interpretation that the SER is derived from the ER or ribosome-associated rER and then applied to the nuclear envelope. The data from the study of these callus sieve elements suggest, instead, that the SER is derived from the nuclear envelope membranes as the nucleus is broken down, and that it has, therefore, a separate origin from the pre-existing ER system. Once formed from the nuclear envelope, the SER migrates to the periphery of the cell where it becomes associated with the plasma membrane (Fig. 9.10). This interpretation, in agreement with that of Srivastava and O'Brien (1966), suggests that the original ER of the cell, mostly in the form of rER with ribosomes attached, degenerates during sieve-element differentiation, marking the end of normal protein synthesis in the cell. The SER can be viewed, therefore, not as a replacement for the rER, but instead, as a separate membrane system with a function and morphology unique to the sieve element.

Muscle cells of animals also have a distinct membrane system, the sarcoplasmic reticulum (SR). This membrane system, like the SER, is unique to a differentiated cell type and lacks ribosomes. The function of the SR in muscle cells is to sequester calcium ions and to regulate the contraction and relaxation of the muscle (Jørgensen and Campbell 1984). The SER of phloem cells also appears to serve as a calcium storage site. When the histochemical technique for calcium localization using pyroantimonate is applied to the *S. tortuosus* callus sieve elements (Sjolund and Jensen 1985), dense deposits are localized over the SER (Fig. 9.11). This technique also demonstrates the storage of calcium in the vacu-

Fig. 9.10. Sieve-element reticulum *(SER)* membranes developing from the nuclear envelope. The nucleus has degenerated and the cytoplasm of the sieve element is filled with free ribosomes. No endoplasmic reticulum membranes with ribosomes attached remain in the cell. The *SER* membranes are also becoming associated with the plasma membrane next to the cell wall *(CW)*. × 47500. **Fig. 9.11.** A callus sieve element *(SE)* and two surrounding callus parenchyma *(P)* cells as seen following a histochemical procedure for the localization of calcium. The dense deposits seen in the SER of the sieve element and in the vacuoles *(V)* of the parenchyma cells are precipitates of calcium and antimony. No deposists are seen in the parenchyma cytoplasm. × 14000

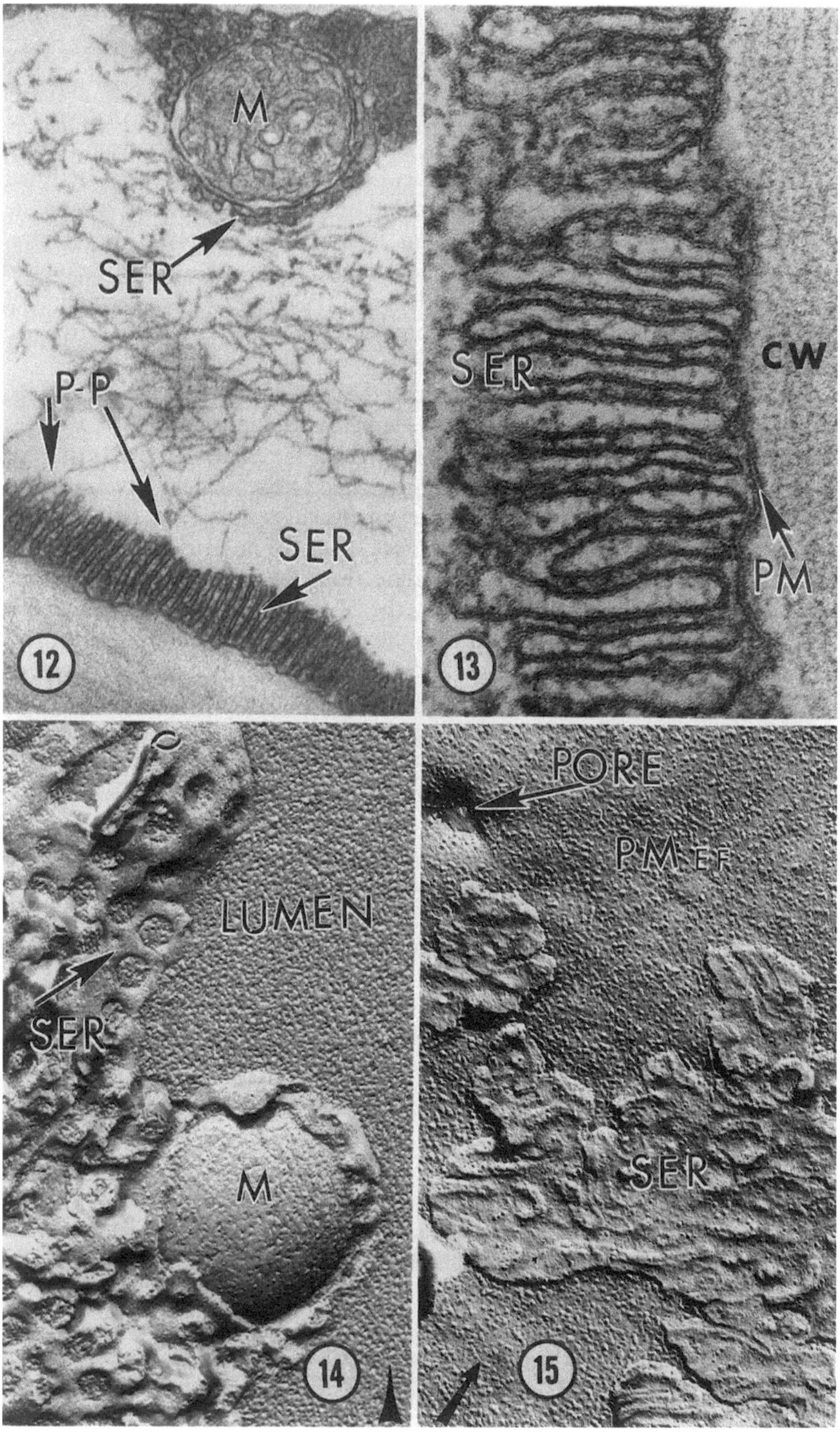
M
SER
P-P
SER
12
SER
CW
PM
13
LUMEN
SER
M
14
PORE
PM EF
SER
15

oles of the adjacent callus parenchyma cells, but little calcium is detected in the parenchyma cytoplasm. These observations were extended by Arsanto (1986) to include X-ray analysis of the SER, confirming the sequestering of calcium by these membranes. This role for the SER, distinct from the role of the rER in protein synthesis in the undifferentiated sieve elements, also suggests that the SER deserves a special designation, distinct from that of the ER. Since the vacuole is lost during sieve-element differentiation, the SER formation may be required to provide a calcium sequestering membrane system in the sieve elements. Surprisingly, the calcium sequestering protein, calsequestrin, normally associated with animal muscle cell SR membranes (Campbell et al. 1983) has recently been isolated from the *S. tortuosus* callus cells (Krause et al. 1989). It is important to determine if the calsequestrin protein is associated with the SER membrane system; these experiments are in progress (Chou and Sjolund, unpublished results).

The morphology of the SER membranes is well documented for the *S. tortuosus* callus phloem (Sjolund and Shih 1983 a). The SER forms stacks of membranes which may be parallel to (Fig. 9.11) or at right angles to (Figs. 9.12, 9.13, 9.17) the sieve-element plasma membrane. Other areas of the sieve element may be enclosed in single membranes of the SER system, closely appressed to the plasma membrane (Figs. 9.8, 9.15). The SER membranes also enclose the mitochondria which persist in the mature sieve elements (Figs. 9.12, 9.14). The unstacked, single cisterna regions of the SER associated with the plasma membrane are fenestrated (Fig. 9.14), providing points of continuity between the plasma membrane of the sieve element and the lumen of the cell (Sjolund and Shih 1983 a). The SER may function, therefore, as an anchoring system, keeping the mitochondria and the ATP produced by them from being swept along in the

Fig. 9.12. Thin-sectioned view of the *SER* membranes in a callus sieve element of *S. tortuosus*. The *SER* surrounds the mitochondrion *(M)* on one side of the cell and is seen at right angles to the plasma membrane on the other side. The P-protein *(P-P)* is anchored to the *SER* membranes at several sites. This cell was processed without injury or cutting of the callus nodule. × 72000. **Fig. 9.13.** A higher magnification of a region of the SER-plasma membrane *(PM)* interface next to the cell wall *(CW)* of a callus sieve element. The two membranes are closely associated, but remain as distinct, separate membranes. × 165000 (Sjolund and Shih 1983 a). **Fig. 9.14.** A region of the *SER* membrane system from an unstacked area surrounding a mitochondrion. Compare this freeze-fracture image to the area labeled *SER* which surrounds the mitochondrion in Fig. 9.12. The *SER* membrane forms a continuous sheet that is fenestrated *(arrow)*. The plasma membrane has been removed during fracture and the lumen of the sieve element is seen through the fenestrations of the *SER*. Similar fenestrations are visible in the *SER* region around the mitochondrion in Fig. 9.12. A mitochondrion *(M)* is seen in this freeze-fracture view. × 62000 (Sjolund and Shih 1983 a). **Fig. 9.15.** A region of the sieve element plasma membrane *(PM)* seen in a freeze-fracture replica as the exterior half of the membrane *(EF)*. The *SER* membranes are in close contact with the plasma membrane and torn edges are visible where the *SER* membranes that projected at right angles into the cell have been sheared off. A developing pore is seen in this view as a sheath of plasma membrane that extends out from the cell, into the pore. This pore is less well developed than that seen in Fig. 9.8. × 100000 (Sjolund and Shih 1983 b)

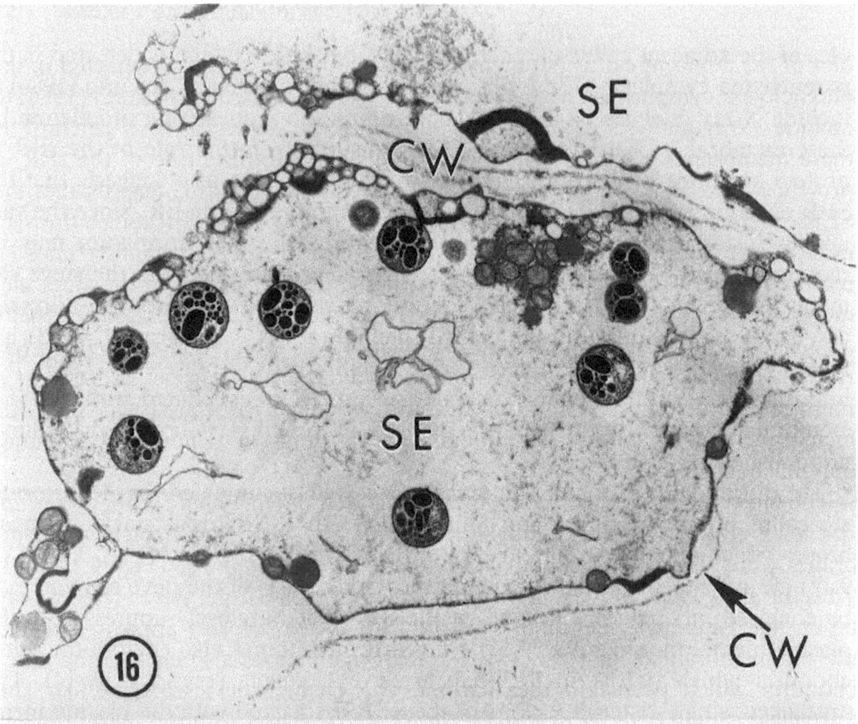
SE
CW
SE
CW
16

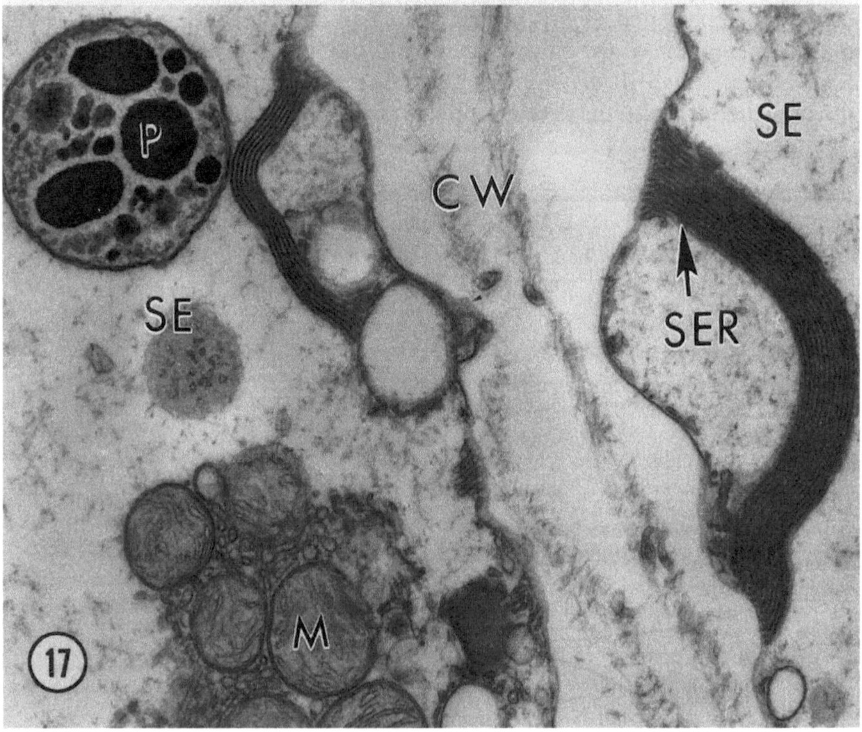
P
SE
CW
SE
SER
M
17

assimilate translocation stream. This anchoring function is also seen in the attachment sites for the P-protein filaments (Figs. 9.6, 9.12) which may keep the P-protein network from being swept along in the assimilate stream.

The association between the SER and the sieve element plasma membrane is also of interest in light of the calcium storage role played by these membranes. The regulation of callose synthesis by enzymes located in the plasma membrane has been shown to be activated by increased cytoplasmic calcium levels (Fink et al. 1987), suggesting that the SER may play an important role in the regulation of callose synthesis in the sieve-element cell wall, especially the rapid deposition of callose which follows injury (Sjolund et al. 1983). The potential role of the calsequestrin protein in this process needs to be examined.

9.8 Isolation and Partial Purification of Callus Sieve Elements

Since callus phloem exists as separate, isolated islands surrounded by parenchyma cells, we have explored the separation of the phloem cells from these cultures. The islands of callus phloem are less complex than the veins of leaves isolated by van Bel and Koops (1985). The cell walls of callus sieve elements from *S. tortuosus* cultures are greatly increased in thickness compared to those of the surrounding callus parenchyma cells (Fig. 9.3). Histochemical evidence suggests that the sieve-element wall has a modified chemical composition (Lucas and Franceschi 1982). The differences between the sieve-element walls and the walls of the parenchyma cells have been exploited through the use of cell wall-digesting enzymes. These experiments, initially begun to allow the rapid freezing of isolated sieve elements for freeze-fracture analysis (Sjolund and Shih 1983 b), have resulted in the separation of sieve elements from callus parenchyma cells.

In *S. tortuosus* callus tissue, the sieve-element wall is more resistant to digestion than is that of the parenchyma cells, especially the walls which contain sieve pores and link sieve elements together. Following a 20-min digestion with pectinase and cellulase enzymes, the sieve element complexes (sieve tubes) remain joined together, retain wall material, and are ovoid to elongate in shape. The *S. tortuosus* parenchyma cells, on the other hand, form true, single proto-

Fig. 9.16. Callus sieve elements *(SE)* separated from the callus tissue following enzymatic digestion of the cell walls. The sieve elements retain their mitochondria, plastids, and *SER* membranes. Some of the cell wall material *(CW)* persists around the sieve elements. × 8500.
Fig. 9.17. A higher magnification view of the *S. tortuosus* callus sieve elements seen in Fig. 9.16. The association between the plasma membrane and the *SER* membranes has persisted during the digestion and separation process. Fibrils of cell wall *(CW)* material are visible in this sieve plate area, and membrane-lined channels are seen in the wall. The sieve-element mitochondria *(M)* and plastids *(P)* appear normal. × 28000

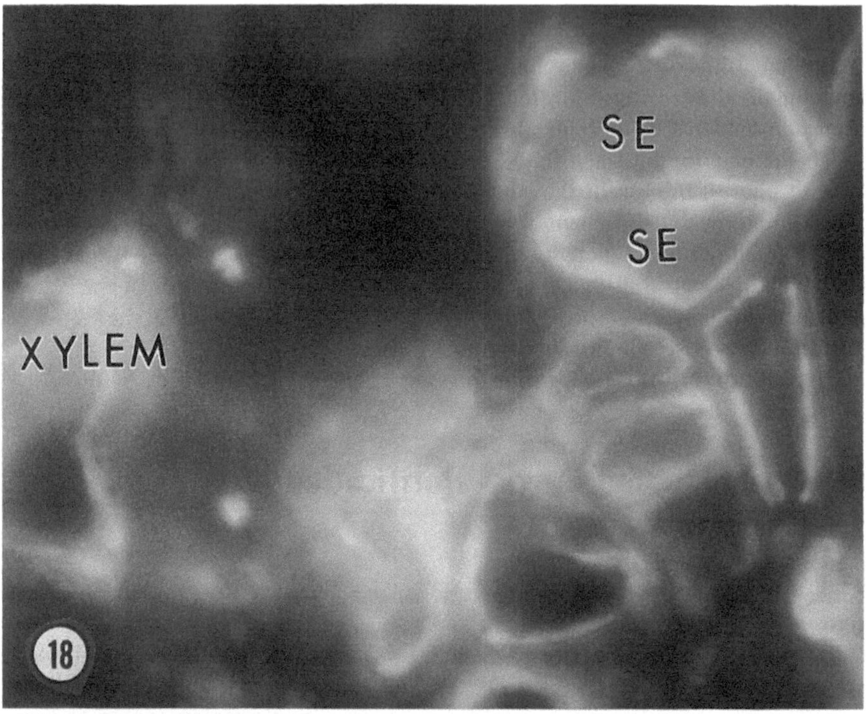
SE
SE
XYLEM
18

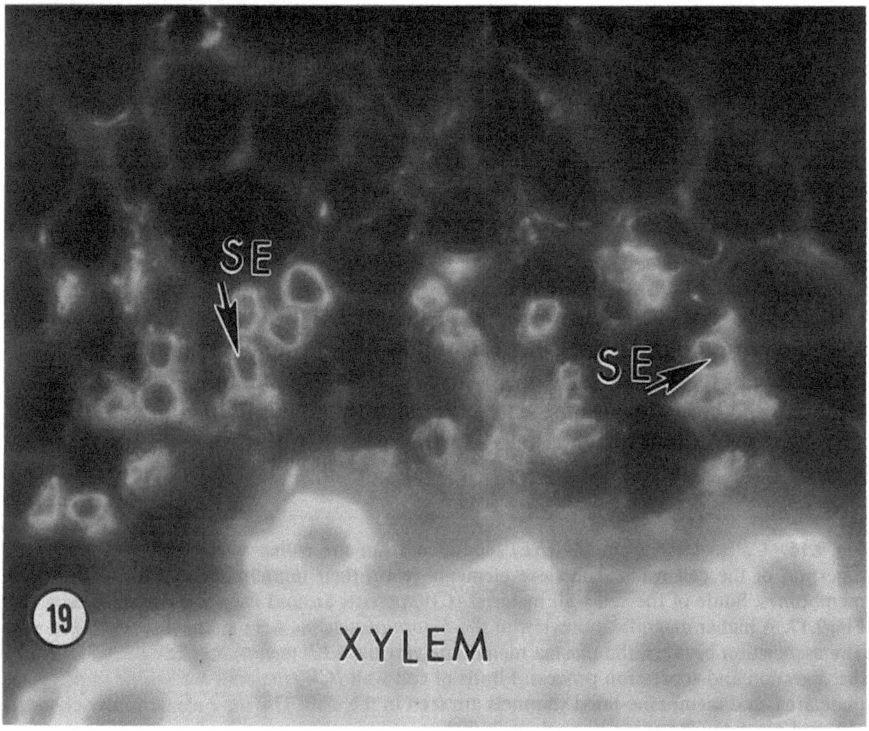
SE
SE
XYLEM
19

plasts, round and cell wall-free. These two cell classes can be separated on the basis of their size differences through the use of nylon membranes. The small parenchyma protoplasts pass through a 70-μm filter, while the linked sieve elements are retained on the filter. The sieve-element plasma membrane, the SER, mitochondria and the persistent sieve-element plastids are retained in the isolated sieve elements (Figs. 9.16, 9.17). The close association between the sieve-element plasma membrane and the SER membrane system also persists (Fig. 9.17) during the isolation procedure. The fibrils of wall material which surround the isolated sieve elements are especially dense in the walls that contain sieve-area pores (Figs. 9.16, 9.17). Preliminary evidence, based on differential enzymatic digestion (Weber and Sjolund, unpublished results), suggests that the fibrils are cellulose. Since the sieve element wall is subjected to high levels of hydrostatic pressure following phloem loading, the thickened layers of wall material may provide resistance to expansion. It is of interest to note that few companion cells are seen associated with the isolated sieve elements (Fig. 9.16) although they are present as part of the callus phloem. Since the thickness of the cell wall of the companion cell is more similar to that of the parenchyma cells (Fig. 9.3) and is chemically different from that of the sieve elements (Lucas and Franceschi 1982), the enzymatic digestion procedure appears to remove the companion cells from the sieve elements.

9.9 Antibody Formation Against Callus Sieve Elements

Since it is possible to prepare fractions enriched in *S. tortuosus* callus phloem cells (Figs. 9.16, 9.17), these phloem-rich fractions have been injected into mice in order to raise anti-phloem antibodies. Sera containing polyclonal antibodies formed against the injected cells were obtained. The goal of this study is to identify phloem cells through the use of the anti-phloem antibodies and an indirect immunofluorescence technique.

◁

Fig. 9.18. Immunofluorescence following treatment of a frozen section of *S. tortuosus* callus tissue with serum from a mouse injected with callus phloem cells similar to those in Fig. 9.16. The *yellow* fluorescence is due to autofluorescence by a xylem element. The *yellow-green* fluorescence is due to the binding of a goat anti-mouse IgG antibody conjugated with fluorescein (Organon Teknika-Cappel) to the mouse primary antibodies. The antibodies are located over the callus sieve elements, primarily over the cytoplasm and not the cell walls. Surrounding callus parenchym cells show little fluorescence. From a slide photographed at × 1000. **Fig. 9.19.** A frozen section of a stem of an intact *S. tortuosus* plant treated with the antibodies made against callus sieve elements. The autofluorescence of xylem is *yellow*, and the localization of the polyclonal mouse antibodies is seen as the *yellow-green* fluorescence from the fluorescein-conjugated secondary antibodies. The *label* appears over phloem sieve elements in the stem sections, indicating a cross-reaction between the antibodies formed against callus sieve elements and sieve elements of the intact plant from which the callus was obtained. From a slide photographed at × 440

Liquid nitrogen was used to freeze *S. tortuosus* callus tissue and frozen sections of the callus were cut with a cryostat. The sections were mounted on glass slides, incubated with the diluted mouse sera (control or anti-phloem) for 45 min, washed with cold buffer, and then incubated with a goat anti-mouse IgG antibody conjugated with fluorescein (Organon Teknika-Cappel, West Chester, PA). After washing, the sections were examined in an ultraviolet, epifluorescent microscope to determine the localization of the antibodies. Xylem elements in the callus tissue are seen as bright yellow cells in these preparations as a result of their autofluorescence. The goat anti-mouse antibodies are recognized by their yellow-green fluorescence (Fig. 9.18), marking the location of the primary, mouse antibodies. The polyclonal antibodies formed in mice following the injection of isolated phloem cells are found to be predominantly localized over the callus sieve-tube members (Fig. 9.18) in the sections of callus tissue. The localization is greatest in the cytoplasm of the sieve elements and little label is seen in the phloem cell walls. Sections incubated with control sera lack the phloem localization.

When sections of the intact plant *S. tortuosus* stems are cut and treated with the antibodies raised against the callus sieve elements, as described above, the mouse polyclonal antibodies are also seen to be localized over the phloem sieve-tube members (Fig. 9.19). In this case the antibodies raised against the cells formed in vitro can be seen to cross-react with the phloem cells formed in vivo. Since these polyclonal antibodies are raised against whole phloem cells, the mouse sera contain many different antibodies. The enrichment of the sieve elements used for injection appears to produce, however, a corresponding enrichment in the level of phloem-specific antibodies formed by the mice. This procedure is also applicable to the formation and identification of phloem-specific monoclonal antibodies (King and Morrow 1988). Monoclonal antibodies that can recognize sieve-element plasma membrane markers could be used to scale up the isolation of sieve elements using fluorescence-activated cell sorting techniques (Alexander et al. 1985). Such antibodies would also be useful in the study of phloem development in cell cultures and in vivo.

Acknowledgments. The author wishes to thank K. Jensen, M. Chou, B. Cho, and E. Komor for their assistance and valuable suggestions and Academic Press for permission to reprint figures from the *Journal of Ultrastructure Research*. Portions of this work were supported by NSF grant PCM 83-13507. Financial assistance from the Organon Teknika Corp.-Cappel (West Chester, PA 19380) supported the cost of printing the color plate of the antibody localization.

References

Alexander R, Cocking E, Jackson P, Jeff J (1985) The characterization and isolation of plant heterokaryons by flow cytometry. Protoplasma 128: 52-58

Aloni R (1987) Differentiation of vascular tissues. Anu Rev Plant Physiol 38: 179-204

Anderson R, Cronshaw J (1970) Sieve element pores in *Nicotiana* pith culture. J Ultrastruct Res 32: 458-471

Arsanto J (1986) Ca^{2+}-binding sites and phosphatase activities in sieve element reticulum and P-protein of chick-pea phloem. A cytochemical and X-ray microanalysis survey. Protoplasma 132: 160-171

Barnett J (1978) Fine structure of parenchymatous and differentiated *Pinus radiata* callus. Ann Bot (London) 42: 367-373

Bornman C, Abbott J, Noel A (1977) *Nicotiana tabacum* callus studies VII. Quantitative hormonal and sucrose effects on sieve element dimension. Z Pflanzenphysiol 83: 207-211

Braun E (1977) A freeze-etch and thin-section study of mycoplasmas in *Vinca rosea* phloem. J Ultrastruct Res 60: 44-51

Butenko R (1964) Plant tissue culture and plant morphogenesis. Acad Sci USSR, Moscow. (Translated from Russian (1968) by Artmann M, Isr Program Sci Transl Jerusalem)

Campbell K, MacLennan D, Jørgensen A, Mintzer M (1983) Purification and characterization of calsequestrin from canine cardiac sarcoplasmic reticulum and identification of the 53,000 dalton glycoprotein. J Biol Chem 258: 1197-1204

Cronshaw J, Anderson R (1971) Phloem differentiation in tobacco pith culture. J Ultrastruct Res 34: 244-259

Daie J (1989) Phloem loading of sucrose: update and opportunities in molecular biology. Plant Mol Biol Rep 7: 106-115

Eleftheriou E (1986) Ultrastructural studies on protophloem sieve elements in *Triticum aestivum L.* Nuclear degeneration. J Ultrastruct Mol Struct Res 95: 47-60

Esau K (1972) Changes in the nucleus and the endoplasmic reticulum during differentiation of a sieve element in *Mimosa pudica L.* Ann Bot (London) 36: 703-710

Esau K (1978) Developmental features of the primary phloem in *Phaseolus vulgaris L.* Ann Bot (London) 42: 1-13

Esau K, Gill R (1972) Nucleus and endoplasmic reticulum in differentiating root protophloem of *Nicotiana tabacum.* J Ultrastruct Res 41: 160-175

Falconer M, Seagull R (1988) Xylogenesis in tissue culture III: Continuing wall desposition during tracheary element differentiation. Protoplasma 144: 10-16

Fink J, Jeblick W, Blaschek W, Kauss H (1987) Calcium ions and polyamines activate the plasma membrane-located 1,3-β-glucan synthase. Planta 171: 130-135

Fukuda H, Komamine A (1980) Establishment of an experimental system for the study of tracheary element differentiation from single cells isolated from the mesophyll of *Zinnia elegans.* Plant Physiol 65: 57-60

Fukuda H, Komamine A (1985) Cytodifferentiation. In: Vasil IK (ed) Cell culture and somatic cell genetics of plants, vol 2. Cell growth, nutrition, cytodifferentiation, and cryopreservation. Academic Press, New York London Tokyo, pp 149-212

Gautheret R (1959) La culture des tissus végétaux. Techniques et réalisations. Masson, Paris

Hanson A, Edelman J (1970) Phloem in carrot calluses. Planta 93: 171-174

Jeffs R, Northcote D (1966) Experimental induction of vascular tissue in an undifferentiated plant callus. Biochem J 101: 146-152

Johnson RPC (1973) Filaments but no membranous transcellular strands in sieve pores in freeze-etched, translocating phloem. Nature (London) 244: 464-466

Jørgensen A, Campbell K (1984) Evidence for the presence of calsequestrin in two structurally different regions of myocardial sarcoplasmic reticulum. J Cell Biol 98: 1597-1602

King S, Morrow K (1988) Monoclonal antibodies produced against antigenic determinants present in complex mixtures of proteins. BioTechniques 6: 856-861

Kohlenbach H, Schmidt B (1975) Cytodifferenzierung in Form einer direkten Umwandlung isolierter Mesophyllzellen zu Tracheiden. Z Pflanzenphysiol 75: 369-374
Komor E, Thom M, Maretzki A (1981) The mechanism of sugar uptake by sugarcane suspension cells. Planta 153: 181-192
Krause K, Chou M, Thomas M, Sjolund R, Campbell K (1989) Plant cells contain calsequestrin. J Biol Chem 264: 4269-4272
Lucas W, Franceschi V (1982) Organization of the sieve-element walls of leaf minor veins. J Ultrastruct Res 81: 209-221
Sjolund R, Cho B (1986) Phloem formation in suspension cultures of *Streptanthus*. In: Cronshaw J, Lucas W, Giaquinta R (eds) Plant biology vol 1, Phloem transport. Liss, New York, pp 161-167
Sjolund R, Jensen K (1985) EGTA-soluble antimonate precipitates are formed in phloem sieve element reticulum (SER) membranes. Plant Physiol 77 Suppl: 147
Sjolund RD, Shih CY (1980) Freeze-fracture studies of membranes in developing sieve elements in a plant tissue culture. Proc Electron Microsc Soc Am 38: 636-637
Sjolund RD, Shih CY (1983 a) Freeze-fracture analysis of phloem structure in plant tissue cultures I. The sieve element reticulum. J Ultrastruct Res 82: 111-121
Sjolund RD, Shih CY (1983 b) Freeze-fracture analysis of phloem structure in plant tissue cultures II. The sieve element plasma membrane. J Ultrastruct Res 82: 189-197
Sjolund RD, Shih CY, Jensen K (1983) Freeze-fracture analysis of phloem structure in plant tissue cultures III. P-protein, sieve area pores, and wounding. J Ultrastruct Res 82: 198-211
Spanner D (1978) Sieve-plate pores, open or occluded? A critical review. Plant Cell Environ 1: 7-20
Srivastava L, O'Brien T (1966) On the ultrastructure of cambium and its vascular derivatives II. Secondary phloem of *Pinus strobus L.* Protoplasma 61: 277-293
Stanzel M, Sjolund R, Komor E (1988) Transport of glucose, fructose and sucrose by *Streptanthus tortuosus* suspension cells. I. Uptake at low sugar concentrations. Planta 174: 201-209
Thorsch J, Esau K (1981) Changes in the endoplasmic reticulum during differentiation of a sieve element in *Gossypium hirsutum*. J Ultrastruct Res 74: 183-194
van Bel A, Koops A (1985) Uptake of [^{14}C]sucrose in isolated minor-vein networks of *Commelina benghalensis L.* Planta 164: 362-369
Wetmore R, Rier J (1963) Experimental induction of vascular tissues in callus of angiosperms. Am J Bot 50: 418-430
Wooding F (1968) Fine structure of callus ploem in *Pinus pinea*. Planta 83: 99-110

10 Wound-Sieve Elements

ALEXANDER SCHULZ

10.1 Introduction

The survival of higher plants is highly dependent upon the phloem path linking assimilate sources and sinks. In an individual plant the elongation, capacity and function of this path have to be finely adjusted to internal and environmental conditions. This means, at the cellular level, that the initiation of new sieve elements, the timing of their differentiation and the start of their function are subject to the changing requirements of the surrounding tissue.

In intact plants, the causal relations of phloem development and function are difficult to investigate, since initiation and differentiation of sieve elements occur continuously and asynchronously in adjacent, (pro-) cambial cell files. Phloem research has therefore increasingly focused on conditions where new elements are experimentally induced, either in tissue cultures (see Chap. 9), or by joining stock and scion (see Chap. 11) or by severing single or all vascular bundles.

Severing the vascular system leads to the development of wound-phloem and -xylem elements which bridge the vascular discontinuities and reconnect the interrupted bundles within a few days (Figs. 10.1–10.3). Since the experimental injury determines the beginning and the conditions of regeneration, the investigation of wound phloem offers chances to follow up the absolute timing of sieve-element differentiation and to elucidate the structure and contents of cells that become capable of translocating assimilates. In contrast to sieve elements differentiated within tissue cultures, wound-sieve elements are functional (cf. Schulz 1987) and - unlike with graft phloem - incompatibility reactions do not occur.

10.2 Tissue Changes During Wound-Phloem Development

10.2.1 Tissue of Origin

The general ability to produce wound phloem seems to be restricted to plants with secondary vascular development, e. g., to dicotyledons. In some monocotyledons the high number of regular anastomoses is apparently sufficient to shunt assimilates from cut to intact bundles, thus making wound phloem unnecessary (Behnke and Sukkri 1971). In other monocotelydons, only very young regions close to the apical meristem seem able to react to bundle severance by developing wound phloem and wound xylem (Aloni and Plotkin 1985). Monocotyledons with anomalous secondary growth have not yet been tested for wound phloem.

Depending on the method of injury, different plants and tissues of the dicotyledons have been reported to be capable of wound-phloem development:

If the entire longitudinal phloem of tree stems is interrupted by the decapitation of the apex, the cambial products forming after the injury change their polarity. Transverse bridges of sieve tubes develop which have members of cambial origin, but some of them arise from ray initials also (Neeff 1914).

If single vascular bundles of primary shoots in several herbaceous plants are severed, the tissue interface between cortex and pith, the procambial cylinder (Eschrich 1953), provides a favorable region for vascular development. Accordingly, most of the wound-sieve elements interconnecting an interrupted bundle with an intact one in *Coleus* (Fig. 10.1) or *Impatiens* are the products of an early activation and specialization of this tissue, though some elements may arise also

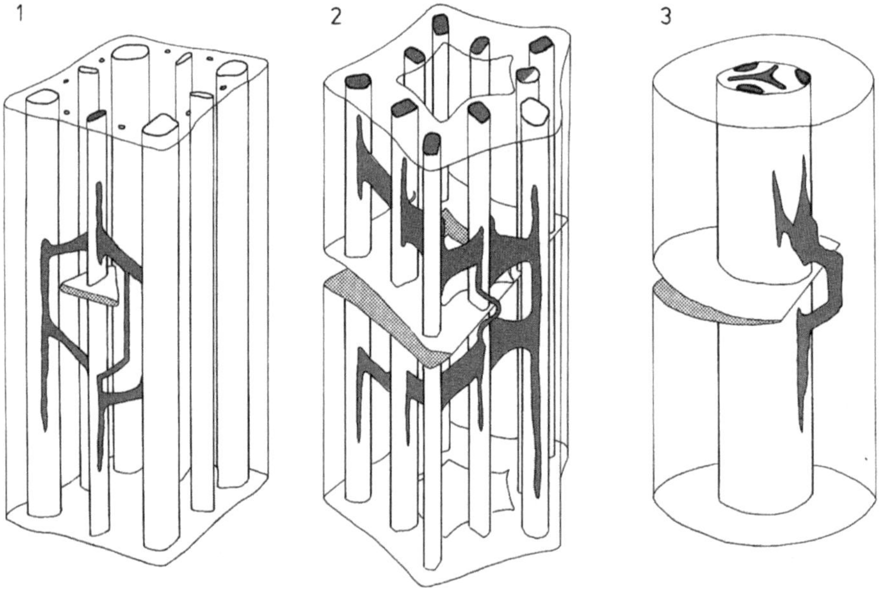

Figs. 10.1–10.3. Induction of wound-phloem in different plants and organs (the severed bundles are *dark screened* in the *upper cross-section*. **Fig. 10.1.** The interruption of a median bundle of the fifth internode of *Coleus* leads to the development of vascular bridges which, within 6 days, pass the wound or contact to an intact corner bundle. **Fig. 10.2.** The transsection of all but one or two bundles in the hypocotyl of *Cucurbita* induces transverse connections of wound phloem (and xylem) between cut and intact bundles within 3 days. **Fig. 10.3.** Following severance of the root stele of *Pisum* a vascular bridge develops in the remaining cortex strip within less than 3 days

from pith or cortex cells that become meristematic by wounding (von Kaan-Albest 1934; Eschrich 1953).

The severance of all but one or two vascular bundles in *Cucurbita* activates cell division in the interfascicular zone bridging the cut and intact bundles (Fig. 10.2) and, subsequently, the production of wound-sieve and -tracheary elements. In contrast to *Impatiens* and *Coleus*, however, the undisturbed *Cucurbita* never develops an interfascicular cambium (Esau 1965). Thus, in *Cucurbita* the role of the procambial cylinder will be taken over by the activation of a parenchymal area at the interface between the bundles.

An even more unusual place for phloem development is the cortex of roots, the stele of which was experimentally severed (Fig. 10.3: *Pisum;* cf. Robbertse and McCully 1979). The wound-sieve tubes leave the interrupted stele, cross the endodermal layer and run through cortical parenchyma in order to bypass the incision. The cortex is recognized by the presence of intercellular spaces (Fig. 10.5). Induced by the injury, the endodermis and cortex, which in intact roots are never involved in the development of vascular tissue, are now activated and become meristematic. Products of this process eventually give rise to wound-sieve elements (Schulz 1986 a).

10.2.2 Developmental Pattern of Wound Phloem

The latitudinal development in regular phloem starts with the first sieve elements in meristematic tissue marking the phloem poles - protophloem - and continues in adjacent cell files - metaphloem - in a centripetal order till eventually the cambium is activated and gives rise to secondary sieve elements (Esau 1969).

In the shoot of the herbaceous *Impatiens* and *Coleus*, the pattern of divisions preparing wound-sieve elements within the interfascicular zone resembles that of the secondary phloem (Behnke and Schulz 1980), i. e., the phloem initials divide once or several times to give rise to a phloem parenchyma cell and a sieve-element/companion-cell complex. Accordingly, one might call this process an early cambial activation of single cells at the site of future wound-sieve tubes. In comparison to the regular secondary development, only the plane of divisions is altered in order to fit the long axis of each member to the new course of the wound-sieve tube.

In the root stele of *Pisum*, the development of wound phloem is comparable to that in *Coleus* and *Impatiens*, since the transitional region between bundle and wound phloem equally shows an early activation of (pro-) cambial tissue (Schulz 1986 b). In the root cortex, however, the situation is different. The first developing wound-sieve tubes provide a new course bypassing the incision and forming one or more phloem strands, adjacent to which - comparable to regular metaphloem - sieve-element differentiation continues. Eventually, a cambium develops between the first formed wound-sieve and wound-tracheary elements adding at the phloem side "secondary" wound-sieve elements (Schulz

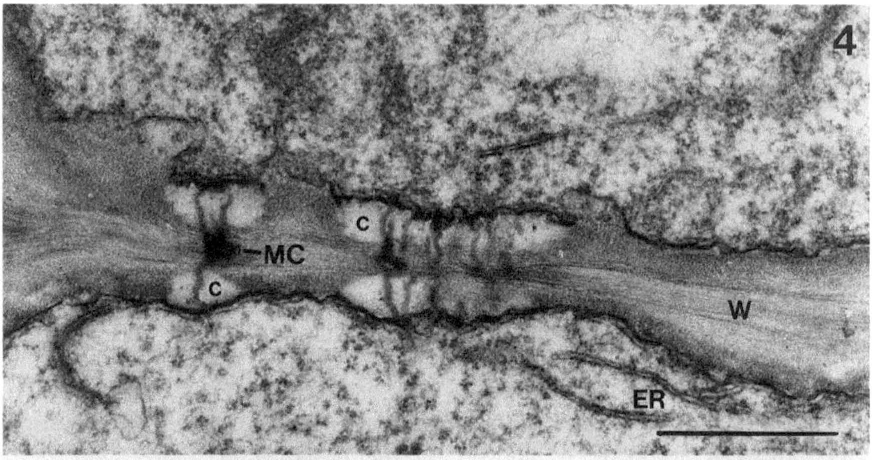
4
MC
c
c
W
ER

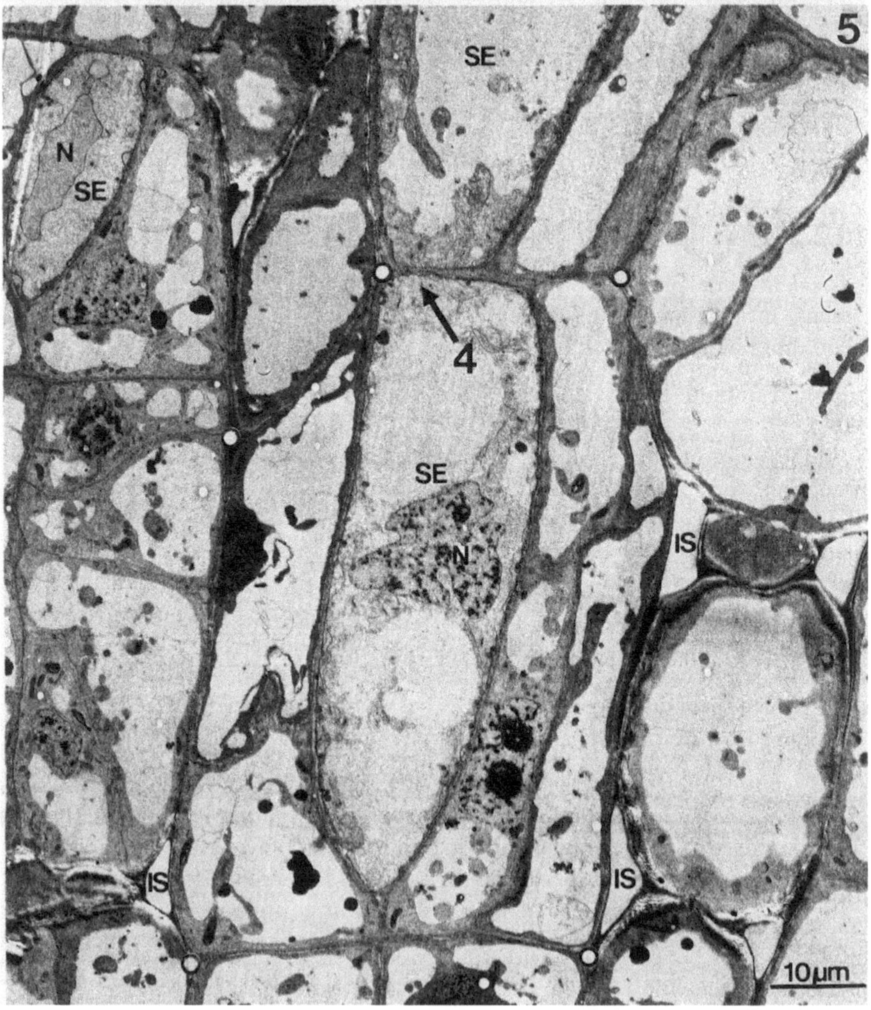
5
SE
N
SE
4
SE
N
IS
IS
IS
10µm

1986c, 1987). Consequently, the phloem regeneration in the cortex may be compared to the entire process of bundle-phloem development including its primary and secondary products.

10.2.3 Size and Shape of Wound-Sieve Elements

Compared to regular sieve elements, the shape of wound-sieve elements is very variable. It is determined by the outline of the original cell and the angle by which this cell is subdivided during the initial, meristematic divisions. Of these, the first one is already subject to the general change of polarity following wounding (Schulz 1988). Due to the predominantly parallel orientation of the second and later divisions, wound-sieve elements form an often arc-like or otherwise bent file of small cells. Sieve plates occur in portions of the original cell walls only (Figs. 10.4, 10.5; cf. Eschrich 1953; Behnke and Schulz 1980; Kollmann et al. 1983; Schulz 1986a, 1987).

The size of a wound-sieve element is limited by the size of its mother cell. In the cortex of pea roots, the first wound-sieve elements are the products of only a few divisions. The minimal number of divisions was reported, according to serial section analysis, to be one. In that case, the original cortex cell (with dimensions of 200 by 45 μm) acted as a sieve-element/companion-cell complex (Schulz 1986a). Accordingly, the wound-sieve element also attained a length of 200 μm and shared the width of the original cell with the companion cell. If not the first, but a later division is the unequal division that gives rise to the sieve element and its companion cell, each further division decreases the size of the sieve element. In general, those members of wound-sieve tubes which are the first to develop after wounding are larger than later ones.

The report of a direct conversion of a cortex cell into a wound-sieve element (Hardham and McCully 1982a) was not confirmed by serial sections. Such a conversion, however, may be achieved by the application of colchicine to regenerating roots (Hardham and McCully 1982b).

◁

Figs. 10.4 and 10.5. Nucleate wound-sieve elements in the root cortex of *Pisum sativum*, 48 h regeneration time (48 h RT). **Fig. 10.4** (*part* of Fig. 10.5): Several plasmodesmata link two wound-sieve elements and are surrounded by callose platelets *(c)*. In the middle of the wall *(W)* they fuse to median cavities *(MC)*. Cisternae of rough ER are still present in the lower element. ×25000. **Fig. 10.5.** The parenchyma cells have been subdivided several times, but their original outline may still be derived *(white circles)* from the location of the intercellular space *(IS)* and the larger thickness of the original walls. Three wound-sieve elements *(SE)* are recognized by callose at plasmodesmata (see *arrow* for higher magnification in Fig. 10.4) and lobed nuclei *(N)*. ×1300

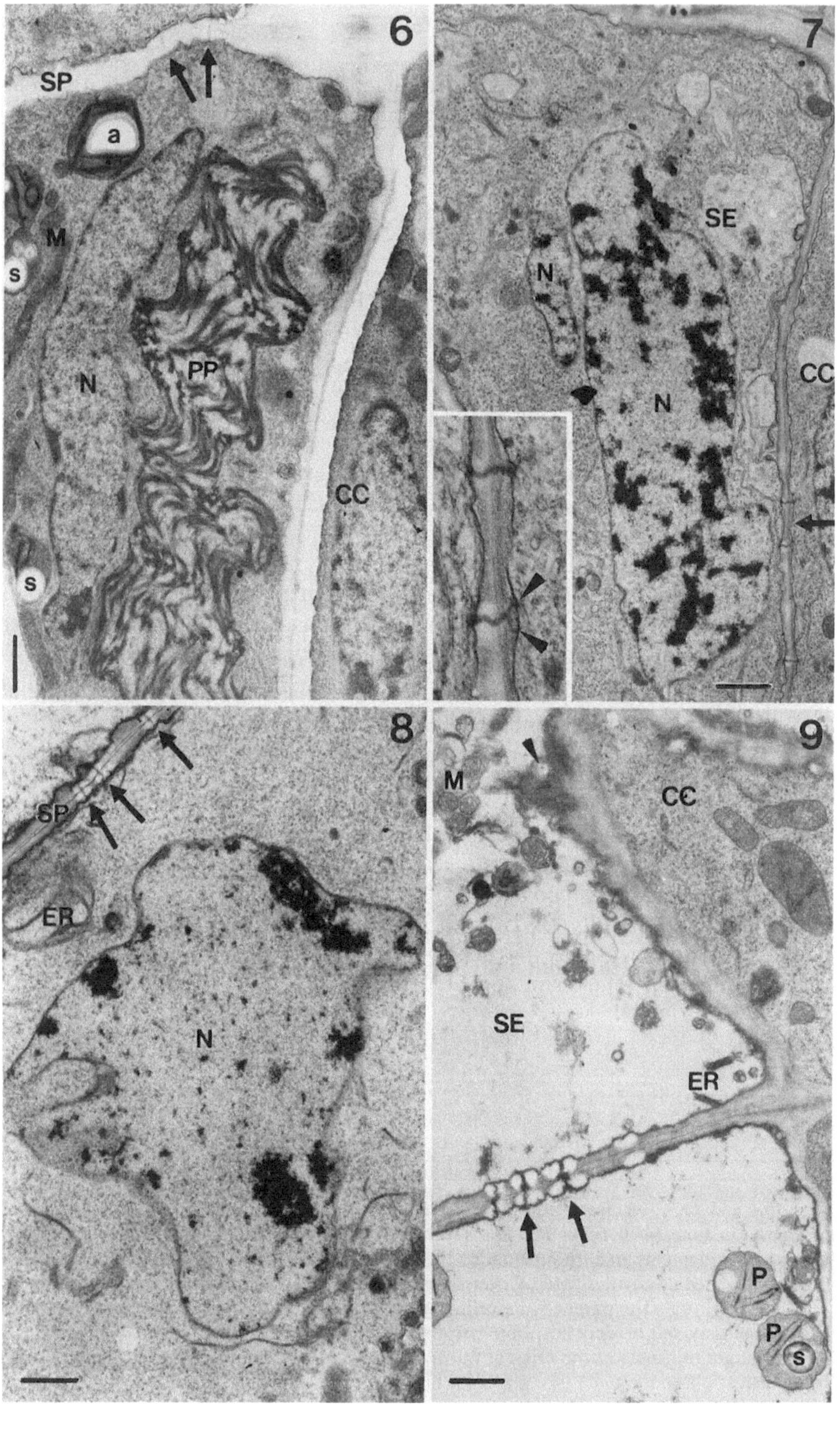
6
SP
a
M
s
N
PP
CC
s
7
SE
N
N
CC
8
SP
ER
N
9
M
CC
SE
ER
P
P
s

10.3 Cytoplasm of Wound-Sieve Elements

In many aspects the *differentiation* of wound-sieve elements follows the pattern established for regular phloem development. Some differences, however, result from the origin of sieve elements from cells with a nonvascular specialization. Compared with procambium and cambium, this specialization is visible in the contents of the plastids, in a high vacuole-to-cytoplasm ratio and in the thickness of the cell wall.

Apart from their origin and shape the *mature* wound-sieve elements are comparable to bundle-sieve elements. Differences in ultrastructure concern the form and amount of P-protein, and the distribution of sieve areas, respectively sieve plates.

10.3.1 Nucleus and Vacuole

A wound-sieve element at the start of its specific differentiation has mostly an ovoid nucleus with few nucleoli and more or less evenly dispersed chromatin. Simultaneous with the appearance of P-protein in the cytoplasm and callose at the plasmodesmata, a lobing of the nucleus occurs, possibly indicating a high level of activity (Figs. 10.5–10.7). Soon afterwards chromatin condenses at a few sites on the envelope which at this stage may or may not become distorted (Fig. 10.8). Successively the chromatin disappears, while portions of the nuclear envelope may persist for longer times. Recognized by their pores, remnants of the nuclear envelope were found among convoluted ER membranes (Schulz 1986a, 1987). Except for such remnants, the mature wound-sieve element (Fig. 10.9) is enucleate.

◁

Fig. 10.6. Nucleate *(N)* wound-sieve element with P-protein body *(PP)* consisting of whirls of filaments, plastids - one with a residual amyloplast-starch grain *(a)* and the others with the spherical starch grains *(s)* of sieve-element plastids - and an elongated mitochondrion *(M)* in dense cytoplasm. The sieve plate *(SP)* is identified by prospective pores *(arrows)*. *CC* companion cell; 144 h RT, *Coleus blumei*. ×8000. **Fig. 10.7.** In this section the lobed nucleus *(N)* of a wound-sieve element *(SE)* is depicted as two portions. The *arrow* points to the position of the inset showing the specific connections between sieve element and companion cell *(CC)* in higher magnification (×26000). On the sieve-element side callose surrounds the plasmatic connection which branches toward the companion cell into two plasmodesmata *(arrowheads)*; 48 h RT, *Pisum sativum*. ×8000. **Fig. 10.8.** Distorted shape of, and chromatin condensation at the envelope indicate the disintegration of the nucleus *(N)* in a wound-sieve element. It has a granular, evacuolate ground plasm, some complexing ER cisternae, and a sieve plate *(SP)* showing paired callose platelets at future sieve pores *(arrows)*; 48 h RT, *Pisum sativum*. ×8000. **Fig. 10.9.** Nearly mature wound-sieve element with widening sieve pores *(arrows)*, plastids *(P)* including thylakoidal remnants and spherical starch grains *(s)*, ER stacks, and mitochondria *(M)* in an electron-lucent ground plasm. Callose at sieve pores and around the connection *(arrowhead)* to the companion cell *(CC)*; 48 h RT, *Pisum sativum*. ×8000

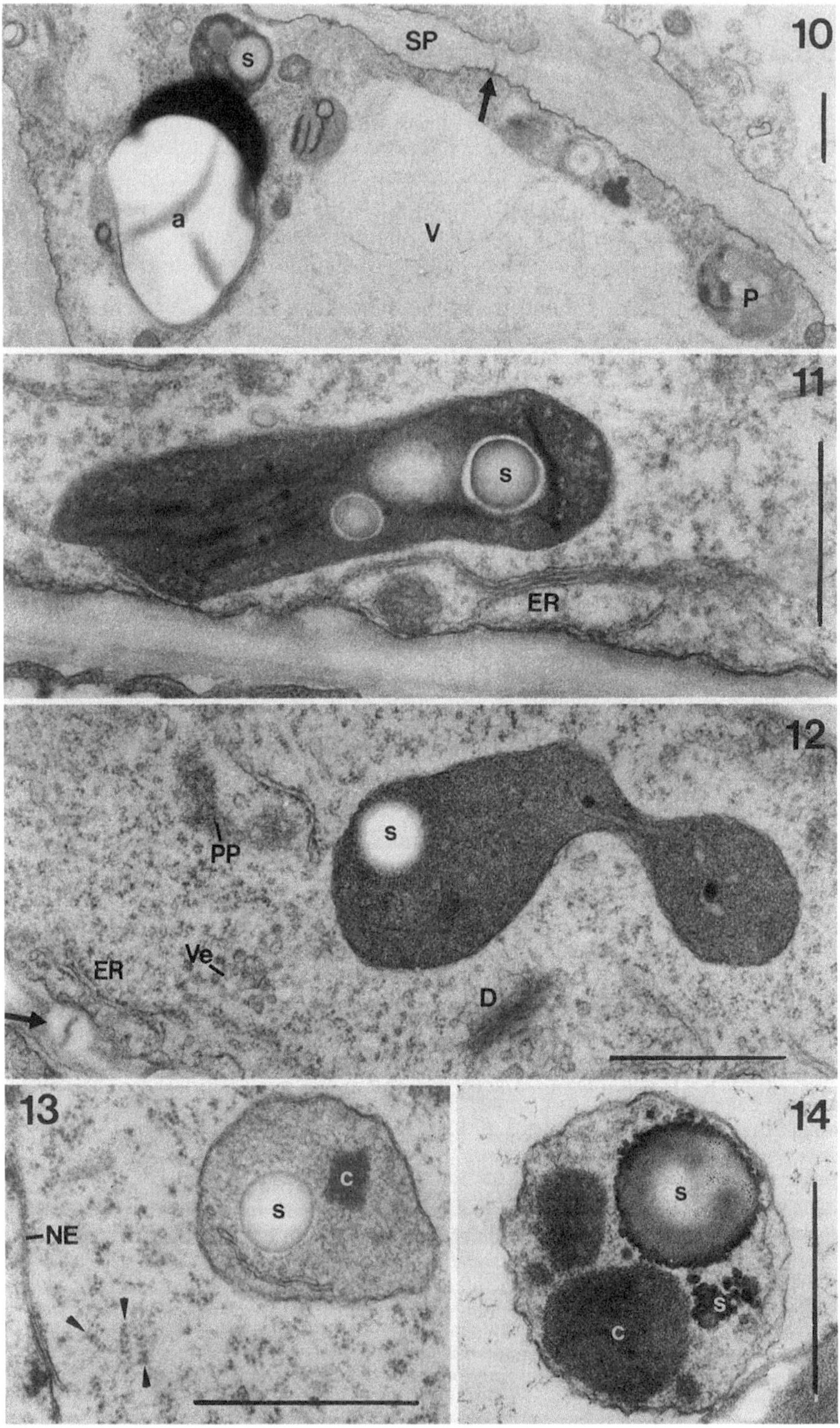
10
SP
s
a
V
P
11
s
ER
12
s
PP
ER
Ve
D
13
c
s
NE
14
s
c
s

During early developmental stages, the vacuoles of wound-sieve elements are larger than those of regular sieve elements and thus reflect the origin of these elements from parenchyma cells. The vacuoles disappear - as far as this can be visualized by electron microscopy - concomitantly with the condensation of the chromatin inside the nuclear envelope. At this stage, the cytoplasmic matrix of a wound-sieve element is granular (Fig. 10.8).

10.3.2 Sieve-Element Plastids

In bundle phloem, sieve-element plastids arise from proplastids and develop species-specific inclusions, e. g., spherical starch grains, protein crystals and protein filaments (Behnke 1981), the functional role of which is not yet known. The indispensability of such inclusions was tested, therefore, by investigating wound-sieve elements that were induced in nonvascular, specialized tissues.

Even when induced in cortical tissue, mature wound-sieve elements exhibited plastids that were comparable in shape and size with those of bundle phloem (Figs. 10.9, 10.13, 10.14). Species-specific starch grains *(Coleus)* or starch grains plus protein crystals *(Pisum)* were documented, although the tissue of origin did not contain proplastids, but chloroplasts with thylakoids or amyloplasts with large starch grains (Behnke and Schulz 1983). More important, however, was the detection of relics of the former plastid type during the course of plastid conversion: Figures 10.6 and 10.10 show starch grains of amyloplast origin in plastids of *Coleus* wound-sieve elements. Several grana were found to persist in plastids of a *Pisum* wound-sieve element (Fig. 10.9). Both types of inclusions disappear before the plastid matrix becomes electron-lucent. The breakdown of the amyloplast-type starch grain always preceded the synthesis of the spherical small starch grains which are typical of sieve-element plastids. The latter are characterized not only by their shape, but also by a reddish staining with iodine and digestion with amylase (Behnke and Schulz 1983).

Fig. 10.10. Wound-sieve element of *Coleus blumei* with vacuole *(V)*, one plastid containing a residual amyloplast-starch grain *(a)* and several sieve-element plastids *(P)* already having spherical starch grains *(s)*, but also thylakoidal remnants. *Arrow* points to pore in future sieve plate *(SP)*; 144 h RT. ×9500. **Fig. 10.11.** Plastid in a cortical wound-sieve element of *Pisum sativum* with dense matrix, thylakoidal grana and spherical starch grains *(s)*. Cisternae of the ER are in initial stages of stacking; 48 h RT. ×25000. **Fig. 10.12.** Stelar wound-sieve element of *Pisum sativum* with precursory P-protein *(PP)*, dicytosome *(D)* with numerous, partly coated vesicles *(Ve)* and a plastid containing a spherical starch grain *(s)* in its dense matrix. The distortion of the plastidal form indicates budding. *Arrow* points to callose at future sieve pore; 18 h RT. ×25000. **Fig. 10.13.** Wound-sieve element of *Pisum sativum* with helical polysomes *(arrowheads)* and a plastid with a spherical starch grain *(s)* and protein crystal *(c)*. *NE* envelope of the disintegrating nucleus; 48 h RT. ×32000. **Fig. 10.14.** Sieve-element plastid in mature element of cortical wound phloem with spherical and particulating starch grains *(s)* and protein crystals *(c)*; 72 h RT, *Pisum sativum*, ×29000 (Behnke and Schulz 1983)

The conversion of amyloplasts or chloroplasts to sieve-element plastids often involves stages associated with the division or budding of plastids (Fig. 10.12). A multiplication of the plastids prior to sieve-element maturation is furthermore indicated by the grouping of new sieve-element plastids with spherical starch grains around one plastid with a residual amyloplast-starch grain (e.g., Figs. 10.6, 10.10; cf. Behnke and Schulz 1983).

Plasmatic and plastidal differentiation seem to occur relatively independently: in a sieve element with an electron-lucent cytoplasmic matrix, plastids with a dense matrix were found adjacent to those with an electron-lucent matrix (Schulz 1986b).

10.3.3 Mitochondria and Dictyosomes

In differentiating wound-sieve elements, the morphology of mitochondria and dictyosomes corresponds to that in regular phloem.

Starting from an elongated amoeboid shape (Fig. 10.6), the mitochondria become spherical (Figs. 10.15, 10.16), concomitantly with the plastids. They persist in mature sieve elements (Fig. 10.9) where their cristae often appear slightly swollen.

Dictyosomes are already numerous in the newly meristematic cortex cells (Schulz 1988). Many vesicles surround them during the nucleated stages of sieve-element development (e.g., Fig. 10.12), possibly indicating a role in the deposition of wall material. An increase of wall thickness becomes visible early after the final division of the precursor cell. The disappearance of the dictyosomes starts with a decrease of the number of vesicles and proceeds with a reduction of the number of cisternae at stages of sieve-element development where the ground plasm is granular and the vacuole is already missing (as, e.g., Fig. 10.8). Dictyosomes do not persist beyond the stage of sieve-pore widening.

10.3.4 Endoplasmic Reticulum and Ribosomes

The changes of the endoplasmic reticulum during differentiation are very similar to those in regular phloem. Accordingly, the rough ER cisternae of nucleate elements contact one another and form parietal stacks of smooth cisternae at later stages of sieve-element differentiation (Fig. 10.9).

There is an alternative for the conversion of rough to smooth ER during sieve-element differentiation: from flat cisternae, the ER may convert into tubular and partly branching forms (Figs. 10.15, 10.16). In mature wound-sieve elements of *Coleus* large networks of tubular ER persisted in mature wound-sieve elements at parietal positions (Schulz 1979).

In earlier stages of sieve-element differentiation, ribosomes are mostly bound to the ER. Later, free helical polysomes occur frequently, sometimes close to precursory P-protein, thus indicating an involvement in P-protein synthesis

(Fig. 10.12; cf. Behnke 1974). Later stages of sieve-element differentiation depict cells with smooth ER, a granular ground plasm, but without a vacuole. The granular appearance of the ground plasm might be due to the dispersed ribosomal remnants. In mature, enucleate sieve elements with widened sieve pores no structure remains which resembles ribosomes (Figs. 10.14, 10.18).

10.3.5 P-Protein and Microtubules

In regular, vascular phloem P-protein (according to a structural definition) may occur in the form of filaments, tubules and crystals (see Chap. 13), the latter being common in Papilionoideae (Behnke and Pop 1981).

Nucleate wound-sieve elements of *Coleus* are characterized by one or more whirl-like bodies assembled from numerous P-protein filaments. They often occur in a central position close to the nucleus (Fig. 10.6). The maturation of the sieve-element protoplast involves the dispersal of the filaments over the cell lumen and their presence in the pores (Behnke and Schulz 1980).

A precursory form of filamentous P-protein appears in the cytoplasm of young wound-sieve elements of *Pisum* when callose becomes first detectable at the plasmodesmata connections. It is aggregated into several small patches and located in close vicinity to ER and polysomes (Figs. 10.12, 10.15). This early form of P-protein was also found in regular phloem of pea ("slime", Bouck and Cronshaw 1965; "amorphous material", Wark and Chambers 1965) and of bean ("fine fibrous precursory material", Esau 1978). It is very probable, that this precursory form is the origin of the P-protein filaments, found in mature wound- (and regular) sieve elements dispersed throughout the lumen (Fig. 10.18), filling also the sieve pores (Fig. 10.21).

Tubular P-protein was not detected in wound phloem of pea roots by the present author. This accords with regular root phloem (Zee and Chambers 1968, 1969), but contrasts with its occurrence in regular, secondary epicotyl phloem (Wark and Chambers 1965; Zee 1968).

P-protein crystals are regularly limited to late metaphloem and secondary phloem of *Pisum* roots (Zee and Chambers 1969; Schulz 1984). Primary wound phloem differentiated in cortical tissue never contained crystals, but only small to moderate amounts of filamentous P-protein (Schulz 1984). In contrast, wound phloem, originating in early activated (pro-) cambial tissue of the stele, exhibited P-protein crystals. Perhaps due to the fixative used, these crystals tended to disperse a little, thereby still reflecting their angular outline (see Figs. 20, 22 in Schulz 1986b; compare Lawton 1978). Wound phloem in the cortex of pea roots was found to form P-protein crystals only after the start of secondary development, i.e., when a new cambium is activated between the primary (first formed) wound-phloem and -xylem elements (see Sect. 10.2.2).

Microtubules were found in nucleate, but not in mature wound-sieve elements of *Coleus* (Schulz 1979) or *Pisum* (Schulz 1984), exactly corresponding to their occurrence in regular phloem (see Chap. 6).

10.4 Symplastic Connections of Wound-Sieve Elements

For the reconnection of the broken phloem continuum, wound-sieve elements have to develop in mature, nonvascular tissue, often perpendicular to the original polar axis. As in regular phloem, they are interconnected by typical sieve pores (Behnke and Schulz 1980; Schulz 1986a, 1987). Considering the entire process of phloem regeneration, the most interesting points in question are concerned with the origin of the new (wound-) sieve pores, and with the linkage between newly developed wound-sieve tubes and the former existing, experimentally severed bundle phloem.

10.4.1 Connections Between Sieve Elements

The investigation of wound phloem of *Coleus, Cucurbita* and *Pisum* showed that sieve pores develop from plasmodesmata and their differentiation follows a pattern similar to that in bundle phloem (Behnke and Schulz 1980; Schulz 1986a, 1987 and unpublished material by the present author). In pea roots, remnants of median cavities were found in early stages of this differentiation, strongly suggesting that most of the plasmodesmata formerly belonged to primary pit fields (Schulz 1987). Moreover, the random position and distribution of these pit fields seems to determine whether they will develop into either simple or compound sieve plates. In *Pisum,* a compound sieve plate developed from three adjacent pit fields (see Fig. 5 in Schulz 1986a), although regular pea phloem exhibits only simple sieve plates.

Pre-existing plasmodesmata are not only the origin of wound-sieve pores, but seem also to be initially involved in determining the course of a developing wound-sieve tube. The progressive addition of sieve elements occurs as a result of the division of cells at the end of a developing sieve tube. This can be determined from the analysis of several cases of incomplete wound phloem in pea roots (Schulz 1984, 1986a). The choice which of several cells at the tip of this

Figs. 10.15–10.18. Sieve plates of cortical *Pisum sativum*-wound phloem of different stages of differentiation. ×30000. **Fig. 10.15.** Two vacuolate *(V)* wound-sieve elements are linked by future sieve pores. Pore sites are marked by callose *(c)* around plasmodesmata and association of single ER cisternae *(arrowheads).* Rough ER-cisternae *(ER)* show transitions to smooth tubular forms and branching *(arrow); M* mitochondrion; 48 h RT. **Fig. 10.16.** Future sieve plate with remnant of a median cavity *(MC)* of the former primary pit field (now occluded by electron-dense material) and callose *(c)* at future sieve pores. Rough and smooth ER occur in the upper element, the latter in form of branching tubules *(arrows); M* mitochondrion; 48 h RT. **Fig. 10.17.** Glancing section through prospective sieve plate depicting callose *(c)* and ER cisternae *(arrowheads)* at future sieve pores, and remnants of the former pit cavity *(arrows)* in the middle of the wall *(W);* 49 h RT. **Fig. 10.18.** The widened pores *(arrows)* in a sieve plate between two mature wound-sieve elements are filled with *ER* and/or P-protein filaments. Dispersed P-protein filaments at *arrowheads; (c)* callose 72 h RT

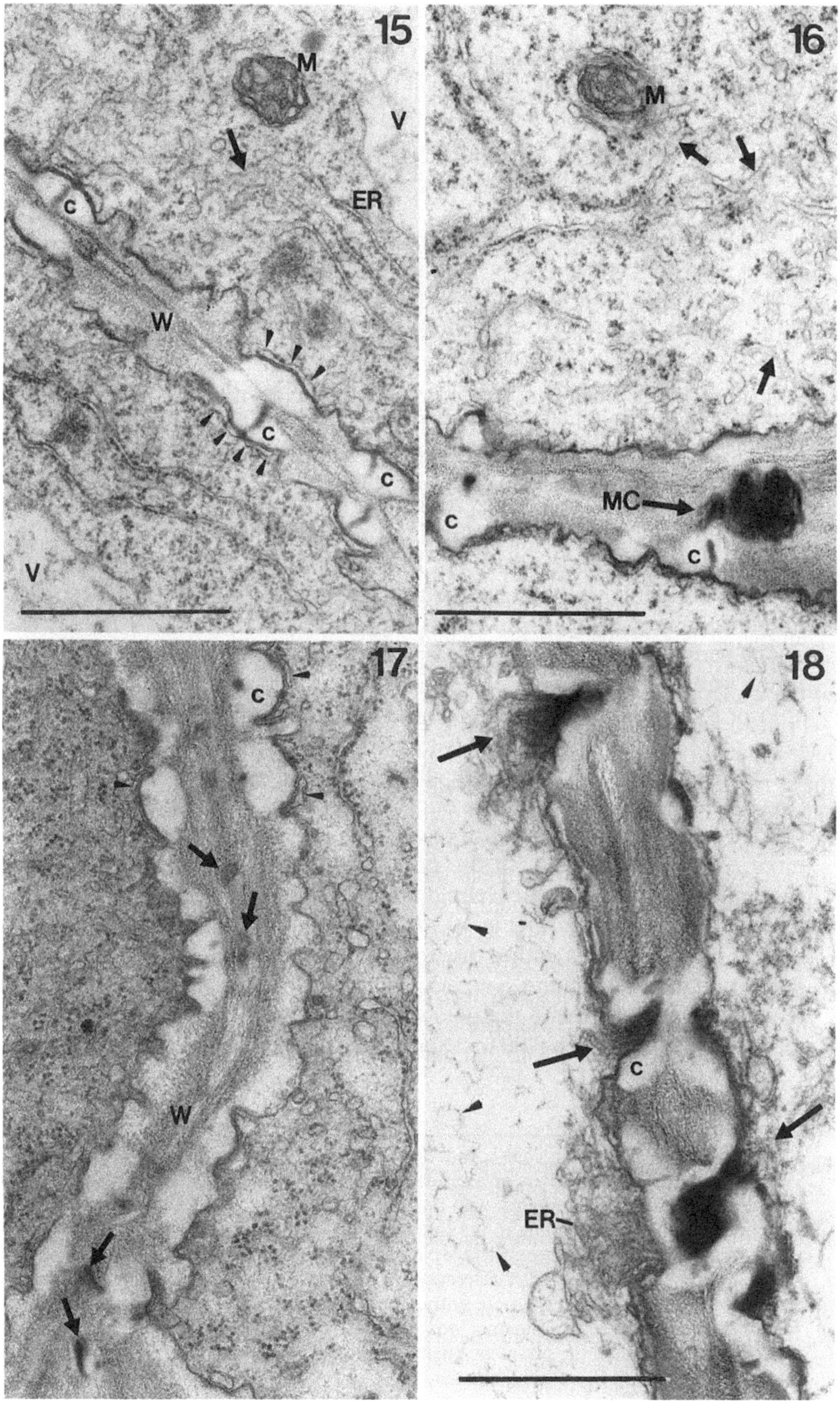
15
M
V
ER
c
W
c
c
V
16
M
MC
c
c
17
c
W
18
c
ER

wound-sieve tube is to be connected seems to be made on the basis of the highest number of plasmodesmata. Thus it is possible that two cells, both connected with the youngest sieve element via pit fields, will develop into sieve elements. The frequent branching of a wound-sieve tube is the result of such a situation (Schulz 1986b; see also "three-poled" wound-sieve elements in *Impatiens;* Eschrich 1953).

The course of wound-sieve tubes which bypass the wound is seldom straight. This is easily explained by the absence of pit fields in some positions. Although direct evidence is not available, all results to date indicate that there is no de novo formation of plasmodesmata at places, where - for the shortest course - sieve pores would be advantageous.

Primary pit fields have two ultrastructural characters making them recognizable in young wound-sieve elements: they are thinner than surrounding wall portions (Schulz 1986a) and their plasmodesmata fuse in the middle of the wall to median cavities (Figs. 10.4, 10.19). Both characters disappear in the course of sieve-element maturation, since wall material is also deposited at the noncallosic portions of the former pit field during the general increase in wall thickness (compare Fig. 10.15 with Figs. 10.17 and 10.18), and the remnants of the cavity, although sometimes persisting beyond maturity, are then occluded by unknown material (Fig. 10.16; cf. Schulz 1987).

The conversion of a pit-field plasmodesma to a sieve pore involves changes of the wall, the plasmalemma and the ER.

Prior to any vascular development, an activation of the original cortex cells takes place in wounded pea roots (Schulz 1988). The plasmodesmata between these cells have desmotubules that are contiguous with ER cisternae (Fig. 10.19). The first appearance of callose platelets at the plasmodesma identifies the cell as a future wound-sieve element and occurs where ER cisternae are more closely associated with the pore sites (Figs. 10.15, 10.16, 10.17). When, subsequently, the symplastic channel starts to widen, it is difficult to observe the fate of the desmotubule (Fig. 10.20). The median cavities, formerly linking several plasmodesmata, either disappear completely or are found to form lacunae within the wall without symplastic contact (Schulz 1987). Individual ER cisternae may persist close to the pore site (Fig. 10.20). Eventually, the pore lumen reaches its regular width (100 to 200 nm in *Pisum*). Figure 10.21 shows a pore of

Figs. 10.19–10.21. Plasmatic connections in developing, cortical wound phloem. *Pisum sativum.* × 100000. **Fig. 10.19.** Part of a primary pit field between activated cortex cells. Rough cisternae of the *ER* are continuous with the desmotubules filling the plasmodesmatal channel. Extensions of the plasmalemma in the middle of the wall are a portion of the median cavity *(MC);* 48 h RT. **Fig. 10.20.** Prospective sieve pore between two wound-sieve elements showing first indications of widening. *ER* cisternae, still apposed to the plasmalemma at callosic wall parts *(c)* are removed from the pore entry, but may occur also free in cytoplasm; 69 h RT. **Fig. 10.21.** Sieve pore between mature wound-sieve elements after 72 h regeneration time. Callose collars *(c)* surround the plasmatic channel which is bordered by the plasmalemma *(arrowheads)* and filled with P-protein filaments

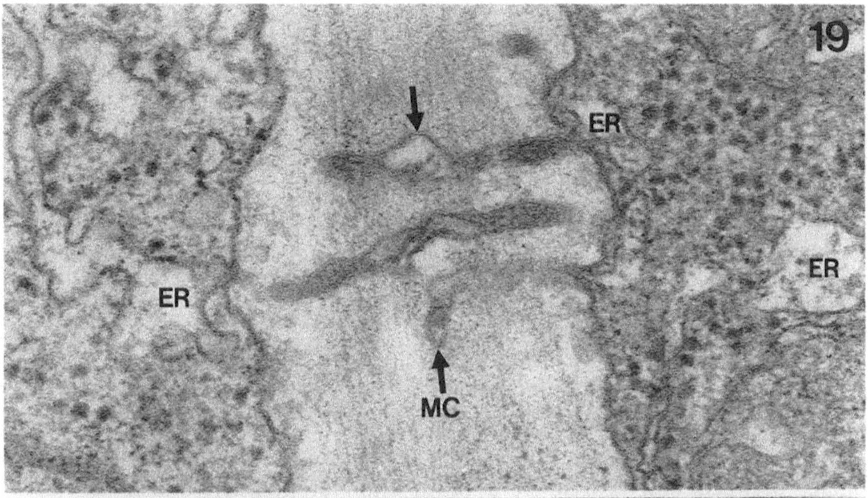
19
ER
ER
ER
MC

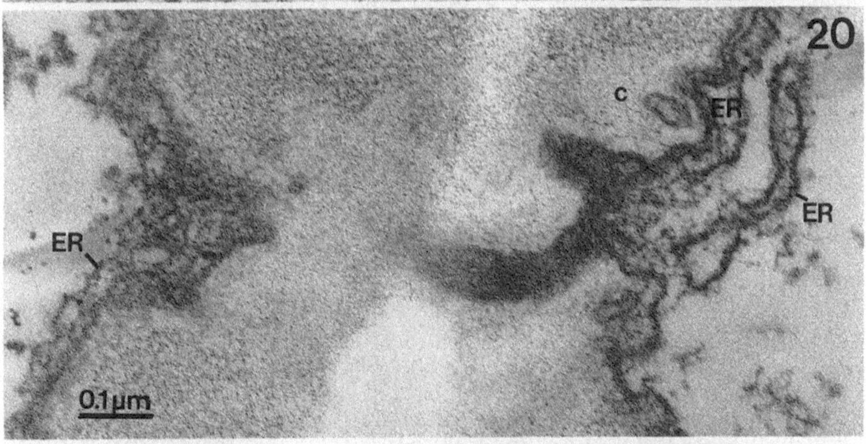
20
c
ER
ER
ER
0.1µm

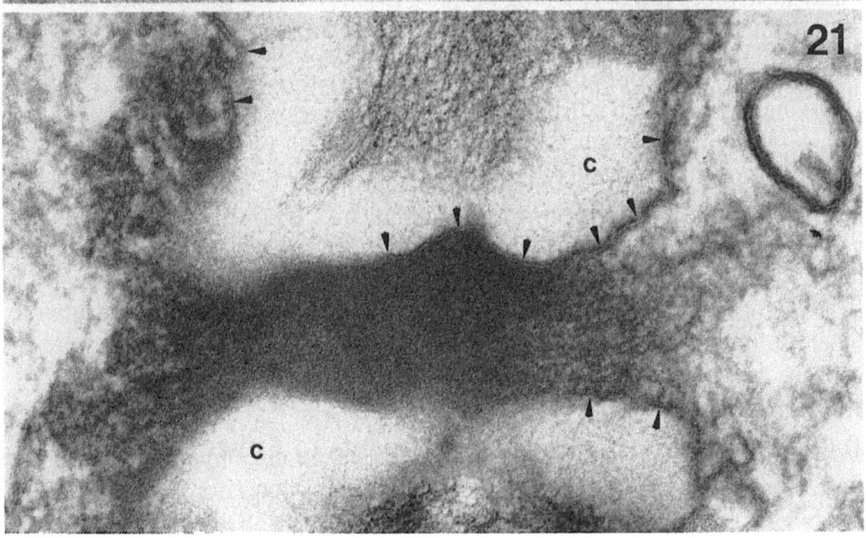
21
c
c

wound phloem 3 days after wounding. It is filled with P-protein and/or ER, even though the wound-sieve tubes of this age are just *starting* to translocate ^{14}C-labeled sucrose in pea roots (Schulz and Gersani, unpublished). The failure to detect ultrastructurally "open" sieve pores in regular and wound phloem of *Pisum* (Schulz 1984) might be explained either by their plugging due to pressure release or their constriction due to callose production (see Hughes and Gunning 1980), both of which are induced by the fixation procedure and facilitated by the small width of the pore lumen.

Sieve elements can only be linked symplastically when they are in about the same, nucleate stage. The secondary phloem in trees is an example where sieve plates and sieve areas develop exclusively in radial walls, since the tangential walls separate cells of different ages. It was therefore interesting to analyze the contact between new wound-sieve tubes and cut bundle phloem. Eschrich already showed in 1953 that wound-sieve tubes are not connected directly to bundle-sieve tubes, but run rather long distances parallel to them (see also Benayoun et al. 1975; Behnke and Schulz 1980). As a consequence, the restriction of the phloem conduit to newly formed elements would mean that, following severance of the entire root stele, the existing bundle-sieve tubes cannot participate in translocation and have to be replaced. In stems, the wound-sieve tubes would have to run at least to the next nodes (after a parallel course alongside the bundles) where they could possibly fuse with still immature, transitory phloem of, e. g., undeveloped axillary buds.

In order to settle this important point, attempts were made to follow all the sieve tubes in a cut phloem trace of pea roots interrupted 2 days earlier and to look at their connections with wound-sieve tubes, using serial ultrathin sections (Schulz 1986b). It could be shown, that - similarly to the relations in stems - wound-sieve tubes run over a long distance (500 μm and more) parallel to existing phloem. Some symplastic contact, however, was provided by lateral sieve areas and single pores which were distributed over this distance. They were connected to mediating sieve elements that, again via lateral pores, eventually were symplastically continuous with those bundle-sieve tubes which were cut by wounding. Starting with a distance of at least three sieve elements from the incision, some sieve plates of cut sieve tubes were found to show indications of a removal from the pores of plugging material, i. e., P-protein and callose. The mediating sieve elements arose from procambial cell tiers, which before wounding were still nucleate. Thus, via lateral contacts and mediating sieve elements, a direct symplastic contact was established between wound phloem and bundle phloem, although again sieve pores developed only when the cells were at about the same developmental stage (see also Schulz 1986b).

10.4.2 Connections from Sieve Elements to Other Cell Types

Wound-sieve elements are symplastically well linked to companion cells by the typical branched sieve-pore plasmodesmata connection (Fig. 10.7; cf. Behnke and Schulz 1980). Obviously, this specialization is part of the genetic program which has to be followed when the sieve-element/companion-cell differentiation is initiated.

Callose was always found to surround those plasmodesmata that were directed toward other cell types but were not continuous through the wall. It is presumed, therefore, that all these connections are occluded during the course of sieve-element differentiation.

10.5 Companion Cells

A consistent feature of wound-sieve elements is their association with companion cells. The latter are not only identified by their close connections to sieve elements (see Sect. 10.4.2), but also by numerous mitochondria and a specific density of ribosomes. In *Coleus, Cucurbita* and *Pisum* no principal difference in size between a wound-sieve element and its companion cell is evident. In the cortex of pea roots, the large companion cells of first-formed wound-sieve tubes have respectively large vacuoles. Wound-companion cells in pea roots are distinguished from surrounding parenchyma by a higher activity of the enzyme succinate-dehydrogenase (Schulz 1984) which is in accordance with a similar histochemical localization in regular phloem (cf. Lehmann 1973).

10.6 Comparison Between Wound- and Bundle-Sieve Elements

The intensive studies of the last decade have documented that several events typical for bundle phloem of angiosperms also occur in wound phloem. Starting with the *unequal division* of a cell that gives rise to the sieve element and its companion cell, these events include, for the sieve element, the *disintegration* and *loss* of the nucleus, vacuoles, Golgi apparatus, ribosomes and cytoskeletal elements as well as the *formation* of typical sieve-element plastids, P-protein and sieve pores. These characteristics, and the association of a companion cell, appear to be essential for cells in order to be able to translocate assimilates.

A difference, such as the small amount of P-protein in wound phloem of pea roots, emphasizes the importance of the time factor in development. The genetic program for "sieve-element differentiation", valid in cells of any origin, might, for example, switch on and off the P-protein synthesis according to the time available, and not to a programmed, final amount. In pea roots there is obviously only a little time available after the complete severance of the stele, since the root part cannot long survive when isolated (see Fig. 10.3). When the phloem is reconnected and the assimilate translocation normalized, the differentiation of sieve elements can continue at the regular, lower speed. Crystals of P-protein, found only in the secondary wound phloem, might in addition be indicative of a correlation between P-protein amount (=time available for its synthesis) and the resulting form. The three choices are (1) only filaments (root protophloem, primary wound phloem); (2) filaments plus a crystal (late metaphloem and secondary regular, and wound phloem in roots); and (3) filaments, one crystal and tubuli (secondary phloem of shoots only).

In conclusion, the comparison of wound and regular phloem, as shown for P-protein above, yields additional information concerning the decisive steps in sieve-element specialization and, thereby, facilitates the analysis of the relationship between structure and function in this important tissue.

Acknowledgment. Supported in part by the Deutsche Forschungsgemeinschaft.

References

Aloni R, Plotkin T (1985) Wound-induced and naturally occurring regenerative differentiation of xylem in *Zea mays* L. Planta 163: 126-132

Behnke H-D (1974) Comparative ultrastructural investigations of angiosperm sieve elements: aspects of the origin and early development of P-protein. Z Pflanzenphysiol 74: 22-34

Behnke H-D (1981) Sieve-element characters. Nord J Bot 1: 381-400

Behnke H-D, Pop L (1981) Sieve-element plastids and crystalline P(hloem)-protein in Leguminosae: micromorphological characters as an aid to the circumscription of the family and subfamilies. In: Polhill RM, Raven PH (eds) Advances in legume systematics. Academic Press, New York London, pp 707-715

Behnke H-D, Schulz A (1980) Fine structure, pattern of division, and course of wound phloem in *Coleus blumei.* Planta 150: 357-365

Behnke H-D, Schulz A (1983) The development of specific sieve-element plastids in wound phloem of *Coleus blumei* (S-type) and *Pisum sativum* (P-type), regenerated from amyloplast-containing parenchyma cells. Protoplasma 144: 125-132

Behnke H-D, Sukkri B (1971) Anastomoses in the internode of *Dioscorea:* their frequency, structure, and function. Z Pflanzenphysiol 66: 82-92

Benayoun J, Aloni R, Sachs T (1975) Regeneration around wounds and the control of vascular differentiation. Ann Bot (London) 39: 447-457

Bouck GB, Cronshaw J (1965) The fine structure of differentiating sieve tube elements. J Cell Biol 25: 79-95

Esau K (1965) Plant anatomy, 2nd edn. John Wiley & Sons, New York London Sidney

Esau K (1969) The phloem. In: Zimmermann W, Ozenda P, Wulff HD (eds) Encyclopedia of plant anatomy, vol 5, pt 2. Borntraeger, Berlin Stuttgart, 505 pp

Esau K (1978) Developmental features of the primary phloem in *Phaseolus vulgaris* L. Ann Bot (London) 42: 1-13

Eschrich W (1953) Beiträge zur Kenntnis der Wundsiebröhrenentwicklung bei *Impatiens holsti.* Planta 43: 37-74

Hardham AR, McCully ME (1982a) Reprogramming of cells following wounding in pea (*Pisum sativum* L.) roots. I. Cell division and differentiation of new vascular elements. Protoplasma 112: 143-151

Hardham AR, McCully ME (1982b) Reprogramming of cells following wounding in pea (*Pisum sativum* L.) roots. II. The effects of caffeine and colchicine on the development of new vascular elements. Protoplasma 112: 152-166

Hughes JE, Gunning BES (1980) Glutaraldehyde-induced deposition of callose. Can J Bot 58: 250-258

Kollmann R, Dörr I, Schulz A, Behnke H-D (1983) Funktionelle Differenzierung der Assimilatleitbahnen. Ber Dtsch Bot Ges 96: 117-132
Lawton DM (1978) P-protein crystals do not disperse in uninjured sieve elements in roots of runner bean *(Phaseolus multiflorus)* fixed with glutaraldehyde. Ann Bot (London) 42: 353-361
Lehmann J (1973) Zur Lokalisierung der Dehydrogenasen des Energiestoffwechsels im Phloem von *Cucurbita pepo* L. Planta 111: 187-198
Neeff F (1914) Über Zellumlagerung. Ein Beitrag zur experimentellen Anatomie. Z Bot 6: 465-547
Robbertse PJ, McCully ME (1979) Regeneration of vascular tissue in wounded pea roots. Planta 145: 167-173
Schulz A (1979) Über Feinbau und Entwicklung von Wundphloem bei *Coleus blumei* (Lamiaceae). Staatsexamensarbeit, Fak Biol, Univ Heidelberg
Schulz A (1984) Licht- und elektronenmikroskopische Untersuchungen zur Entwicklung und Funktion von Wundphloem bei *Pisum sativum* L. (Fabaceae). Thesis, Ruprecht-Karls-Univ, Heidelberg
Schulz A (1986a) Wound phloem in transition to bundle phloem in primary roots of *Pisum sativum* L. I. Development of bundle-leaving wound-sieve tubes. Protoplasma 130: 12-26
Schulz A (1986b) Wound phloem in transition to bundle phloem in primary roots of *Pisum sativum* L. II. The plasmatic contact between wound-sieve tubes and regular phloem. Protoplasma 130: 27-40
Schulz A (1986c) Die Entwicklung von Wundleitbündeln in der Wurzelrinde von *Pisum sativum* L. In: Botanikertagung Hamburg, Abstracts, p 184
Schulz A (1987) Sieve-element differentiation and fluoresceine translocation in wound-phloem of pea roots after complete severance of the stele. Planta 170: 289-299
Schulz A (1988) Vascular differentiation in the root cortex of peas: premitotic stages of cytoplasmic reactivation. Protoplasma 143: 176-187
von Kaan-Albest A (1934) Anatomische und physiologische Untersuchungen über die Entstehung von Siebröhrenverbindungen. Z Bot 27: 1-94
Wark MC, Chambers TC (1965) Fine structure of the phloem of *Pisum sativum*. I. The sieve element ontogeny. Aust J Bot 13: 171-183
Zee S-Y (1968) Ontogeny of cambium and phloem in the epicotyl of *Pisum sativum*. Aust J Bot 16: 419-426
Zee S-Y, Chambers TC (1968) Fine structure of the primary root phloem of *Pisum*. Aust J Bot 16: 37-47
Zee S-Y, Chambers TC (1969) Development of the secondary phloem of the primary root of *Pisum*. Aust J Bot 17: 119-214

11 Sieve Elements of Graft Unions

RAINER KOLLMANN AND CHRISTL GLOCKMANN

11.1 Introduction

In grafted plants the various relations between scion and stock are based on exchange of heterogeneous substances between the partners; many of them being organic compounds which are translocated in the symplast (see reviews by Brabec 1965 and Carr 1976). Although symplastic continuity between cells in the graft interface has often been postulated, reliable experimental proof was difficult to obtain because of the uncertainty in localizing the exact boundary between the partners at the cellular level. This problem has been overcome by investigating heterografts with species-specific cell markers (Kollmann and Glockmann 1985; Kollmann et al. 1985). Secondarily formed plasmodesmata have been recognized in the fusion walls interconnecting the partner cells in these heterogeneous systems. However, of main importance for the success, i. e. for a long-lasting survival of the graft, a connection of the vascular bundles of scion and stock has to be achieved (Vöchting 1892; Simon 1908, 1930; Küster 1925; Funk 1929; Hayward and Went 1939; see also the reviews by Krenke 1933 and Brabec 1965). Vascular connections are formed within a region of mixed tissue derived from both partners, which we define as the graft union in the strict sense. While xylem contact has been recognized with certainty and convincingly described in the early literature (Vöchting 1892; Simon 1908; Crafts 1934; Silberschmidt 1935/36); continuous phloem connections have always been difficult to demonstrate unambiguously (Herse 1908; Crafts 1934) on account of the small size and irregular shape of the sieve elements (see also the review by Krenke 1933).

Translocation of various substances between graft partners has been demonstrated several times (de Stigter 1961, 1966, 1971 a, b; review by Carr 1976). Many authors, however, have described the graft union as a barrier resulting in an accumulation between scion and stock (Silberschmidt 1933, 1935/36; de Stigter 1971 a). Differences in this accumulation effect exist between the various grafting systems studied so far (Funk 1929; Moore and Walker 1981). This is obvious especially with heterografts (grafts between different species) as compared with homografts (grafts between plants of the same species) and autografts (grafts within one and the same plant), respectively (for further details of definition see Brabec 1965). There are several complications encountered when grafts, especially in heterogeneous systems, are established. The question arises whether the reason for these complications are inherent phenomena of incompatibility between the graft partners based on a species-specific recognition sys-

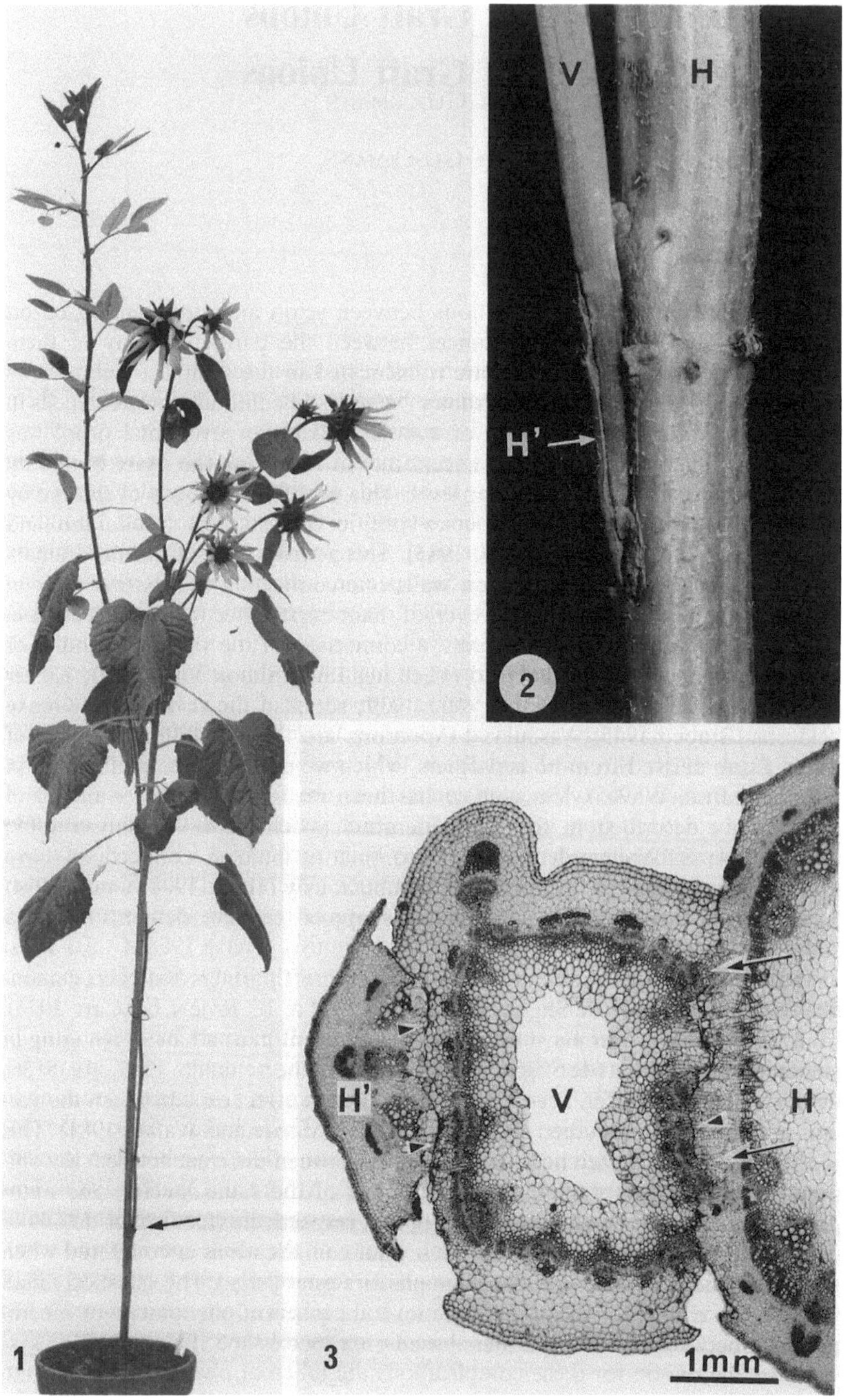
V
H
H'
2
H'
V
H
1mm
1
3

tem (Yeoman et al. 1978; Yeoman 1984) or whether a more general physiological and/or anatomical disharmony must be taken into account (Funk 1929; Silberschmidt 1935/36; Deloire and Hébant 1982; Feucht et al. 1983; Moore 1983, 1984 a, b; Schmid et al. 1984; Schmid and Feucht 1986).

11.2 Grafting Procedure

The success in grafting often depends on the technique applied. Favored methods are referred to as cleft-grafting and side-grafting, the first leaving a stump of the stock with all the vascular bundles cut, the latter having the scion inserted in a longitudinal incision of the stock (Figs. 11.1–11.3) with only a few of the stock bundles disrupted. An exact matching of the various tissues of stock and scion is of great importance for the graft (Crafts 1934; Esau 1965, 1977; Kollmann et al. 1985). The positioning of vascular tissue resulting in superimposed cambia evidently promotes the grafting process.

11.3 Histology and Cytology of the Graft Union

11.3.1 General Aspects of Development

Callus Formation. During the development of a graft union a sequence of several distinct steps can be described (see reviews by Krenke 1933; Esau 1965, 1977; McCully 1983). The first structural event is the formation of callus by hypertrophy and proliferation of living parenchyma cells at the wound surface of scion and stock (Stoddard and McCully 1979; Moore 1982, 1983, 1984 a). Tissues near the cambium are the most active in this process of wound restitution.

Isolation Layer. Depending on the technique and plant material used, breakdown products of cells destroyed by cutting usually form an isolation layer (necrotic layer or contact layer) at the graft interface (Muzik and La Rue 1954;

◁

Fig. 11. 1. A 3-month-old *Vicia faba/Helianthus annuus* graft; *arrow:* position of the grafted area. × 0.1 (Photograph courtesy of J. Kornmacher). **Fig. 11. 2.** Close-up of grafted area with part of the scion *Vicia (V)* inserted between main part *(H)* and flap *(H')* of the stock *Helianthus.* × 2 (Photograph courtesy of J. Kornmacher). **Fig. 11. 3.** Transection of 16-day-old side-graft union with the scion *Vicia (V)* between main part *(H)* and flap *(H')* of the stock *Helianthus; arrows* mixed *V/H* tissue; *arrowheads* isolation layer. Note the position of *Vicia* and *Helianthus* vascular bundles

st
sc
4
100 µm

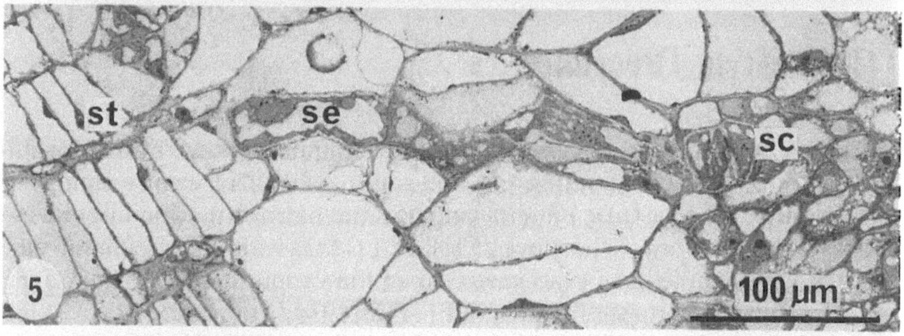
st
se
sc
5
100 µm

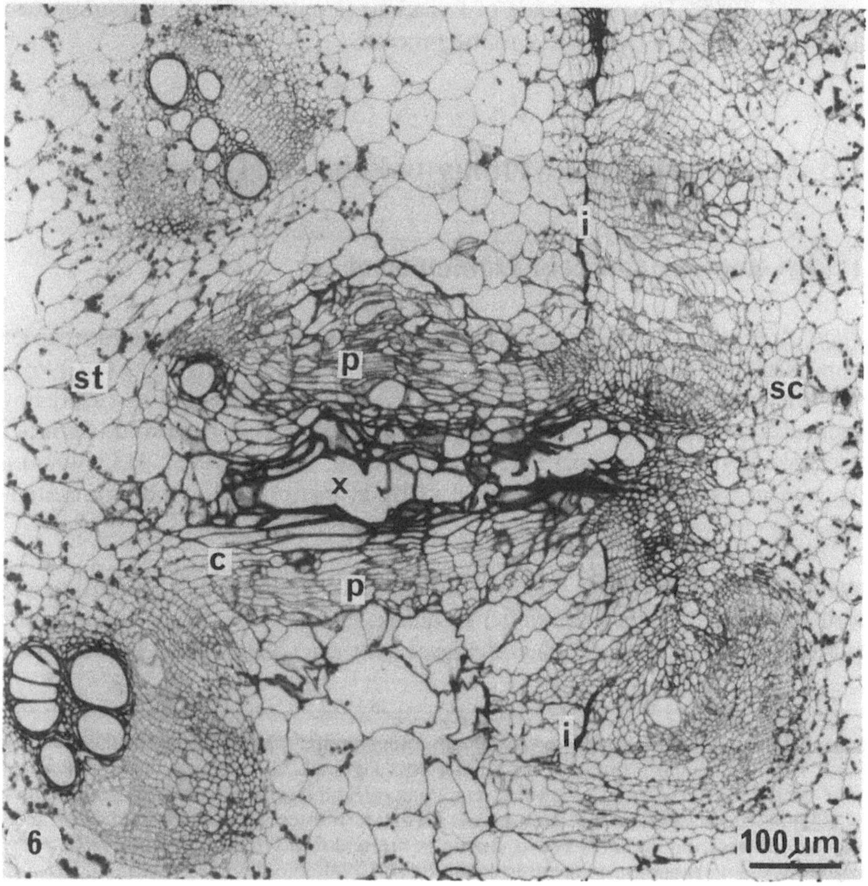
i
p
st
sc
x
c
p
i
6
100 µm

Stoddard and McCully 1979; Deloire and Hébant 1982; Moore and Walker 1981; Moore 1982; McCully 1983).

Mixed Callus of the Graft Partners. During further development and expansion of the calluses the isolation layer is disrupted and/or dissolved. While invading the open space between the partners, callus cells interdigitate, forming a homogeneously mixed tissue derived from scion and stock. In most grafts studied to date, the whole process lasts about 2-4 days following grafting (Stoddard and McCully 1979, 1980), heterografts as a rule being somewhat slower than homo- and autografts, respectively. In the mixed callus the partner cells become interconnected symplastically by plasmodesmata which are secondarily formed in the fusion walls (Kollmann and Glockmann 1985; Kollmann et al. 1985).

Wound Repair Xylem. From day 3 on tracheids and tracheary elements differentiate either directly from callus cells or after several subdivisions (Simon 1930; Crafts 1934; Silberschmidt 1935/36; Hayward and Went 1939; Lindsay et al. 1974; Stoddard and McCully 1979, 1980; Jeffree and Yeoman 1983; see also the review by McCully 1983). These first differentiated tracheids correspond to the wound type of xylem elements usually formed after the severance of vascular strands (Hardham and McCully 1982a, b; Lipetz 1970). Continuous wound xylem strand bridging the graft interface and interconnecting the cut vascular bundles of scion and stock are found in *Lycopersicon* autografts as early as 4 days after grafting and in *Coleus* autografts from day 6 on (Stoddard and McCully 1980).

Wound Repair Phloem. The first-formed sieve elements, as a rule, differentiate in graft unions from callus cells (Crafts 1934) somewhat later than xylem elements. Scattered single sieve elements occur in *Lycopersicon* autografts at day 4, and more extended single phloem strands are observed from day 6 on after grafting. This is in contrast to other wound situations where phloem is restored prior to xylem (Lipetz 1970; Robbertse and McCully 1979).

Generally, the development of the first sieve elements in graft unions corresponds to that described by other authors on phloem restitution after wounding (Eschrich 1953; Behnke and Schulz 1980; Hardham and McCully 1982a, b; Kollmann et al. 1983; Schulz 1986a, b; Tiedemann 1986, 1989). Sieve elements and their companion cells are formed by several consecutive divisions of paren-

◁

Fig. 11.4. Phloem bridge crossing the graft union and interconnecting the vascular bundle of scion *(sc)* and stock *(st)*; longitudinal section; aniline blue fluorescence of sieve-plate callose; heterograft: *Impatiens walleriana/Impatiens oliveri.* **Fig. 11.5.** Wound repair phloem: Young sieve elements *(se)* redifferentiating from callus cells in the graft union; *st* stock; *sc* scion. 10-day-old heterograft: *Cucumis/Cucurbita.* (Specimen courtesy of R. Tiedemann). **Fig. 11.6.** Vascular bundle crossing the graft union interconnecting the transectioned bundles of stock *(st)* and scion *(sc)*. Cambium *(c)* and phloem *(p)* continuity between vascular bridge and bundles of stock and scion; *x* xylem; *i* isolation layer; 21-day-old heterograft: *Cucumis/Cucurbita* (Specimen and photograph courtesy of R. Tiedemann)

chymatic callus cells (Fig. 11.5). The individual sieve elements often are of small size and irregular shape, and, typically, the sieve tubes show a meandering course within the graft union. Connection with the phloem of severed vascular bundles occurs via sieve elements simultaneously derived from vascular cambium (Eschrich 1953; Behnke and Schulz 1980; Schulz 1986 b).

About 6-9 days after grafting the reorganization of phloem bridging the partner bundles (Fig. 11.4) is completed (Stoddard and McCully 1979, 1980; Kollmann et al. 1983; Tiedemann 1986, 1989). The fine structure of the wound sieve elements is in accordance with the phenomena described for regular sieve element differentiation (Behnke and Schulz 1980). Sieve pores develop from plasmodesmata scattered or grouped in primary pit fields of the initial parenchyma cell (Behnke and Schulz 1980; for further details, see Chap. 10); again, wound sieve elements in graft unions show the same peculiarities.

Vascular Cambium. During the early development of the wound repair phloem and xylem in the graft union a cambium is not involved (Simon 1930; Crafts 1934; Kollmann et al. 1983; Tiedemann 1986, 1989; see also review by McCully 1983). A more elaborate vascular system interconnecting the graft partners, however, is established by the activity of a vascular cambium; it develops in the mixed callus of the graft union between the first formed xylem and phloem and in line with the cambia of the vascular bundles of stock and scion (Sharples and Gunnery 1933; Silberschmidt 1935/36; review by Esau 1965, 1977). Cambia crossing the graft union give rise to new phloem and xylem strands, thus forming complete vascular bundles which interconnect the vascular bundles of the graft partners (Fig. 11.6). In herbaceous grafts examined so far most vascular connections develop in this way within 2-3 weeks after grafting (Crafts 1934; Stoddard and McCully 1979; Moore and Walker 1981; Deloire and Hébant 1982; Moore 1984 b; Tiedemann 1986, 1989; Kollmann and Glockmann, unpublished results). Unlike wound repair phloem and xylem cells, the conducting elements derived from a cambium have a more regular, elongated shape and are often arranged in tiers.

11.3.2 Sieve-Element Contact at the Graft Interface

While there is no discrepancy in the literature on the general structure and development of the sieve elements in the graft union, a main, and until quite recently open, question was concerned with the nature of the cell contacts at the

Figs. 11.7.-11.11. Specific cell markers of *Vicia* and *Helianthus* sieve elements. **Fig. 11.7.** S-type plastid only with starch *(s)* of *Helianthus* sieve element. **Fig. 11.8.** P-type plastid with starch *(s)* and protein *(pr)* of *Vicia* sieve element. **Fig. 11.8 a.** Protein crystals in *Vicia* plastid (from Fig. 11.8, enlarged). **Fig. 11.9.** Tubular P-protein in *Helianthus* sieve tubes. **Fig. 11.10.** Filamentous P-protein in *Vicia* sieve tubes. **Fig. 11.11.** P-protein body in *Vicia* sieve element

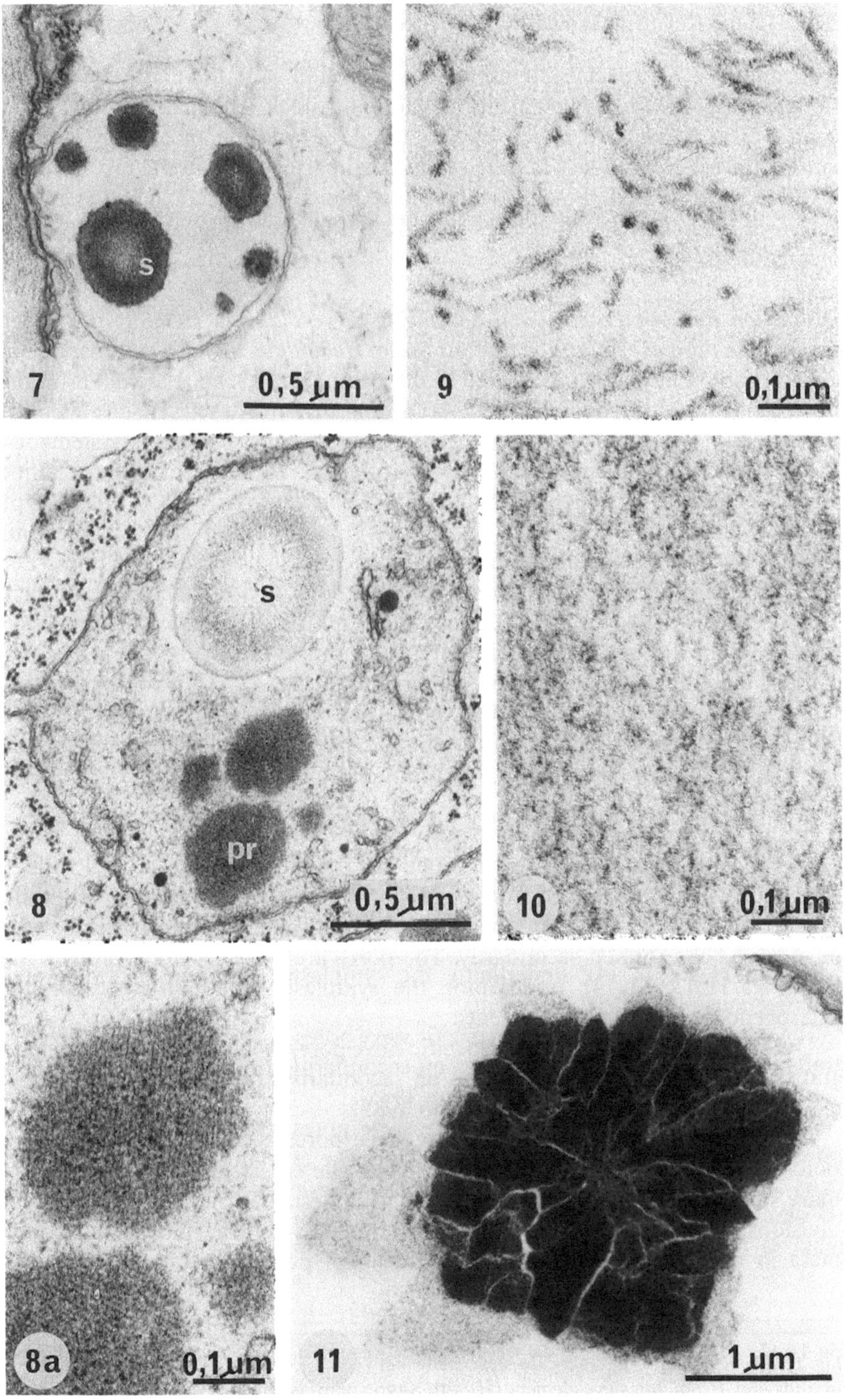
7
s
0,5 µm
9
0,1µm
8
s
pr
0,5 µm
10
0,1 µm
8a
0,1 µm
11
1 µm

graft interface. Of special interest was the question as to the existence of open sieve pores between sieve elements of scion and stock. The main reason for the controversial debate (see review by McCully 1983) was the uncertainty in determining the exact boundary between individual partner cells in the mixed callus. As already mentioned, these difficulties have now been overcome by the investigation of heterografts with species-specific cell markers (Kollmann and Glockmann 1985).

Distinction Between Primary and Secondary Sieve Plates in Heterografts. Using the combination of *Vicia faba* grafted on to *Helianthus annuus,* cell markers in the graft union have been identified in the parenchymatic cells as well as in the sieve elements of the mixed callus. According to Behnke (1981) sieve element plastids and P-protein provide a reliable character in seed-plant systematics. *Vicia* (Fabaceae) contains specific P-type plastids with starch and crystalline protein contents (Figs. 11.8, 11.8 a), while *Helianthus* (Asteraceae) has S-type plastids with starch only (Fig. 11.7). The P-protein of *Vicia* shows filamentous (Fig. 11.10) or, occasionally, also tubular structures in its dispersed form (Zee 1969); furthermore, one crystalline P-protein body is a characteristic of each *Vicia* sieve element (Fig. 11.11). Sieve elements of *Helianthus,* on the other hand, contain dispersed P-protein only in a tubular form (Fig. 11.9). The cell markers mentioned above are the same in both the sieve elements of intact plant shoots and in the graft union. Also, no differences have been found between callus-derived wound-repair and cambium-derived sieve elements. Thus, there are reliable markers determining whether a sieve element belongs to the scion or to the stock.

In this way it was possible to identify interspecific sieve plates in the *Vicia/Helianthus* heterograft (Fig. 11.12). Such sieve plates must have been formed secondarily in a wall derived from the fusion of two adjacent partner cell walls. The secondary sieve plates are perforated by many sieve pores which likewise must have been formed secondarily. The pores are traversed by P-protein and ER (Figs. 11.13, 11.14), underlining the symplastic continuity between the heterogenous partner sieve elements.

The experimental proof of interspecific sieve pores in *Vicia/Helianthus* heterografts is in accordance with findings on secondarily formed interspecific plasmodesmata (Kollmann and Glockmann 1985).

Actually, as is the case with primary sieve pores, secondarily formed sieve pores are also derived from plasmodesmata. The crucial difference is that primary sieve pores develop from primary plasmodesmata in walls formed by cell divisions, while secondary sieve pores differentiate from secondary plasmodesmata in non-division walls. The mechanism of pore formation in secondary

Fig. 11.12. Secondary, interspecific sieve plate between *Vicia* sieve element *(V)* with P-type plastids and *Helianthus* sieve element *(H)* with S-type plastids. Heavy callose deposition at the sieve pores

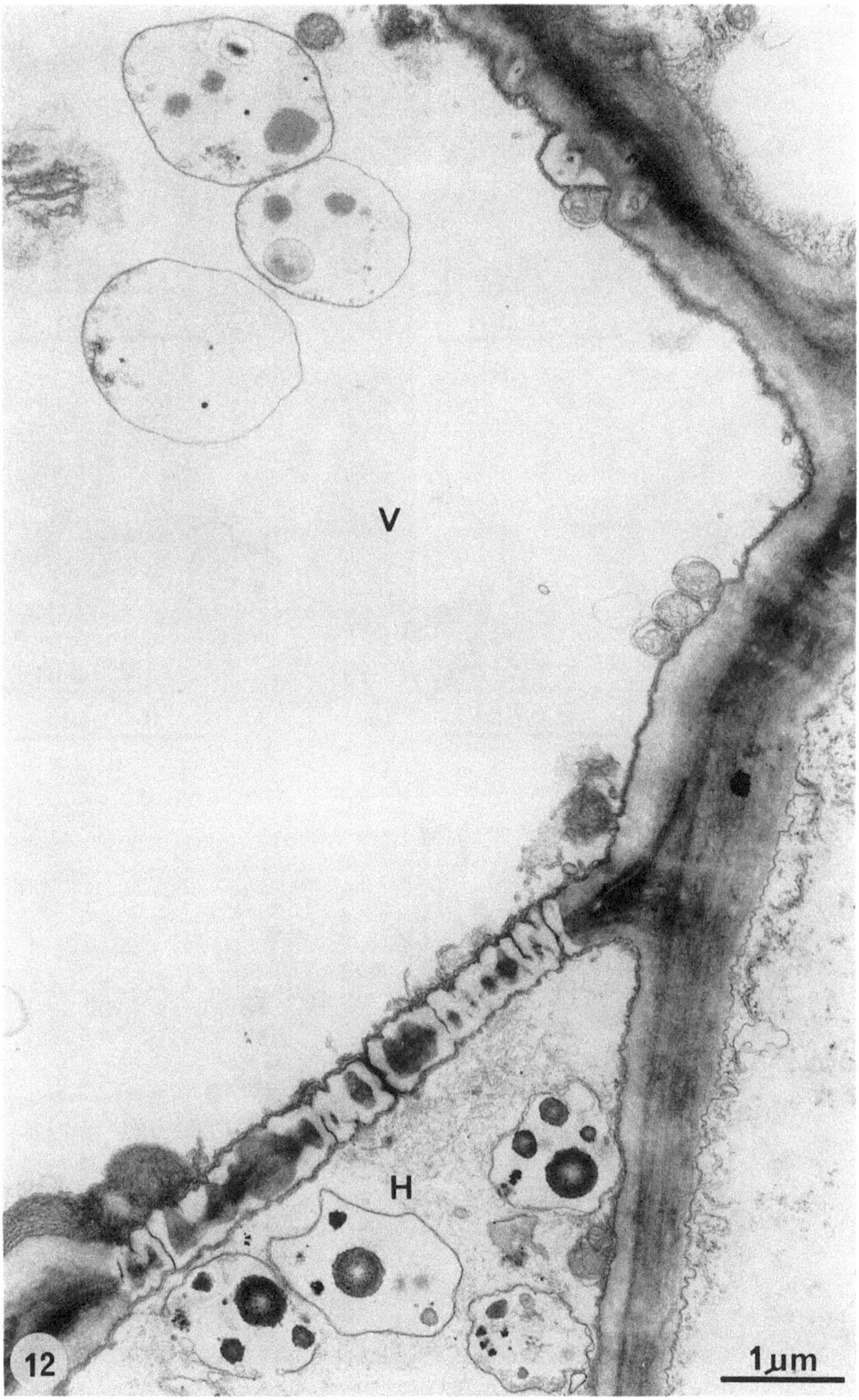
V
H
12
1 µm

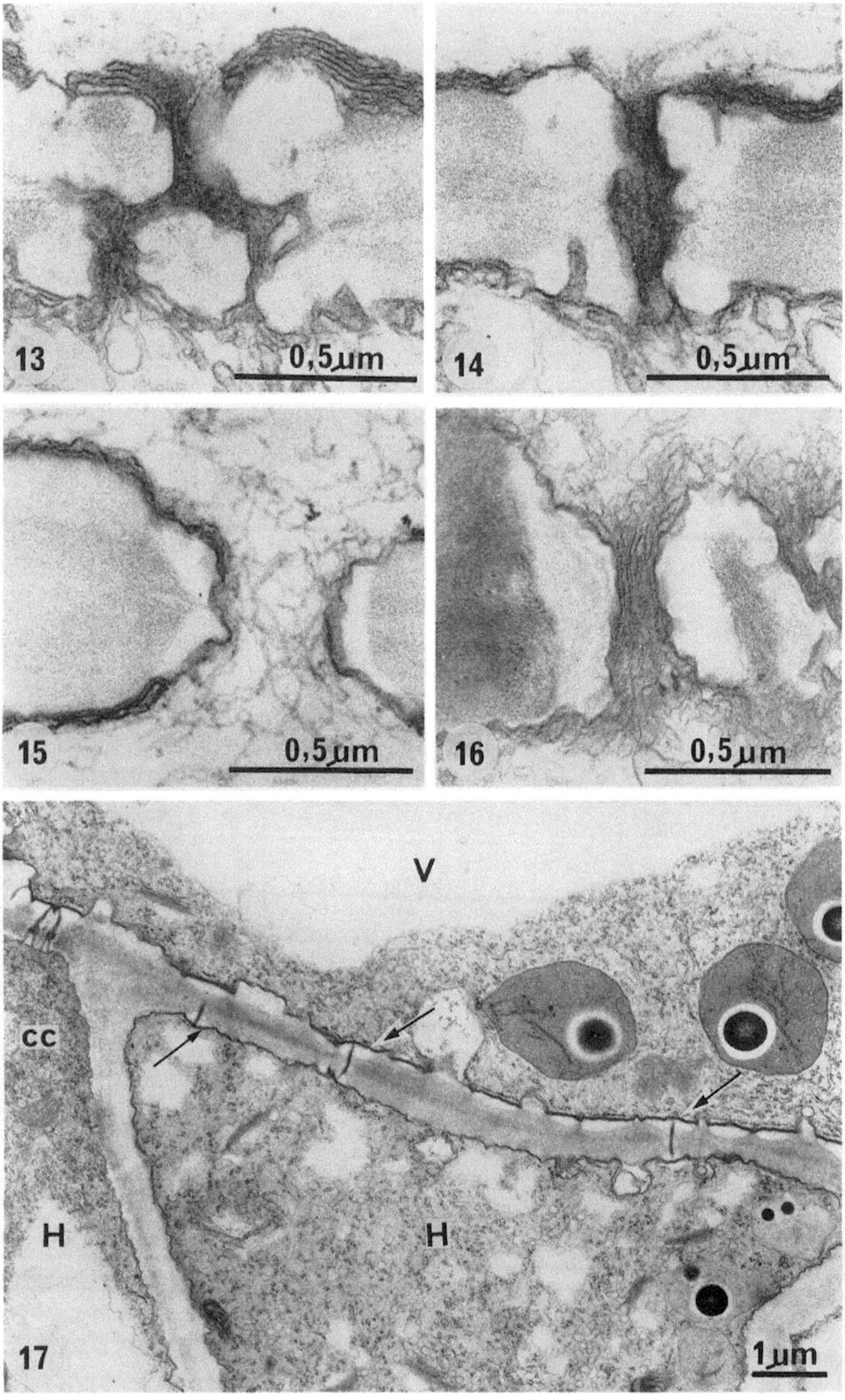
13
0,5 µm
14
0,5 µm
15
0,5 µm
16
0,5 µm
V
cc
H
H
17
1 µm

sieve pores most likely follows the same principle as described for primary sieve pores (Esau et al. 1962): callose platelets are involved in the transformation of plasmodesmata into sieve pores (Fig. 11.17) in a way that is still unknown. As a striking characteristic of secondary sieve plates when compared with primary ones, branched sieve pores occasionally occur (Fig. 11.13). The secondarily formed branched sieve pores resemble to some extent sieve pores of lateral sieve areas in woody dicotyledons (Evert et al. 1971) and in conifers (Kollmann 1975). Most probably, branched sieve pores derive from branched plasmodesmata, while single unbranched plasmodesmata give rise to simple sieve pores (Fig. 11.14). Another typical feature of the interspecific sieve plates (as compared to primary ones) in the *Vicia/Helianthus* graft (Figs. 11.15, 11.16) is heavy callose deposits which restrict the pore openings to such a degree that the fine-structure of the P-protein and ER tubules can only rarely be discerned. The question of whether or not this phenomenon represents a pathological effect or if it indicates some kind of incompatibility between the graft partners is open to debate.

Identification of Secondary Sieve Plates in Autografts. Lacking partner-specific cell markers, localization of the graft interface in autografts and most heterografts is more difficult than in the *Vicia/Helianthus* graft described above. Therefore, only an indirect method of identifying the sieve element contact between scion and stock can be applied. By means of complete serial sectioning for light and electron microscopy the exact course of phloem bridges (Fig. 11.18) has been followed and three-dimensionally reconstructed step-by-step (Figs. 11.20 a, b). Within this series each single sieve plate of a connecting sieve tube has to be examined with the electron microscope (Fig. 11.19) to determine the state of its sieve pores (Fig. 11.21). By this method Monzer and Kollmann (1986), Tiedemann (1986, 1989) and Kollmann and Glockmann (unpublished results) have proved, in various auto- and heterografts, symplastic continuity of sieve tubes interconnecting the graft union. Most of the sieve pores were simple and had callose deposits of varying thickness depending on the graft examined. Occasionally, sieve plates have been found in the graft union which, besides simple pores, had branched connections (Fig. 11.21). These structures have been interpreted, in accordance with the findings in *Vicia/Helianthus* grafts, as secondary sieve plates and sieve pores, respectively, formed between the partner cells at the graft interface. Branching in general might be a typical feature of secondarily formed intercellular connections.

Fig. 11.13. Secondarily formed interspecific, branched sieve pore in *Vicia/Helianthus* graft. **Fig. 11.14.** Secondarily formed interspecific, unbranched sieve pore in *Vicia/Helianthus* graft. **Fig. 11.15.** Primarily formed intraspecific *Helianthus* sieve pore. **Fig. 11.16.** Primarily formed intraspecific *Vicia* sieve pore. **Fig. 11.17.** Young interspecific sieve plate between *Vicia* sieve element *(V)* and Helianthus sieve element *(H)*; callose deposition at the site of pore formation *(arrows)*; interspecific one-sided branched plasmodesmata between *Helianthus* companion cell *(cc)* and *Vicia* sieve element

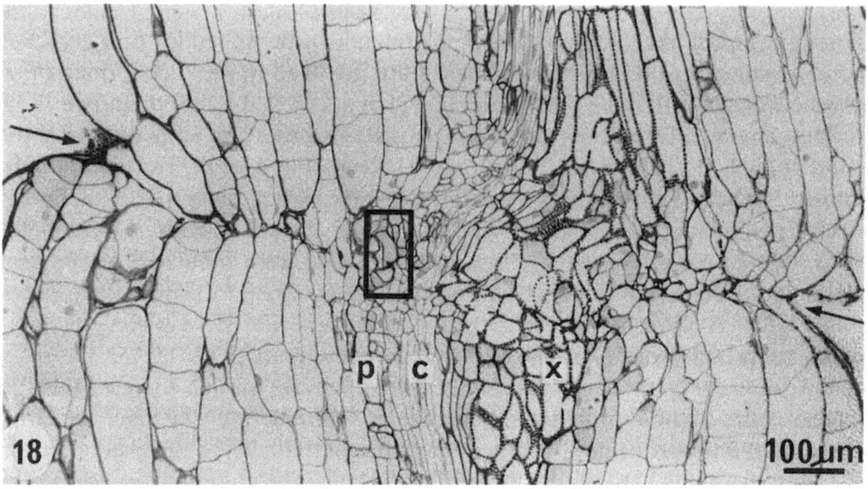
p
c
x
18
100 µm

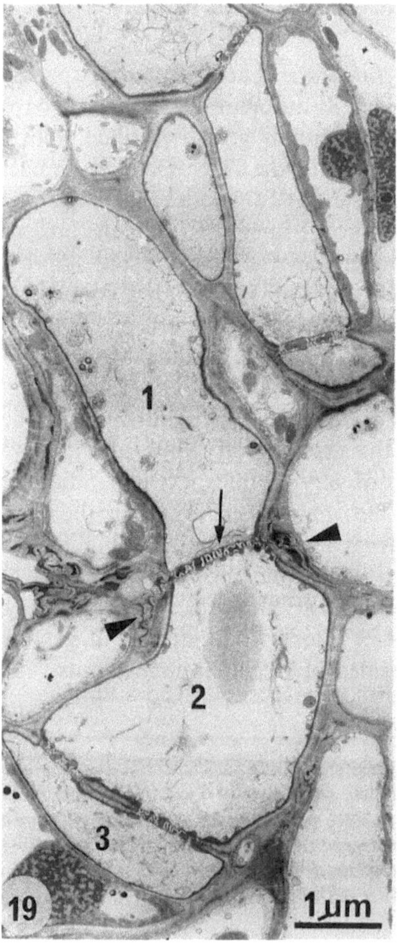
1
2
3
19
1 µm

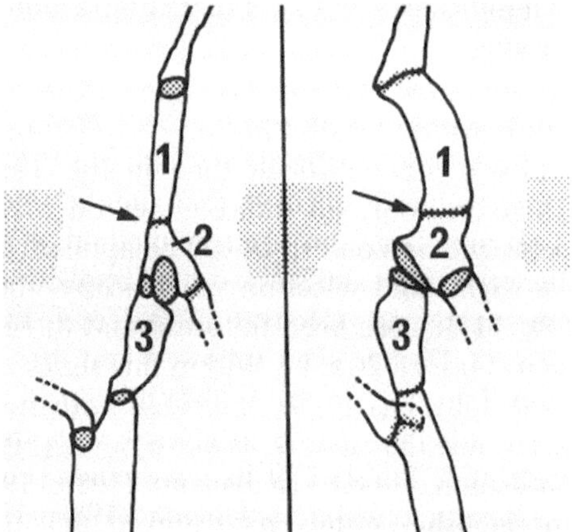
1
2
3
1
2
3
20
b
a

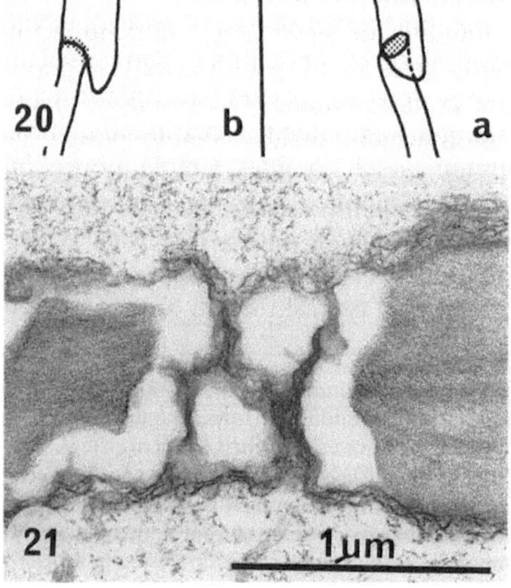
21
1 µm

There is, however, one essential difference between inter- and intraspecific secondary sieve plates: while in autografts the diameter of the pores in all sieve plates is nearly identical, in heterografts the interspecific pores generally are narrower than intraspecific ones. This distinction is independent of variations in callose deposits.

Sieve Element-Companion Cell Communication in Heterografts. As is common in angiosperms, sieve elements of graft unions are always associated with companion cells. However, in contrast to the situation in regular phloem, sieve elements and their companion cells are not always and exclusively derived from a division of one mother cell; a secondary connection of both cell types is also possible. This has been proved with the *Vicia/Helianthus* graft, where the individual partner cells have species-specific markers. In Fig. 11.17 a companion cell of *Helianthus* is associated with young sieve elements of both *Helianthus* and *Vicia,* the latter being interconnected by an undifferentiated, interspecific sieve plate. The protoplasts of the *Helianthus* companion cell and the *Vicia* sieve element are joined by plasmodesmata. These interspecific symplastic connections must have been formed secondarily in the non-division wall between the companion cell and the sieve element. It is noteworthy that these secondary cell connections show the same fine-structure as the primarily formed cell bridges. They represent the usual one-sided, branched plasmodesmata joined by a median nodule to a single sieve pore (Esau 1969).

From this observation we can conclude that an interaction between a sieve element and its companion cell is not absolutely dependent on an ontogenetic relationship of the partners involved. Furthermore, it is obvious that cells of unrelated plants must be able to cooperate in fundamental processes as a structural and functional unit.

Fig. 11.18. Longitudinal section of a 20-day-old *Vicia*-autograft union. A vascular bundle with xylem *(x)*, cambium *(c)* and phloem *(p)* crosses the graft interface *(arrows)*; *frame* area enlarged in Fig. 11.19. **Fig. 11.19.** Electron micrograph of a serial section of the framed area in Fig. 11.18. File of three sieve elements *(1, 2, 3)* with a secondary sieve plate between cell *1* and *2 (arrow)*; residues of the isolation layer (compressed wall material) at both sides of the secondary sieve plate *(arrowheads)*. **Fig. 11.20.** Reconstruction of the course of the single sieve tube as shown in Figs. 11.18 and 11.19 (sieve elements *1–3*) crossing the graft interface *(stippled marginal areas)*; sieve plates are marked by *dense stippling*; *arrow* secondary sieve plate. Specimen was analyzed by consecutive serial sections. *a* Surface view in the plane of the section of Figs. 11.18 and 11.19. *b* Reconstructed profile perpendicular to the plane of sectioning of Figs. 11.18 and 11.19 (Reconstruction and drawing by J. Monzer). **Fig. 11.21.** Branched secondary sieve pores with heavy callose deposits in a secondary sieve plate; serial section to Fig. 11.19

11.4 Function of Phloem Connections in Graft Unions

The transport of assimilates in grafts has been shown in the past by de Stigter (1961, 1966, 1971 a, b) with autografts *(Silene armeria)* as well as with heterografts (muskmelon/*Cucurbita*) and more recently by Tiedemann (1986) with *Cucumis/Cucurbita* and by Rachow-Brandt (1987) with *Lycopersicon/Solanum* and *Vicia/Helianthus*. The principle of a solute exchange between scion and stock was demonstrated with the autoradiographic technique using ^{14}C-photosynthates.

The main transport described by these authors proceeded from scion to stock; occasionally transport occurred in the opposite direction from stock to scion on the order of up to 10% of the downward stream. Many different factors, especially the strength of various sinks may influence the rate and direction of translocation in a grafted system.

Recently, a quantitative relationship between assimilate transport and phloem differentiation was investigated by Tiedemann (1986) and Rachow-Brandt (1987) in three heterografts and their respective autografts, *Cucumis/Cucurbita, Lycopersicon/Solanum* and *Vicia/Helianthus*. Of these *Lycopersicon/Solanum* was the most successful graft, while *Vicia/Helianthus* was more difficult to achieve.

The transport across the graft interface was determined by measuring the amount of ^{14}C-activity in the stock as a percent of the total activity exported from a $^{14}CO_2$-fed scion leaf. The state of the phloem development in the corresponding graft union was defined by the number of interconnecting sieve tubes as analyzed by light-microscopy using serial sections.

As was expected, a close interrelation between transport and phloem restitution in the graft union could be demonstrated. During the first 6 days following grafting only a very low translocation of assimilates into the stock was detectable. As continuous phloem bridges were not established at that time (see Sect. 11.3.1), only a transport in parenchyma cells seems likely. In the *Lycopersicon/Solanum* system, the import of photosynthates into the stock started at day 6 and reached a maximum rate about 9 days after grafting. Due to the prominent sink, more than 80% of the assimilates exported from a scion leaf were transported into the stock, while with the autograft *Lycopersicon/Lycopersicon* only about 60% were imported. Likewise, the heterograft *Cucumis/Cucurbita* showed the first detectable import into the stock at day 7 with a maximum at day 20; at that time only 6% of the assimilates exported from the fed leaf were imported into the stock. The *Cucumis/Cucumis* autograft showed a maximum of transport 10 days earlier than the corresponding heterograft; about 50% of the scion assimilates were imported into the stock. The interfamiliar heterograft *Vicia/Helianthus* turned out to be the most sluggish system. Import into the stock did not commence before day 11 and increased up to day 14. Only 2% of the photosynthates exported from a fed scion leaf were detectable in the stock while the corresponding autografts reached about 30% *(Helianthus/Helianthus)* and 40% *(Vicia/Vicia)*, respectively.

The physiological data support the anatomical observations. The first assimilate import coincides with the development of wound repair phloem, 6-9 days after grafting (see Sect. 11.3.1). The phloem bridges derived from either callus or cambium (depending on the system investigated) are completed 2-3 weeks after grafting, at the same time that transport into the stock has reached a maximum rate.

Finally, differences in the transport rate parallel differences in the number of sieve tubes crossing the graft interface. While in the *Cucumis/Cucurbita* system at 3 weeks after grafting, about 20 sieve tubes per graft union are established, in the autograft *Cucumis/Cucumis*, about 90 sieve tubes interconnect the partner vascular systems. Thus the number of sieve tubes in a graft union determines the transport rate from scion to stock (Tiedemann 1989). The importance of phloem restitution between graft partners is evident and, furthermore, the graft union is not only a mechanical barrier to translocation as was suggested by many authors in the past.

11.5 Questions Concerning the Mechanism of Sieve-Element Formation in Graft Unions

It is generally accepted that in multicellular organisms intercellular communications have vital functions for the cooperation between the various cells, tissues and organs. In grafted plants two or several partners of homogeneous or heterogeneous origin interact as a structural and functional unit. The cooperation is established by intercellular short-distance transport between various parenchymatous tissues at the graft interface and, as is the topic of the present chapter, by long-distance transport in a more elaborate vascular system.

It has been shown that for both transport processes symplastic intercellular connections (plasmodesmata and sieve pores) are developed between the partner cells. These findings are in accordance with the physiological data on grafts (see the review by Carr 1976) but they are in contrast to the observations on parasitic angiosperms often compared with grafted plants (see Kollmann and Dörr 1987 and Chap. 12).

Depending on the origin of scion and stock, the cell contacts across the graft interface are intra- or interspecific; they are, in all cases, secondarily formed by adjacent partner cells in non-division walls. It is noteworthy that secondary formation of cell contacts may even occur between sieve elements and companion cells of different genotypes.

The mechanism of formation of secondary cell bridges is unknown up to date (Jones 1976; Kollmann et al. 1985) and needs further investigation. The phenomenon, however, raises the question as to the existence of special recognition factors involved in the achievement of cell contacts between adjacent cells (Yeoman 1984; Yeoman et al. 1978). So far, no experimental proof has been presented.

Success or failure of a grafting process depend on different factors. With unsuccessful grafts characteristic symptoms in the graft union have often been interpreted as indications of incompatibility between the partners. Besides cell necrosis and degeneration of the vascular system (see review by Moore 1983) a lack of vascular restitution is primarily responsible for the failure of a graft. However, it is open to discussion whether or not these phenomena are due to a genetically determined, inherent system of cellular incompatibility between the grafted plants.

Most important for a successful graft is an exact anatomical matching of scion and stock at the graft interface. In particular, mismatching of the vascular systems and respective cambia often results in an unsuccessful or at least delayed grafting process (Funk 1929; Crafts 1934; Silberschmidt 1935/36; Esau 1965, 1977; Brabec 1965; Kollmann et al. 1985). Differences in phloem differentiation and the rate of translocation as described in various homo- and heterografts are most likely the consequence of this anatomical mismatching between the partners (for further details see Tiedemann 1989). Evidently, the first restitution of a functioning vascular system between the damaged bundles in herbaceous grafts proceeds like a wound restitution within a few days. The question arises as to the mechanism of induction of these fast differentiation processes.

With regard to the wound restitution, it has often been experimentally proved that various morphogenetic substances such as phytohormones (especially auxin and gibberellins) and sugars induce and control the regeneration of cambia and vascular tissues (von Kaan-Albest 1934; LaMotte and Jacobs 1962, 1963; Wetmore and Rier 1963; Thompson and Jacobs 1966; Thompson 1967; Lipetz 1970; Benayoun et al. 1975; Shimomura and Fuzihara 1977; Stoddard and McCully 1980).

Furthermore, it has been shown that a positive correlation exists between the extent of phloem and xylem regeneration and the number of severed bundles (LaMotte and Jacobs 1962, 1963; Benayoun et al. 1975). A leakage of morphogenetic substances from the severed vascular system into adjacent tissues, as suggested by several authors (Wetmore and Rier 1963; Benayoun et al. 1975; Robbertse and McCully 1979), may induce dedifferentiation of parenchymatous callus cells and their differentiation into vascular elements. As to the situation in grafts, the pattern and course of differentiation in the graft union may be "controlled by the ratio of concentrations of two diffusible morphogens" such as auxin and sucrose, as proposed by Warren Wilson and Warren Wilson (1981). If this hypothesis proves true, the stem anatomy and especially the disposition of vascular bundles indeed have a decisive influence on graft-specific phloem- and xylem-restitution in the graft union.

Acknowledgments. The authors are much indebted to J. Kornmacher, Botanical Garden, University of Kiel for skilfully establishing most of the grafts dealt with in this paper. For carefully reading the manuscript and some helpful suggestions we are grateful to J. Monzer. The work was supported by the Deutsche Forschungsgemeinschaft.

References

Behnke HD (1981) Sieve-element characters. Nord J Bot 1: 381-400

Behnke HD, Schulz A (1980) Fine structure, pattern of division, and course of wound phloem in *Coleus blumei.* Planta 150: 357-365

Benayoun J, Aloni R, Sachs T (1975) Regeneration around wounds and the control of vascular differentiation. Ann Bot (London) 39: 447-454

Brabec F (1965) Pfropfung und Chimären unter besonderer Berücksichtigung der entwicklungsphysiologischen Problematik. In: Ruhland W (ed) Handbuch der Pflanzenphysiologie, vol 15/2. Springer, Berlin Heidelberg New York, pp 388-498

Carr DJ (1976) Plasmodesmata in growth and development. In: Gunning BES, Robards AW (eds) Intercellular communication in plants: studies on plasmodesmata. Springer, Berlin Heidelberg New York, pp 243-288

Crafts AS (1934) Phloem anatomy in two species of *Nicotiana,* with notes on the interspecific graft union. Bot Gaz 95: 592-608

Deloire A, Hébant C (1982) Peroxidase activity and lignification at the interface between stock and scion of compatible and incompatible grafts of *Capsicum* on *Lycopersicum.* Ann Bot (London) 49: 887-891

de Stigter HCM de (1961) Translocation of ^{14}C-photosynthates in the graft muskmelon/*Cucurbita ficifolia.* Acta Bot Neerl 10: 466-473

de Stigter HCM de (1966) Parallelism between the transport of ^{14}C-photosynthates and the flowering response in grafted *Silene armeria* L. Z Pflanzenphysiol 55: 11-19

de Stigter HCM de (1971 a) Some aspects of the physiological functioning of the graft muskmelon/*Cucurbita ficifolia.* I. Melon/*Cucurbita* versus cucumber/*Cucurbita:* differences in behaviour due to differences in degree of compatibility. Z Pflanzenphysiol 65: 223-231

de Stigter HCM de (1971 b) Some aspects of the physiological functioning of the graft muskmelon/*Cucurbita ficifolia.* III. Recovery in presence of limited quantities of stock foliage. Z Pflanzenphysiol 65: 281-295

Esau K (1965) Plant anatomy, 2nd edn. John Wiley & Sons, New York

Esau K (1969) The phloem. In: Zimmermann W, Ozenda P, Wulff HD (eds) Encyclopedia of plant anatomy, vol 5, pt. 2. Borntraeger, Berlin Stuttgart, 505 pp

Esau K (1977) Anatomy of seed plants, 2nd edn. John Wiley & Sons, New York

Esau K, Cheadle VI, Risley EB (1962) Development of sieve-plate pores. Bot Gaz 123: 233-243

Eschrich W (1953) Beiträge zur Kenntnis der Wundsiebröhrenentwicklung bei *Impatiens Holsti.* Planta 43: 37-74

Evert RF, Deshpande BP, Eichhorn SE (1971) Lateral sieve-area pores in woody dicotyledons. Can J Bot 49: 1509-1515

Feucht W, Schmid PPS, Christ E (1983) Kompatibilität bei *Prunus avium/Prunus cerasus* - Veredlungen während der Verwachsungsphase. I. Die Struktur des Phloems einschließlich der Siebröhren. Gartenbauwissenschaft 48: 45-50

Funk R (1929) Untersuchungen über heteroplastische Transplantationen bei Solanaceen und Cactaceen. Beitr Biol Pflanzen 17: 404-468

Hardham AR, McCully ME (1982 a) Reprogramming of cells following wounding in pea *(Pisum sativum* L.*)* roots. I. Cell division and differentiation of new vascular elements. Protoplasma 112: 143-151

Hardham AR, McCully ME (1982 b) Reprogramming of cells following wounding in pea *(Pisum sativum* L.*)* roots. II. The effects of caffeine and colchicine on the development of new vascular elements. Protoplasma 112: 152-166

Hayward HE, Went FW (1939) Transplantation experiments with peas. II. Bot Gaz 100: 788-801
Herse F (1908) Beiträge zur Kenntnis der histologischen Erscheinungen bei der Veredelung der Obstbäume. Landwirtschaftl Jahrb 37: 71-136
Jeffree CE, Yeoman MM (1983) Development of intercellular connections between opposing cells in a graft union. New Phytol 93: 491-509
Jones MGK (1976) The origin and development of plasmodesmata. In: Gunning BES, Robards AW (eds) Intercellular communication in plants: studies on plasmodesmata. Springer, Berlin Heidelberg New York, pp 81-103
Kollmann R (1975) Sieve element structure in relation to function. In: Aronoff S, Dainty J, Gorham PR, Srivastava LM, Swanson CA (eds) Phloem transport. Plenum, New York, pp 225-242
Kollmann R, Dörr I (1987) Parasitische Blütenpflanzen. Naturwissenschaften 74: 12-21
Kollmann R, Glockmann C (1985) Studies on graft unions. I. Plasmodesmata between cells of plants belonging to different unrelated taxa. Protoplasma 124: 224-235
Kollmann R, Dörr I, Schulz A, Behnke HD (1983) Funktionelle Differenzierung der Assimilatleitbahnen. Ber Dtsch Bot Ges 96: 117-132
Kollmann R, Yang S, Glockmann C (1985) Studies on graft unions. II. Continuous and half plasmodesmata in different regions of the graft interface. Protoplasma 126: 19-29
Krenke NP (1933) Wundkompensation, Transplantation und Chimären bei Pflanzen. Springer, Berlin
Küster E (1925) Pathologische Pflanzenanatomie, 3rd edn. Fischer, Jena
LaMotte CE, Jacobs WP (1962) Quantitative estimation of phloem regeneration in *Coleus* internodes. Stain Tech 37: 63-73
LaMotte CE, Jacobs WP (1963) A role of auxin in phloem regeneration in *Coleus* internodes. Dev Biol 8: 80-98
Lindsay DW, Yeoman MM, Brown R (1974) An analysis of the development of the graft union in *Lycopersicon esculentum*. Ann Bot (London) 38: 639-646
Lipetz J (1970) Wound healing in higher plants. Int Rev Cytol 27: 1-28
McCully ME (1983) Structural aspects of graft development. In: Moore R (ed) Vegetative compatibility responses in plants. Baylor Univ Press, Waco, Texas, pp 71-88
Monzer J, Kollmann R (1986) Vascular connections in the heterograft *Lophophora williamsii* Coult. on *Trichocereus spachianus* Ricc. J Plant Physiol 123: 359-372
Moore R (1982) Graft formation in *Kalanchoe blossfeldiana*. J Exp Bot 33: 533-540
Moore R (1983) Physiological aspects of graft formation. In: Moore R (ed) Vegetative compatibility responses in plants. Baylor Univ Press, Waco, Texas, pp 89-105
Moore R (1984a) Graft formation in *Solanum pennellii* (Solanaceae). Plant Cell Rep 3: 172-175
Moore R (1984b) The role of direct cellular contact in the formation of compatible autografts in *Sedum telephoides*. Ann Bot (London) 54: 127-133
Moore R, Walker DB (1981) Studies of vegetative compatibility - incompatibility in higher plants. I. A structural study of a compatible autograft in *Sedum telephoides* (Crassulaceae). Am J Bot 68: 820-830
Muzik TJ, La Rue CD (1954) Further studies on the grafting of monocotyledons. Am J Bot 41: 448-455
Rachow-Brandt G (1987) Vergleichende transportphysiologische Untersuchungen an Pfropflingen unterschiedlicher Kompatibilität. Thesis, Univ Kiel
Robbertse PJ, McCully ME (1979) Regeneration of vascular tissue in wounded pea roots. Planta 145: 167-173
Schmid PPS, Feucht W (1986) Carbohydrates in the phloem of *Prunus avium/Prunus cerasus* graftings and of homospecific controls. Angew Bot 60: 201-208
Schmid PPS, Bartscherer HC, Feucht W (1984) Ultrastructural localization of polyphenols in the sieve tubes of *Prunus avium* L. by ferric chloride. Sci Hortic 22: 105-111
Schulz A (1986a) Wound phloem in transition to bundle phloem in primary roots of *Pisum sativum* L. I. Development of bundle-leaving wound-sieve tubes. Protoplasma 130: 12-26

Schulz A (1986 b) Wound phloem in transition to bundle phloem in primary roots of *Pisum sativum* L. II. The plasmatic contact between wound-sieve tubes and regular phloem. Protoplasma 130: 27-40

Sharples A, Gunnery H (1933) Callus formation in *Hibiscus Rosa sinensis* L. and *Hevea brasiliensis* Müll. Arg. Ann Bot (London) 47: 837-840

Shimomura T, Fuzihara K (1977) Physiological study of graft union formation in cactus. II. Role of auxin on vascular connection between stock and scion. J Jpn Soc Hortic Sci 45: 397-406

Silberschmidt K (1933) Beiträge zur Kenntnis der Stoffwechselgemeinschaft zwischen Pfropfpartnern. Planta 19: 729-780

Silberschmidt K (1935/36) Die Abhängigkeit des Pfropferfolges von der systematischen Verwandtschaft der Partner. Z Bot 29: 65-137

Simon S (1908) Experimentelle Untersuchungen über die Entstehung von Gefäßverbindungen. Ber Dtsch Bot Ges 26: 364-396

Simon S (1930) Transplantationsversuche zwischen *Solanum melongena* und *Iresine Lindeni.* Jahrb Wiss Bot 72: 137-160

Stoddard FL, McCully ME (1979) Histology of the development of the graft union in pea roots. Can J Bot 57: 1486-1501

Stoddard FL, McCully ME (1980) Effects of excision of stock and scion organs on the formation of the graft union in *Coleus:* a histological study. Bot Gaz 141: 401-412

Tiedemann R (1986) Struktur und Funktion der Phloemverbindungen zwischen Reis und Unterlage beim Pfropfling *Cucumis sativus* auf *Cucurbita ficifolia.* Thesis, Univ Kiel

Tiedemann R (1989) Graft union development and symplastic phloem contact in the heterograft *Cucumis sativus* on *Cucurbita ficifolia.* J Plant Physiol 134: 427-440

Thompson NP (1967) The time course of sieve tube and xylem cell regeneration and their anatomical orientation in *Coleus.* Am J Bot 54: 588-595

Thompson NP, Jacobs WP (1966) Polarity of IAA effect on sieve-tube and xylem regeneration in *Coleus* and tomato stems. Plant Physiol 41: 673-682

Vöchting H (1892) Über Transplantation am Pflanzenkörper. Untersuchungen zur Physiologie und Pathologie. Laupp, Tübingen

von Kaan-Albest A (1934) Anatomische und physiologische Untersuchungen über die Entstehung von Siebröhrenverbindungen. Z Bot 27: 1-94

Warren Wilson J, Warren Wilson PM (1981) The position of cambia regenerating in grafts between stems and abnormally-oriented petioles. Ann Bot (London) 47: 473-484

Wetmore RH, Rier JP (1963) Experimental induction of vascular tissues in callus of angiosperms. Am J Bot 50: 418-430

Yeoman MM (1984) Cellular recognition systems in grafting. In: Pirson A, Zimmermann MH (eds) Encyclopedia of plant physiology, vol 11. Cellular interactions. NS. Springer, Berlin Heidelberg New York, pp 453-472

Yeoman MM, Kilpatrick DC, Miedzybrodzka MB, Gould AR (1978) Cellular interactions during graft formation in plants; a recognition phenomenon? Symp Soc Exp Biol 32: 139-160

Zee SY (1969) Fine structure of the differentiating sieve elements of *Vicia faba.* Aust J Bot 17: 441-456

Note

Unpublished results of the authors have been included in the present paper.

12 Sieve Elements in Haustoria of Parasitic Angiosperms[1]

INGE DÖRR

12.1 Introduction

The necessity for conducting elements within haustoria of phanerogamic parasites is obvious, especially since holoparasitic angiosperms lack roots and assimilatory organs. Water as well as organic substances must be taken up, therefore, by specialized haustoria (Kuijt 1969; Kollmann and Dörr 1987). The faster the parasite grows, the more efficient must be the cells developed specifically for nutrient transport.

The tremendously rapid rate of spreading of *Cuscuta* is well known. Accordingly, the *Cuscuta* haustoria have, of the parasites studied to date, the most elaborate haustorial transport system, especially with respect to the extended phloem. This report will mainly focus, therefore, on the haustorial phloem of *Cuscuta odorata*, with a comparison of *Cuscuta* to other parasitic angiosperms in Sects. 12.5 and 12.6.

The cellular systems for water and assimilate transport within the haustoria of *Cuscuta* were discovered at quite different times. Xylem was already described in 1827 by von Mohl, while the haustorial phloem was demonstrated over half a century later by Peirce (1893) and by Thoday (1911). Both the latter scientists presented excellent drawings and descriptions of sieve tubes within the haustorium (Fig. 12.3). Their observations were confirmed by the electron microscopical investigations of Israel et al. (1980). The physiological function of this phloem was shown by Schumacher (1934), who - while using the phloem-mobile marker K-fluorescein - indirectly demonstrated the uptake of translocates from *Pelargonium* and their transport in the parasite. Bennett (1944) further demonstrated phloem function in studies on the transmission of plant viruses by dodder.

12.2 Phloem in the Haustorium of *Cuscuta*

A haustorial complex arises immediately after the *Cuscuta* shoot has twined around organs of the host. Parts of the *Cuscuta* cortex, adjacent to the host surface, develop into the structure called the prehaustorium (Fig. 12.1), an organ

[1] Dedicated to my mother

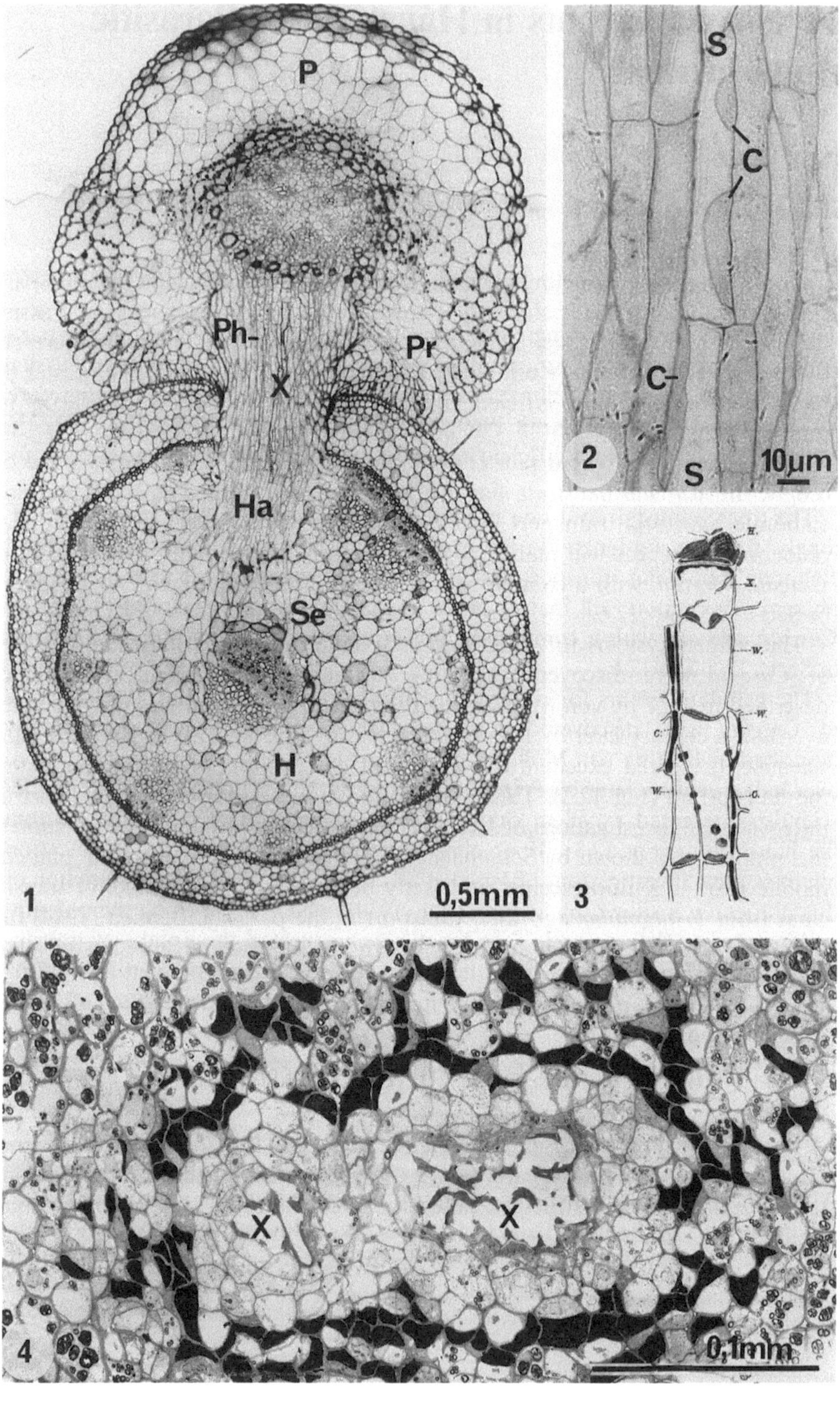
P
Ph-
X
Pr
Ha
Se
H
1
0,5mm
S
C
C
S
2
10µm
3
X
X
4
0,1mm

which connects both plants intimately. The haustorium itself is an endogenously arising organ (Lee 1986) which develops from a disk-like meristem present next to the vascular bundles of the *Cuscuta* shoot. After having penetrated the prehaustorium, the young haustorium enters the host tissue. This cone-like tip of the haustorium stops its growth at a certain depth. Its superficial cells, referred to as "searching hyphae" (Dörr 1968, 1969), proliferate in all directions and establish contact with the vascular bundles of the host. From a phylogenetic point of view it is, even up to now, unclear whether the haustorium represents a modified adventitious root or a unique organ. The internal anatomy shown by a longitudinal section (Fig. 12.1) exhibits some similarities to the histology of roots. Clearly visible is a central xylem which connects the xylem of the *Cuscuta* axis with that of the host. On both sides of the xylem strands sieve elements can be detected. They consist of relatively short cells (Fig. 12.2) separated by transversely orientated sieve plates. One or more typical companion cells, arising from an unequal division and showing dense cytoplasm, are associated with each element (Fig. 12.2). The similarity of modern observations of haustorial sieve elements to the drawing of Thoday (1911) is striking (Fig. 12.3). Near the *Cuscuta* shoot the haustorial phloem elements are extremely short. Isodiametrically shaped sieve elements finally are situated immediately next to the axial sieve elements of the vascular cylinder of the *Cuscuta* shoot so that a phloem continuity in this direction is guaranteed. At the tip of the haustorium a phloem bridge to the assimilate conducting system of the host is achieved by the structures referred to as "absorbing hyphae" (see Sect. 12.4). An uninterrupted phloem connection (Fig. 12.12) between host and parasite, a point of discussion over decades, can be demonstrated by aniline blue fluorescence, revealing a series of fluorescent areas throughout the entire haustorium (Israel et al. 1980).

A transverse section through a haustorium demonstrates the quantity and distribution of phloem elements (Fig. 12.4). A 2-week-old haustorium on a *Primula* petiole, cut transversely at the point of entrance into the host tissue, shows two groups of about 20 xylem elements together in the center of the eliptically shaped haustorium. These elements are surrounded by a large quantity of sieve tubes (about 120 elements in the section depicted in Fig. 12.4 were detected by

◁

Fig. 12.1. Light microscopical survey of a longitudinal section through the haustorium *(Ha)* of the holoparasite *Cuscuta odorata* situated within the petiole of the host *(H) Pelargonium zonale.* The parasitic shoot *(P)* and the host are cut transversely. The prehaustorium *(Pr)* represents the adhesive organ. Xylem *(X)* occurs in the center of the haustorium; *Ph* shows the position of phloem. Searching hyphae *(Se)* traverse the host tissue, occasionally getting contact to the vascular bundles of the host. **Fig. 12.2.** Light micrograph of a longitudinal section of sieve elements *(S)* and companion cells *(C)* within the haustorium of *Cuscuta.* **Fig. 12.3.** Drawing of haustorial sieve elements and companion cells of *Cuscuta* by Thoday in 1911. **Fig. 12.4.** Electron microscopical survey of the central part of a transversely sectioned haustorium *(Cuscuta odorata)* at the point of entrance into the host tissue *(Primula obconica).* The haustorial vascular system is composed of a central xylem *(X),* surrounded by many sieve elements (120 could be detected). Sieve elements are colored *black* for a better demonstration

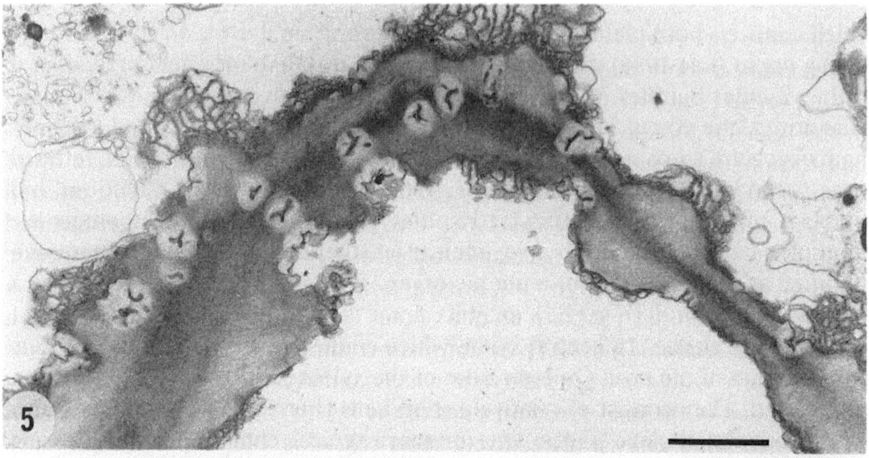
5

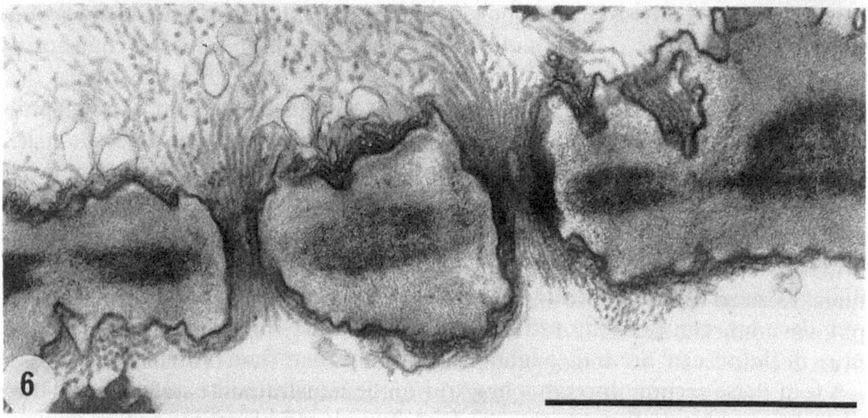
6

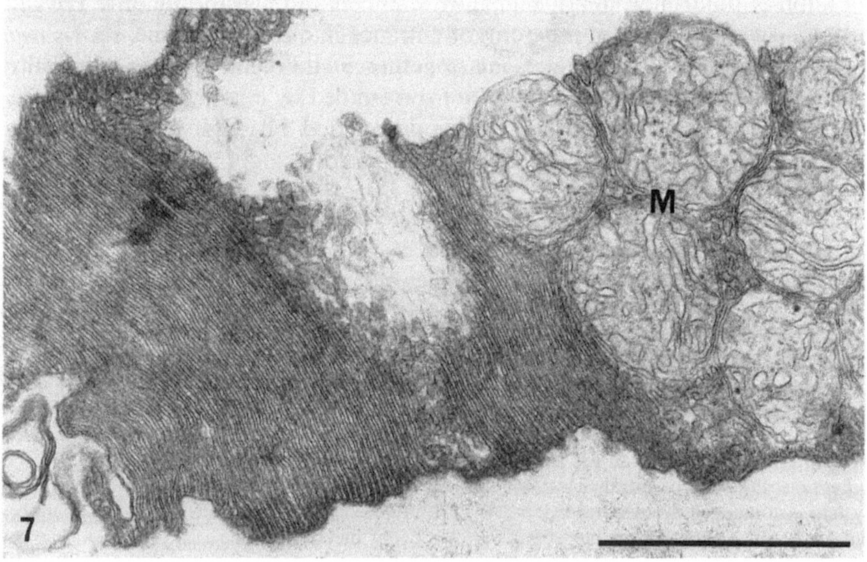
M
7

electron microscopy). This illustrates that far more assimilate conducting elements are present than previously had been assumed. With respect to their fine-structure, the sieve elements within the haustorium show no striking differences from the usual features of angiosperm sieve tubes. An exception to this, however, is seen in the sieve pores; their distribution is more irregular than in normal sieve plates (Figs. 12.5, 12.8). Sieve elements and pores contain P-protein of a tubular nature (Figs. 12.6, 12.15). The walls of the sieve elements are nearly totally covered by sieve-element reticulum (SER), consisting of stacked cisternae (Sjolund and Shih 1983), oriented vertically toward the wall. The distance between the cisternae is very constant (Fig. 12.7). *Cuscuta* has S-type plastids (Behnke 1981).

12.3 Development of Haustorial Sieve Elements

The most conspicuous difference in the development of haustorial sieve elements in *Cuscuta,* in comparison to those of normal plants, is their origin from ordinary parenchyma cells. There are no procambial strands within the haustorium which could serve as precursors for the conducting elements. Being derived from haustorial parenchyma cells, the resulting sieve elements consequently are extremely short, often nearly isodiametrically shaped (Figs. 12.8, 12.10). A prerequisite for the formation of these assimilate conducting elements is at least one cell division, the daughter cells developing into a sieve element and a companion cell (Fig. 12.2). More divisions are possible, in which case phloem parenchyma cells also result (Figs. 12.8, 12.10). Moreover, several sieve tube members, with companion cells and parenchyma cells, may be derived from one parenchyma cell if repeated divisions occur. In any case, the shape of the mother cell is still recognizable (Figs. 12.8, 12.10). This manner of differentiation was also observed (Fig. 12.11) in the haustorium of *Alectra* (Dörr et al. 1979), indicating that this mode of development is common. Identical phloem development has been reported in the regeneration processes occurring during wound phloem development (see Chap. 10) as well as in graft connections (see Chap. 11). The similarity of the origin of these artificially induced assimilate conducting elements to those in haustoria is striking (Kollmann et al. 1983). In contrast, in *Cassytha,* the sieve elements occur only in the upper part (Fig. 12.19) of the haustorium and they originate from a cambial tissue (Damaschke 1982).

◁

Fig. 12.5. Haustorial sieve plate with irregularly distributed sieve pores. **Fig. 12.6.** Sieve pores within the *Cuscuta* haustorium in longitudinal view traversed by P-protein tubules. **Fig. 12.7.** Glancing section through the parietal cytoplasm of a haustorial sieve element. SER cisternae orientated vertically toward the walls are typical. *M* sieve element mitchondria; *Bar lines* = 1 µm

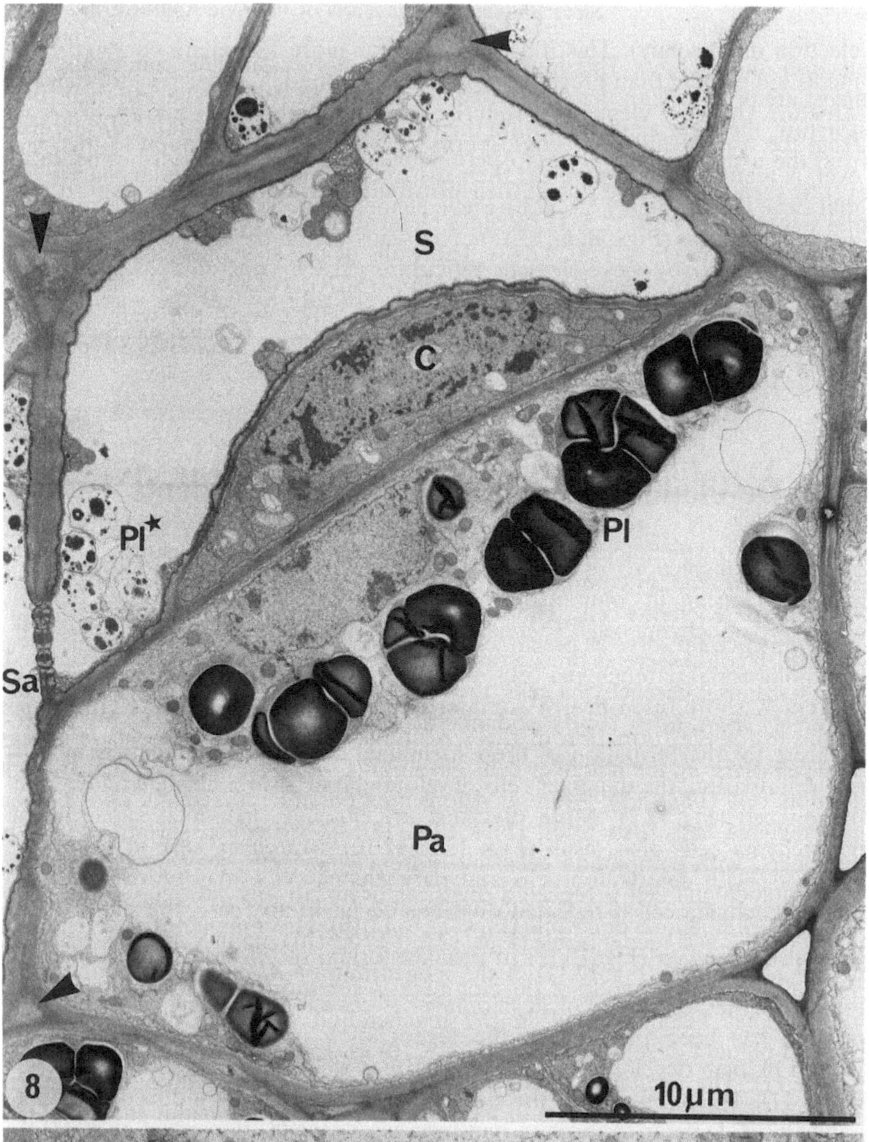
S
C
Pl
Pl
Sa
Pa
8
10µm

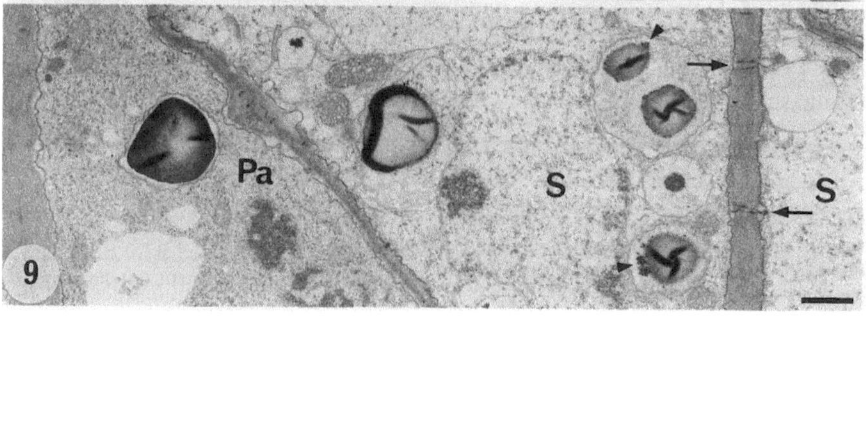
Pa
S
S
9

Intercellular spaces, normally absent in phloem, are found in *Cuscuta* around the newly differentiated sieve tubes. Here they become secondarily filled with wall material, while they remain unaffected around the other division products (Figs. 12.8, 12.10).

The sieve pores are, as previously pointed out, irregularly distributed (Fig. 12.8). This is understandable because they are derived from the plasmodesmata of the mother parenchyma cell, which are aggregated in primary pit fields. This too is in complete accordance with sieve pore development of wound phloem (Schulz 1987) and phloem from graft unions. The formation of sieve pores in the parasite is identical to normal sieve pore development and is characterized by the appearance of small amounts of callose surrounding the plasmodesmata (Fig. 12.9). This callose deposition is helpful in identifying the prospective sieve element at an early stage.

The first indication that a haustorial parenchyma cell is turning into a sieve element is the appearance of typical sieve-tube starch within the plastids (Fig. 12.9). After cell division the straight edges of the amyloplast starch grains take on a more fluffy appearance and small star-shaped starch particles, typical for sieve elements of *Cuscuta,* appear. The process of starch conversion sometimes does not go to completion and residual amyloplast starch grains may remain in the plastids of fully differentiated sieve elements.

The direction of sieve-element differentiation within the haustorium progresses from that part of the haustorium nearest the *Cuscuta* shoot toward the haustorial tip (Dörr 1987). Four days after twining around the host the initial stages of division are observed within the young haustorium, near the stele of the *Cuscuta* shoot. At the same time the first sieve elements are detected. It is also interesting to note that at this time there is not yet any contact established with the conducting elements of the host. This contact does not exist earlier than 6 days after the *Cuscuta* shoot has wrapped itself around the host. It can be concluded that the differentiation of haustorial sieve elements is not correlated with or induced by the host sieve-tube contact. This was confirmed by an investigation of haustoria that developed as free cones without any host tissue. Here, too, haustorial sieve elements are found. It would seem more likely that the growing haustorial tip, acting as a sink and requiring assimilates, induces the differentiation of sieve elements (Dörr, unpublished results).

Fig. 12.8. Haustorial cell of *Cuscuta* divided into a sieve element *(S)*, a companion cell *(C)* and a parenchyma cell *(Pa)*. Correspondingly the elements contain either plastids with sieve tube starch *(Pl*)* or amyloplasts *(Pl)*. The sieve area *(Sa)* is derived from a primary pit field. Intercellular spaces adjacent to the sieve elements *(arrowhead)* are filled, while the others are open. **Fig. 12.9.** Detail of a haustorial parenchyma cell *(Pa)* and two very young sieve elements *(S)*. The starting point of the development of the latter is the appearance of star-like sieve tube starch *(arrowhead)* - often near the amyloplast starch and the callose deposits at the ends of the plasmodesmata *(arrows)*, indicating the future sieve pores; *Bar line*=1 µm

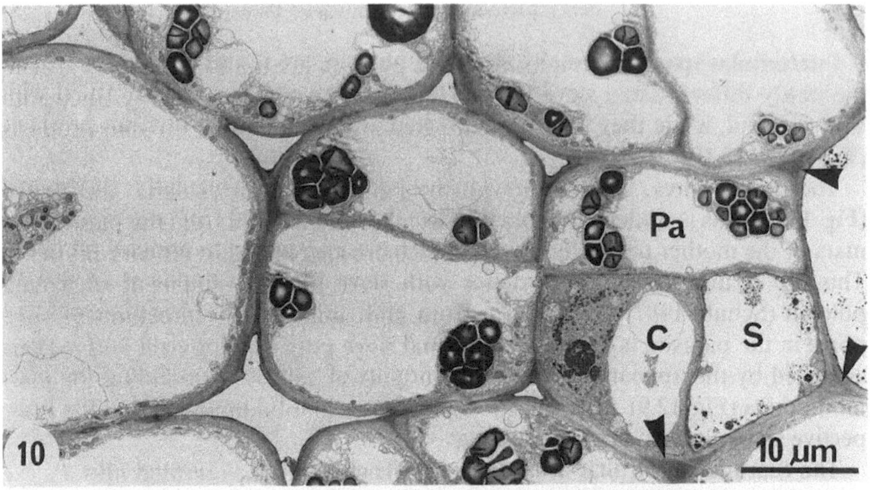

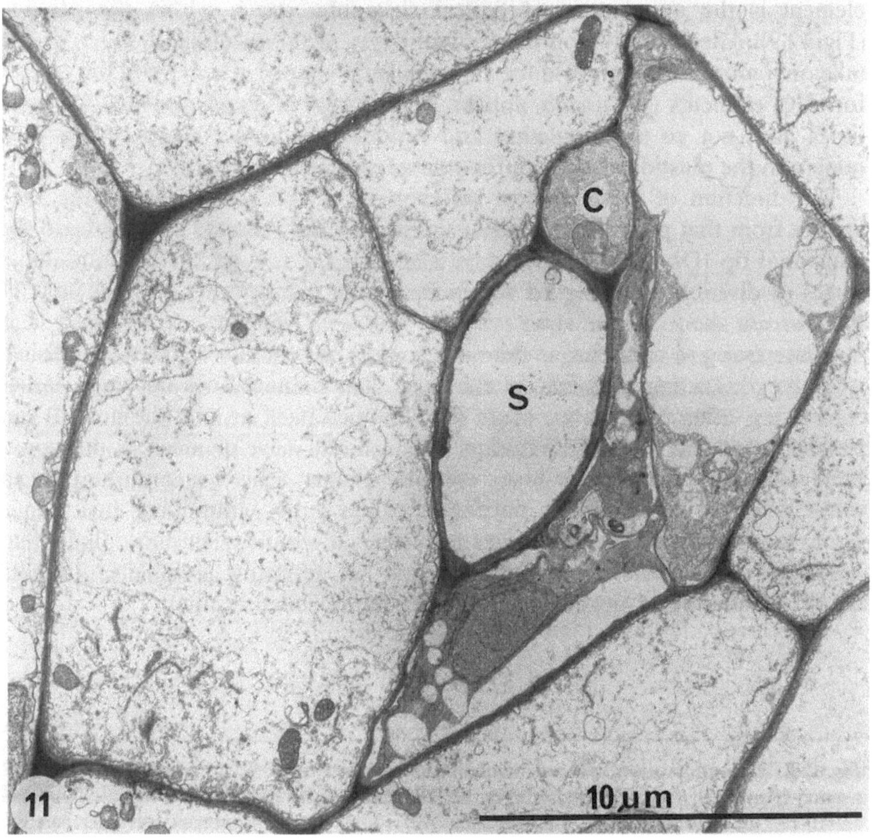

Fig. 12.10. Electron micrograph of transversely sectioned haustorial parenchyma cells of *Cuscuta*. The daughter cells derived from two divisions develop into a sieve element *(S)*, a companion cell *(C)*, and a parenchyma cell *(Pa)*. Intercellular spaces are filled *(arrowhead)* around the sieve element. **Fig. 12.11.** Within the haustorium of *Alectra vogelii* a parenchyma cell has developed (after several divisions) into a sieve element *(S)* and a neighboring companion cell *(C)*, both situated within the area of the old mother cell

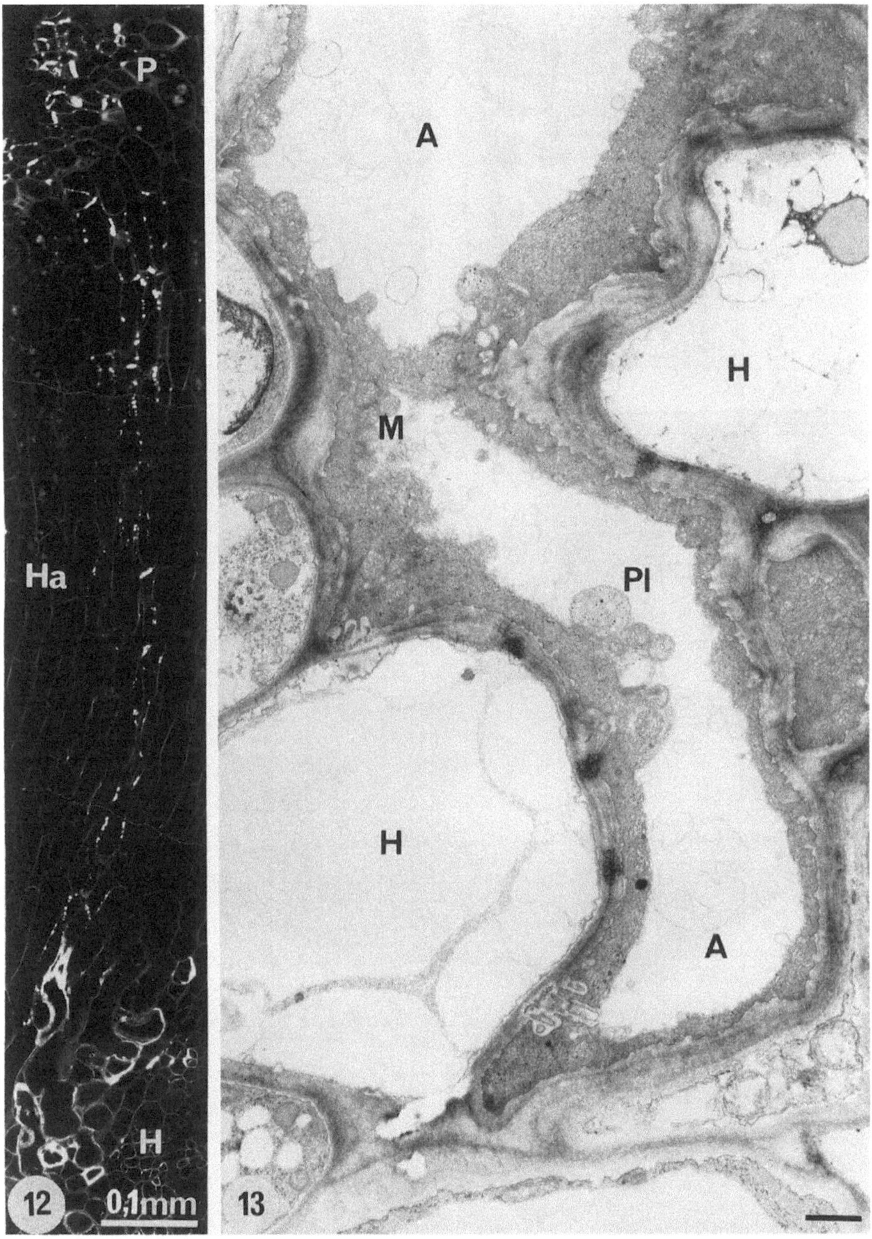

Fig. 12.12. Fluorescence micrograph of a longitudinal section through the *Cuscuta* haustorium *(Ha)*. Bright callose deposits mark the continuity of the sieve elements from the transversely sectioned sieve tubes of the parasite's shoot *(P)* to those of the host tissue *(H)* (Israel et al. 1980). **Fig. 12.13.** Part of the very extended "absorbing" hypha *(A)* situated within the host tissue *(H)*. The parietal protoplast consists of SER, sieve tube plastids *(Pl)* and mitochrondria *(M)*; *Bar line*= 1 μm

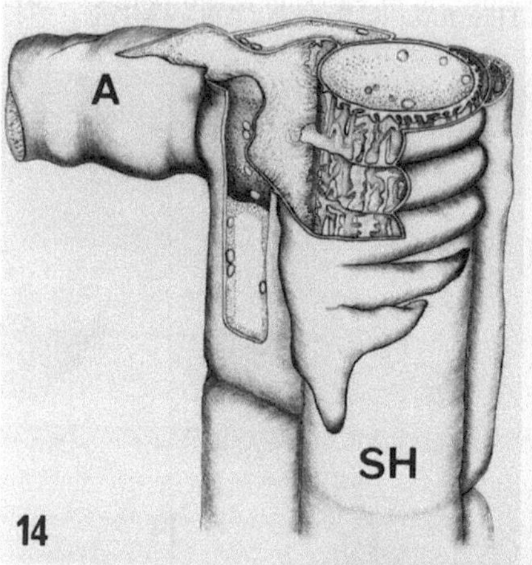

14

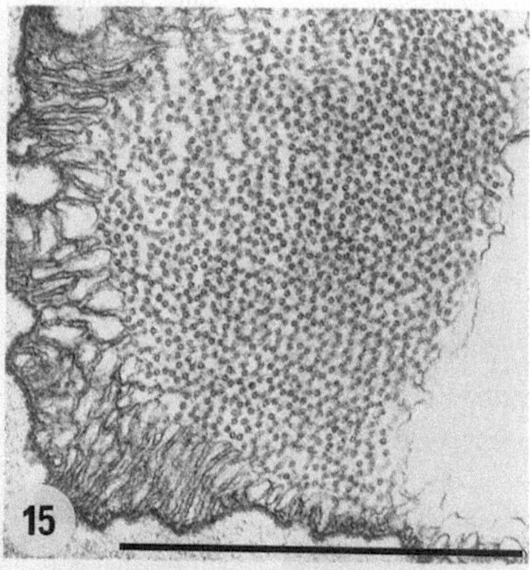
15

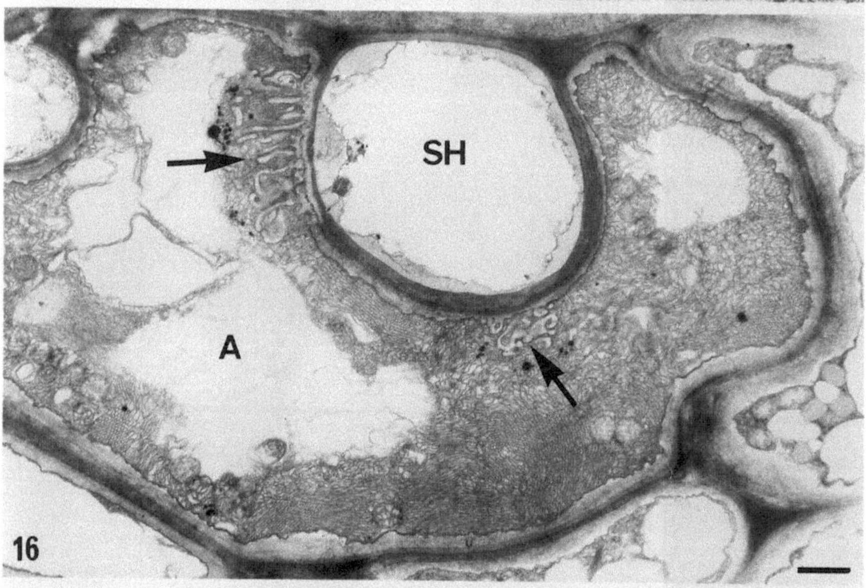

16

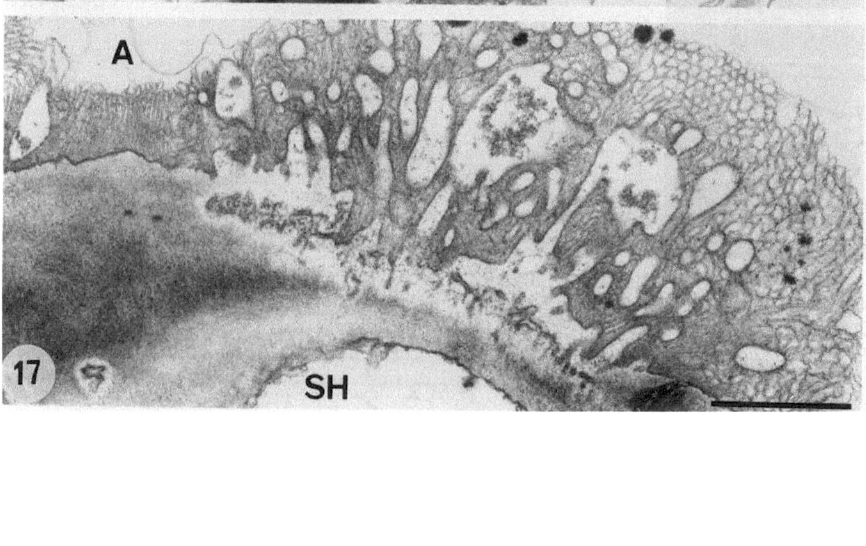

17

12.4 The Contact Hypha of *Cuscuta*

The "absorbing" hypha represents one of the most peculiar and highly differentiated cell types within the plant kingdom. It derives from the so-called "searching" hypha, which switches to assimilate transport upon coming into contact with a sieve element of the host. The "searching" hypha itself is already highly specialized. When leaving the haustorial cone it grows unicellularly, without any cell divisions, like a pollen tube, throughout the host tissue. We observed a total length of these cells of up to 600 μm. An enzymatic ability to dissolve host walls was shown by Nagar et al. (1984). Vascular bundles of the host are reached purely by chance. Within the phloem tissue of the host, growth occurs only in the middle lamellae of the host walls and the tip of the hypha branches repeatedly. A sieve-tube connection to the host is established by finger-like protrusions of the cell which surrounds (Figs. 12.14, 12.16) the foreign, conducting element (Dörr 1972). The conversion of the extended "searching" hypha into an "absorbing" hypha starts at the point of host sieve-tube attachment and advances toward the haustorial organ. So far, we do not know when and where the haustorial sieve elements, differentiating in the opposite direction, join with this system.

The cellular transformation starts with the development of a conspicuous wall labyrinth which greatly enlarges the internal surface of that part of the wall which is common with the host sieve element (Fig. 12.16, 12.17). There are no sieve pores or plasmodesmata in this region so that the uptake of assimilates occurs through the apoplast (Kollmann and Dörr 1969). Simultaneous to the elaboration of wall protuberances, the SER becomes visible within the adjacent cytoplasm. The SER, appearing in extended flat cisternae, often stacked vertically toward the wall, is the dominating structure within the "absorbing" hypha, often totally filling the branched cell tip (Fig. 12.16). The quantity and extent of the SER seen here is much greater than seen in most other plant cells, but its precise function is still unknown. Additional ultrastructural characteristics are identical with those of normal sieve-element development, including the appearance of P-protein, often in large aggregates of protein tubules (Fig. 12.15), the degeneration of the nucleus and the tonoplast, the disappearance of ribosomes, and the typical modification of mitochondria and plastids.

Figs. 12.14.-12.17. "Absorbing" hyphae of *Cuscuta odorata*. **Fig. 12.14.** Schematic drawing of an "absorbing" hypha *(A)* clasping with fingerlike protrusions around a host sieve tube *(SH)*. The common wall shows labyrinthic structures, orientated toward the lumen of the hypha (Dörr 1972). **Fig. 12.15.** Aggregate of tubular P-protein within an "absorbing" hypha. **Fig. 12.16.** Transverse section of a host sieve element *(SH)* surrounded by a longitudinally cut protrusion of the "absorbing" hypha *(A)*. Wall labyrinths *(arrows)* enlarge the surface toward the parasite. SER nearly fills the parasitic cell tip. **Fig. 12.17.** Typical wall labyrinth of the parasitic "absorbing" hypha *(A)* enlarging the wall which is in common with the host sieve tube *(SH)*; *Bar lines* = 1 μm.

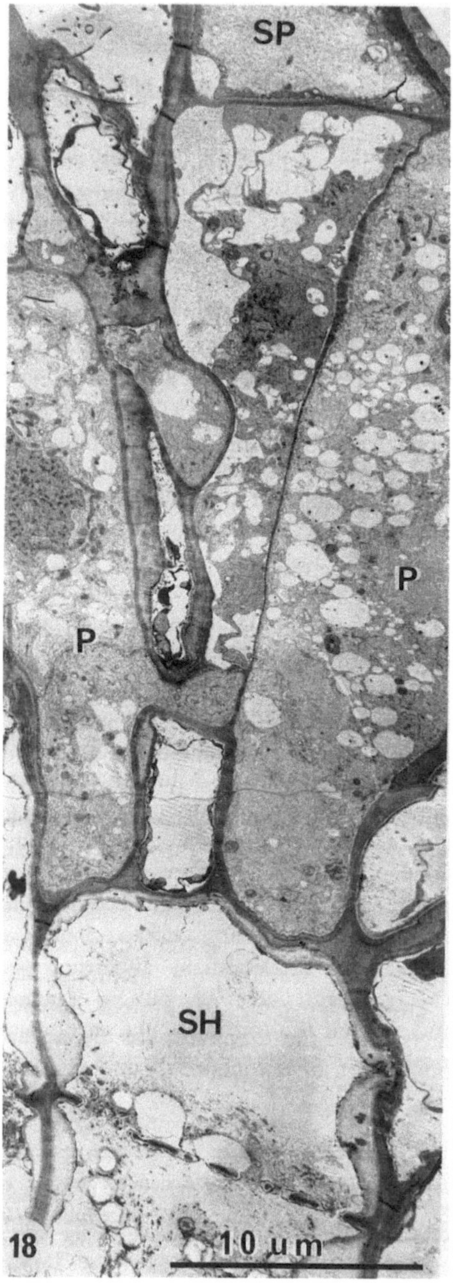
SP
P
P
SH
18
10 um

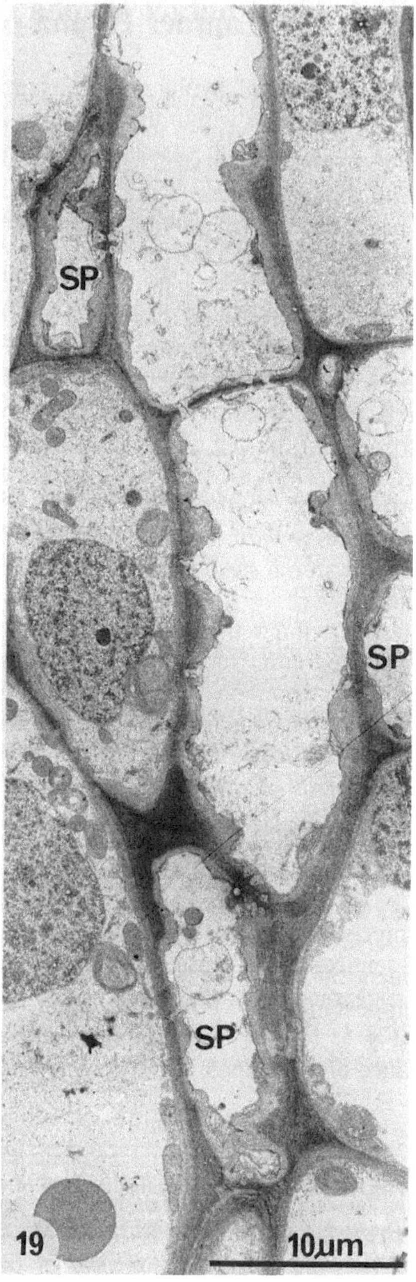
SP
SP
SP
19
10μm

The latter occasionally still contain amyloplast starch besides sieve-tube starch, as previously described for the elements within the haustorium.

A striking difference compared to normal phloem is the absence of a companion cell aside the extremely long "absorbing" hypha (Fig. 12.13). One reason for this might be the previously mentioned lack of cell divisions in the "searching" hypha. Companion cells may also be absent in the protophloem of roots and shoots (Esau 1969) and in the middle of a nodal ganglion (Behnke 1965). Kuijt et al. (1985) reported a lack of companion cells in the endophytic system of *Pilostyles* (Rafflesiaceae).

12.5 Phloem in the Haustoria of Different Parasitic Plants

Studies on the phloem in the haustoria of phanerogamic parasites are quite rare, possibly because their unusual manner of differentiation results in the fact that they cannot easily be recognized by their shape, but must be examined at the ultrastructural level (see Kuijt and Toth 1976).

It is interesting that the heterotrophic parasite *Lathraea* (Scrophulariaceae) has no haustorial phloem, and that it takes up organic substances from the host's bleeding sap during the spring through haustorial xylem elements (Ziegler 1955). In recognition of this special, dual function for the xylem, the term "phloeotracheid" was created by Benson (1910) and seemed justified by studies showing protein and/or starch contents in them, until Fineran et al. (1978) found that these cells are dead at maturity and therefore renamed them graniferous tracheary elements (see also Fineran 1985).

Within the family of Scrophulariaceae *Orobanche* is a plant that lives holoparasitically. Its secondary haustoria have sieve elements which differentiate to within one cell of those of the host. This final cell of the parasite, attached to the sieve tube of the host, is of parenchymatous nature (Fig. 12.18; Dörr and Kollmann 1975). The scrophulariacean hemiparasite *Alectra* develops sieve elements within the primary haustorium which come close to the host phloem, but both systems are always separated by a parenchymatous bridge (Fig. 12.20) consisting of at least eight cells which are, in contrast to *Orobanche*, highly vacuolated (Dörr et al. 1979). Studies on the phloem in *Striga* indicated that the situation was similar to that in *Alectra* (Visser and Dörr 1987; see also Ba 1983).

Fig. 12.18. Contact cells of the parasite *Orobanche ramosa (P)* with highly structured protoplasts next to the longitudinally cut sieve tubes *(SH)* of the host *Cannabis sativa* with sieve plates. A sieve element of *Orobanche (SP)* is situated next to the contact cell (Dörr and Kollmann 1975). **Fig. 12.19.** Longitudinal section through the haustorium of the hemiparasite *Cassytha ciliolata* on *Hibiscus rosa sinensis*. Depicted is the end of the haustorial phloem situated in the upper part of the haustorium (see Fig. 12.21). The final *Cassytha* sieve elements *(SP)* are extremely short (Micrograph: U. Damaschke)

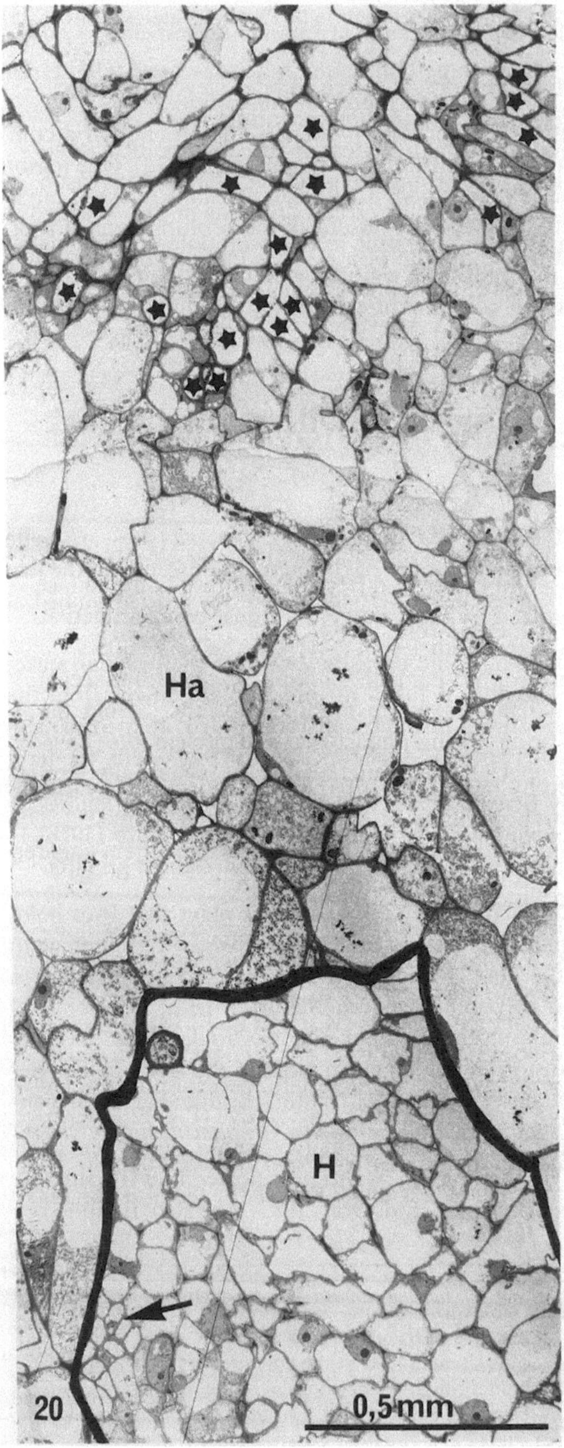
Ha
H
0,5 mm
20

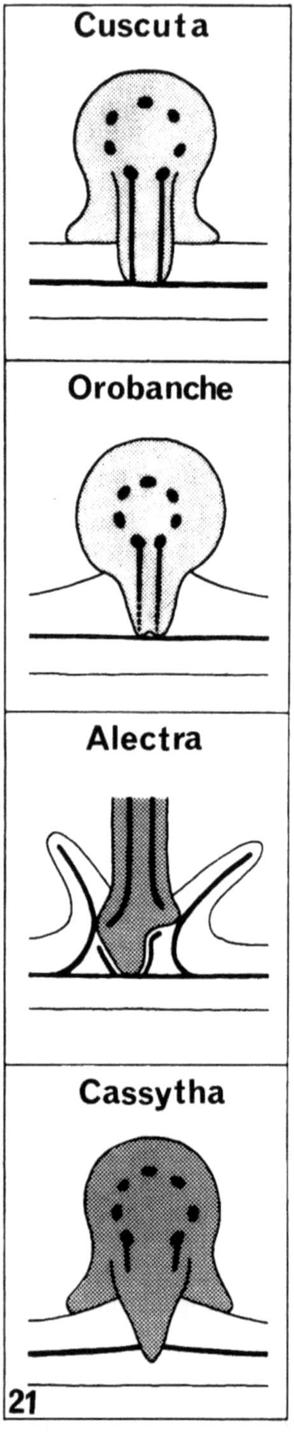
Cuscuta
Orobanche
Alectra
Cassytha
21

Within the haustoria of *Castilleja* (hemiparasitic Scrophulariaceae) phloem elements were recognized by Kuijt and Dobbins (1971). These elements were only found in old haustoria, possibly only during self-parasitism (the host root could not be identified; Dobbins and Kuijt 1973).

This situation is also of interest because of the studies on the haustoria of *Cassytha,* a hemiparasitic member of the Lauraceae (Damaschke 1982). These haustoria show a very reduced phloem system, ending with short sieve elements in the external part of the haustorium. During self-parasitism, however, a continuous phloem system is present throughout the haustorium, fusing with the phloem of the shoot.

The Loranthaceae family apparently has the largest number of species which have been investigated for the occurrence of phloem within haustoria. The most elaborate system is reported for *Phoradendron* by Calvin (1967). He observed by light microscopy some cells with sieve areas which were located only a few cells away from the host phloem. Similar results are described by Sallé (1976, 1979) for the cortical strands of *Viscum album.* Menzies (1954) referred to indistinct phloem in young cortical strands in *Loranthus micranthus.* The holoparasite *Tristerix aphyllus* (Loranthaceae) showed typical phloem elements within multiseriate strands derived from the haustorium (Mauseth et al. 1984, 1985).

12.6 Comparative Aspects

It is not yet possible to make any final conclusions about phloem development within angiosperm haustoria. More structural studies, especially electron microscopical investigations, are needed. This research, based on a plant systematic background would be most informative, but additional, physiological experiments are also needed to determine the amount of transported material in order to develop a clear understanding of parasitism.

In general, it is known that strong sinks require conducting elements for a supply of assimilates. These elements are less necessary to supply weak sinks, since transport via parenchyma cells may be sufficient.

Fig. 12.20. Electron micrograph of a survey of a longitudinal section through the haustorium *(Ha)* of the hemiparasite *Alectra vogelii* on *Voandzeia subterranea (H).* Sieve elements of the parasite are indicated by *asterisks,* those of the host root by an *arrow.* The host-parasite interface is accentuated by the *dark line* drawn on the micrograph. **Fig. 12.21.** Schematic drawing of the occurence of phloem between the holoparasites *Cuscuta* and *Orobanche (light grey)* and the hemiparasites *Alectra* and *Cassytha (dark grey)* and their hosts *(colorless).* The phloem of host and parasite is indicated by *broad black lines. Cuscuta,* a fast growing holoparasite, shows complete phloem connection to its host. *Orobanche,* growing more slowly, shows a haustorial phloem separated from the host by at least one contact cell (symbolized by the *dotted lines*). *Alectra,* living holoparasitically when subterraneous, and hemiparasitically when green, shows several parenchyma cells beween both phloem systems. (*Alectra* induces lateral host roots.) *Cassytha,* throughout its life a hemiparasite, lacks phloem within the inner part of the haustorium

A comparison between four different holo- and hemiparasitically living angiosperms may substantiate this principle (Fig. 12.21). The holoparasite *Cuscuta* represents a strong sink because of its constant growth rate. Its phloem system is continuous throughout the haustorium. The root parasite *Orobanche,* living as a holoparasite, shows a quite different mode of growth. The young tubercle becomes loaded with starch slowly, and the starch is quickly mobilized later on when the shoot is emerging. It is possible that the phloem system between both partners is not nearly as complete as in *Cuscuta*. *Alectra,* a hemiparasite when above ground, is holoparasitic when developing subterraneously. Both stages need to be investigated. *Alectra,* when green and partially autotrophic, has a parenchymatous bridge of several cells between both phloem systems. This might reflect a reduced rate of transport from the host to the parasite (Gouws et al. 1980). *Cassytha* is independent from the host plant throughout its life in terms of assimilates. Accordingly, the haustorial phloem ends rudimentarily in the upper part of the haustorium (Damaschke 1982).

The relationship between the degree of contact the parasite has with the host phloem and the growth rate of the parasite suggest that sink strength may be a factor in determining the kind of haustorial connections that are formed. It should be noted, however, that this concept of "sink strength" may also be applied to the tissues of the haustorium itself and should not be restricted only to other parts of the parasite. The haustorium, which often develops a large mass of tissue, also needs a supply of organic nutrients, and this organ itself may play a role in phloem induction.

Acknowledgment. I am greatly indebted to Christl Glockmann for her help at all times. I thank Margarete Rasch for permanent positive assistance, Jan Monzer for the preparation of the drawing for Fig. 12.21, J. H. Visser (Stellenbosch) for kindly reviewing the English text, Johannes Kornmacher and Claus Clausen for excellent plant cultivation.

References

Ba AT (1983) Biologie du parasitisme chez deux Scrophulariacées tropicales. Thesis, Univ Dakar

Behnke H-D (1965) Über das Phloem der Dioscoreaceen unter besonderer Berücksichtigung ihrer Phloembecken, II. Mitteilung: Elektronenoptische Untersuchungen zur Feinstruktur des Phloembeckens. Z Pflanzenphysiol 53: 214-244

Behnke H-D (1981) Sieve-element characters. Nord J Bot 1: 381-400

Bennett CW (1944) Studies of dodder transmission of plant viruses. Phytopathology 34: 905-932

Benson M (1910) Root parasitism in *Exocarpus* (with comparative notes on the haustoria of *Thesium*). Ann Bot (London) 24: 667-677

Calvin CL (1967) Anatomy of the endophytic system of the mistletoe *Phoradendron flavescens.* Bot Gaz 128: 117-137

Damaschke U (1982) Strukturelle und physiologische Untersuchungen am Stoffaufnahme-System des Hemiparasiten *Cassytha*. Thesis, Christian-Albrechts-Univ Kiel

Dobbins DR, Kuijt J (1973) Studies on the haustoria of *Castilleja* (Scrophulariaceae). II. The endophyte. Can J Bot 51: 923-931

Dörr I (1968) Zur Lokalisation von Zellkontakten zwischen *Cuscuta odorata* und verschiedenen höheren Wirtspflanzen. Protoplasma 65: 435-448

Dörr I (1969) Feinstruktur intrazellulär wachsender *Cuscuta*-Hyphen. Protoplasma 67: 123-137

Dörr I (1972) Der Anschluß der *Cuscuta*-Hyphen an die Siebröhren ihrer Wirtspflanzen. Protoplasma 75: 167-184

Dörr I (1987) The haustorium of *Cuscuta* - new structural results. In: Weber HC, Forstreuter W (eds) Proc 4th Int Symp Parasitic flowering plants, Marburg, FRG, pp 163-170

Dörr I, Kollmann R (1975) Strukturelle Grundlagen des Parasitismus bei *Orobanche*. II. Die Differenzierung der Assimilat-Leitungsbahnen im Haustorialgewebe. Protoplasma 83: 185-199

Dörr I, Visser JH, Kollmann R (1979) On the parasitism of *Alectra vogelii* Benth. (Scrophulariaceae) III. The occurrence of phloem between host and parasite. Z Pflanzenphysiol 94: 427-439

Esau K (1969) The phloem. In: Zimmermann W, Ozenda P, Wulff HD (eds) Encyclopedia of plant anatomy, vol 5, pt. 2. Borntraeger, Berlin Stuttgart, 505 pp

Fineran BA (1985) Graniferous tracheary elements in haustoria of root parasitic angiosperms. Bot Rev 51: 389-441

Fineran BA, Juniper BE, Bullock S (1978) Graniferous tracheary elements in the haustorium of the Santalaceae. Planta 141: 29-32

Gouws J, Visser JH, Grobbelaar N (1980) Some aspects of the bidirectional translocation of ^{14}C-labelled metabolites between *Alectra vogelii* Benth. and *Voandzeia subterranea* (L.) Thou. Z Pflanzenphysiol 99: 225-233

Israel S, Dörr I, Kollmann R (1980) Das Phloem der Haustorien von *Cuscuta*. Protoplasma 103: 309-321

Kollmann R, Dörr I (1969) Strukturelle Grundlagen des zwischenzelligen Stoffaustausches. Ber Dtsch Bot Ges 82: 415-425

Kollmann R, Dörr I (1987) Parasitische Blütenpflanzen. Naturwissenschaften 74: 12-21

Kollmann R, Dörr I, Schulz A, Behnke H-D (1983) Funktionelle Differenzierung der Assimilatleitbahnen. Ber Dtsch Bot Ges 96: 117-132

Kuijt J (1969) The biology of parasitic flowering plants. Univ Cal Press, Berkeley Los Angeles

Kuijt J, Dobbins DR (1971) Phloem in the haustorium of *Castilleja* (Scrophulariaceae). Can J Bot 49: 1735-1736

Kuijt J, Toth R (1976) Ultrastructure of angiosperm haustoria - a review. Ann Bot (London) 40: 1121-1130

Kuijt J, Bray D, Olson AR (1985) Anatomy and ultrastructure of the endophytic system of *Pilostyles thurberi* (Rafflesiaceae). Can J Bot 63: 1231-1240

Lee KB (1986) Studies on the haustorial development of *Cuscuta australis* R. Brown. Thesis, Sung Kyun Kwan Univ, Seoul

Mauseth JD, Montenegro G, Walckowiak AM (1984) Studies of the holoparasite *Tristerix aphyllus* (Loranthaceae) infecting *Trichocereus chilensis* (Cactaceae). Can J Bot 62: 847-857

Mauseth JD, Montenegro G, Walckowiak AM (1985) Host infection and flower formation by the parasite *Tristerix aphyllus* (Loranthaceae). Can J Bot 63: 567-581

Menzies BP (1954) Seedling development and haustorial system of *Loranthus micranthus* Hook. Phytomorphology 4: 397-409

Nagar R, Singh M, Sanwal GG (1984) Cell wall degrading enzymes in *Cuscuta reflexa* and its hosts. J Exp Bot 35: 1104-1112

Peirce GJ (1893) On the structure of the haustoria of some phanerorgamic parasites. Ann Bot (London) 7: 291-327

Sallé G (1976) Le phloème des cordons corticaux du *Viscum album* L. (Loranthacées). Protoplasma 87: 17-25

Sallé G (1979) Relations cyto-physiologiques entre le *Viscum album* L. et le pommier. Bull Soc Bot Fr 126, Actual Bot 79-95

Schulz A (1987) Sieve-element differentiation and fluoresceine translocation in wound-phloem of pea roots after complete severance of the stele. Planta 170: 289-299
Schumacher W (1934) Die Absorptionsorgane von *Cuscuta odorata* und der Stoffübertritt aus den Siebröhren der Wirtspflanzen. Jahrb Wiss Bot 80: 74-91
Sjolund RD, Shih CY (1983) Freeze-fracture analysis of phloem structure in plant tissue culture. I. The sieve element reticulum. J Ultrastruct Res 82: 111-121
Thoday MG (1911) On the histological relation between *Cuscuta* and its host. Ann Bot (London) 25: 655-682
Visser JH, Dörr I (1987) The haustorium. In: Musselman LJ (ed) Parasitic weeds in agriculture, vol 1. *Striga*. CRC, Boca Raton, Fla, pp 89-106
von Mohl H (1827) Über den Bau und das Winden und Ranken der Schlingpflanzen. Preisschr Med Fak Univ Tübingen
Ziegler H (1955) *Lathraea*, ein Blutungssaftschmarotzer. Ber Dtsch Bot Ges 68: 311-318

13 Phloem Proteins

James Cronshaw and Dinkar D. Sabnis

13.1 Introduction

Phloem cells contain specific proteins, some of which may be present in large quantities and visible with both the light and electron microscopes, while others can be demonstrated only with chemical methods such as gel electrophoresis. The first accounts of these substances in the phloem were from observations with the light microscope. In the middle of the nineteenth century Hartig (1854) observed characteristic accumulations of material in phloem, especially on the sieve plates. This material became known as slime and the accumulations as slime plugs. Other early workers recognized that the source of slime in mature sieve elements was from cell inclusions which they called slime bodies and also that the staining reactions of slime indicated a considerable protein content (Wilhelm 1880). In the latter part of the nineteenth century several workers showed that slime plugs resulted from the flow of the contents of sieve elements toward a cut surface as a result of injury (Nägeli 1861; Fischer 1884).

Later, when appropriate techniques became available, studies of slime with the light microscope gave way to studies at the electron microscope level. Difficulties occurred in the early electron microscope studies of phloem tissue because slime was not well preserved by the fixation methods commonly used. It was not until after the introduction of glutaraldehyde as a fixative for plant material and its use for fixing phloem tissue (Bouck and Cronshaw 1965) that detailed studies of slime were made and the term P-protein introduced for this material. The term P-protein was suggested after cytochemical tests had revealed that slime was proteinaceous and electron microscopy had shown it to have a variety of morphological forms, many of which were structurally comparable to cell components known to be proteinaceous (Esau and Cronshaw 1967). Once it was accepted that slime was proteinaceous, and since plant slimes are carbohydrates usually associated with cell walls, continued use of the term for the material in the phloem seemed inappropriate. P-protein has now been generally accepted as the term to describe this material and according to Cronshaw's (1975 a) justification of the use of the term, P-protein is defined as "the proteinaceous material in the phloem which is sufficiently characteristic when observed with the electron microscope to warrant a special term."

Biochemical studies in the 1970's showed that numerous phloem-specific proteins could be identified using sodium dodecyl sulfate polyacrylamide gel electrophoresis and that some of these proteins had hemagglutinating properties. Continuing chemical studies have revealed a wide variety of phloem-specific proteins and suggest that differences in their chemistry may be characteristic of various genera and related to other taxonomic features (Sloan et al. 1976; Sabnis and Hart 1979; Kollmann 1980; McEuen et al. 1981). The unknown relationship between the morphologically distinct P-proteins and the chemically distinct protein subunits fractionated by gel electrophoresis poses problems of terminology. All of these proteins, however, are phloem-specific. Accordingly, we propose to use the term phloem-specific proteins for all of the proteins that are characteristic of the phloem. For the group of these proteins that are morphologically distinct at the electron microscope level and are observed in the cytosol we continue to use the term P-protein.

13.2 P-Protein

In sieve elements at early stages of differentiation P-protein is discernible with the light microscope as discrete accumulations that are called P-protein bodies (Cronshaw 1975 a). Dependent on the species, there may be one or more per cell (Esau 1969). For example, *Nicotiana tabacum* sieve elements have only one large P-protein body per cell, whereas they are numerous in *Cucurbita maxima* sieve elements (Cronshaw and Esau 1967, 1968 a, b). In some species such as *Phaseolus* there may be more than one morphological type of P-protein body (Cronshaw 1975 a). The P-protein bodies are small at first and gradually increase in size as the sieve elements mature. During intermediate stages of differentiation the P-protein bodies disperse in the cytoplasm at about the same time as the autolysis of the nucleus, the reorganization of the endoplasmic reticulum, and the disappearance of the vacuoles, dictyosomes, and ribosomes. During this dispersal, the P-protein bodies appear to swell and if there are several P-protein bodies, may appear to fuse with one another.

Cytochemical studies have demonstrated the proteinaceous nature of P-protein. Early studies showed that mercuric bromphenol blue and ponceau S stain the slime bodies of immature sieve elements and the slime plugs of mature sieve elements (Rouschal 1941; Esau and Cheadle 1965; Cronshaw and Esau 1967, 1968 a), indicating a positive reaction for protein. Yapa and Spanner (1972) showed that dispersed P-protein in sieve tubes of *Tetragonia expansa* and *Lycopersicon esculentum* could be digested by proteolytic enzymes such as pepsin, protease, and pronase confirming the proteinaceous nature. Similar results were obtained by Ilker and Currier (1975) for the P-protein of *Xylosma congestum*. Although some cytochemical tests have demonstrated the presence of lipids and ribonucleic acid in addition to protein in accumulations at sieve plates (Salmon 1946, 1951; Buvat 1963), this is due to the presence of subcellular components other than P-protein that are known to be contained in these accumulations. The absence of a polysaccharide component associated with the filamentous P-protein in *Nymphoides peltata* was demonstrated by Freundlich (1974) using Thiery's periodic acid-thiocarbohydrazide-silver stain. Direct confirmation of the proteinaceous nature of P-protein was provided by Kleinig et al. (1971 b), who solubilized the filamentous structures in *Cucurbita* sieve-tube exudate and showed that they could be reversibly aggregated into filamentous structures which resembled the natural filaments in negative stained preparations.

13.2.1 Distribution

P-proteins have been found in all dicotyledonous and many monocotyledonous species examined and are absent from the gymnosperms and lower vascular plants (see Chap. 3 to 7). Some gymnosperms and seedless vascular plants do contain other phloem-specific proteins however and these will be described in Sect. 13.3.

13.2.2 Morphology

Using the transmission electron microscope P-proteins were shown to have different morphological forms and described to be amorphous, granular, fibrillar, filamentous, tubular or crystalline depending on species and also on the stage of differentiation of the cells. They are most abundant in the sieve elements, but are also found in companion cells and phloem parenchyma cells (Esau 1969; Cronshaw 1975 a).

13.2.3 Filamentous (Fibrillar) P-Proteins

The different morphological forms of P-protein vary considerably in structure and dimension. Filaments of P-protein (often named fibrils, but see Behnke and Dörr 1967) have been observed in all P-protein-containing species that have been studied in detail. These filaments range in width from 2 to 20 nm, and a species may have several morphological types. In many species, such as *Nicotiana* (Cronshaw and Esau 1967), *Acer* (Northcote and Wooding 1966; Wooding 1969; Cronshaw et al. 1973) and *Coleus* (Steer and Newcomb 1969) *Musa* (Behnke 1969) and *Tamus* (Behnke 1969) these filaments show cross-striations with a regular periodicity especially in mature sieve elements (Fig. 13.1). Filamentous P-proteins have been isolated from phloem exudate of several species. Kollmann and co-workers (Kollmann et al. 1970; Kleinig et al. 1971 a) described two types of filaments in exudate from *Cucurbita maxima* and *Nicotiana glauca:* one, 5 nm in diameter had a beaded appearance with regular spacings of about 5 nm, and the other was 9 nm in diameter and the conclusion was that it consisted of two helically wound 4 nm subunits. They were able to solubilize these filaments and reversibly aggregate them into filamentous structures that resembled the native filaments. Cronshaw (1970), and Cronshaw et al. (1973), examined P-proteins in the exudate of *Cucurbita, Cucumis* and *Nico-*

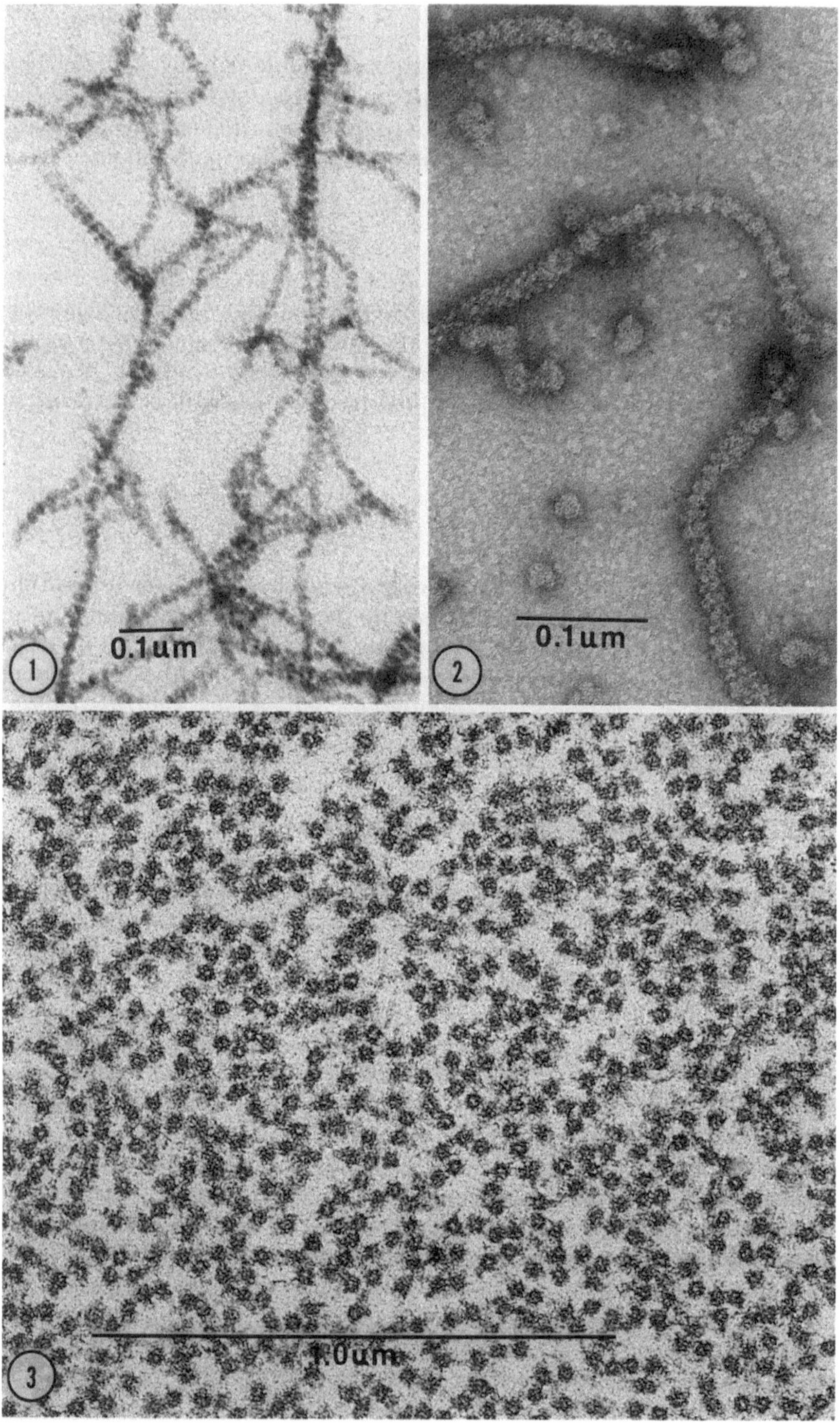
0.1um
1
0.1um
2
1.0um
3

tiana. In *Cucurbita* and *Cucumis*, the bulk of the filaments were 8 nm and formed a network in which these filaments were often organized into larger strands. In *Nicotiana*, they described filaments 30 nm in diameter that consisted of two helically wound subfilaments. These filaments could be disrupted with urea and sodium sulfite, reagents known to disrupt hydrogen bonds and disulfide bonds respectively. Williamson (1972) examined negatively stained exudate from the phloem of *Ricinus communis* and *Cucurbita*. He described filaments 10-27 nm and smaller filaments about 2 nm in diameter in *Cucurbita* exudate. In *Ricinus* exudate, he described two types of filaments, a larger type, 20 nm in diameter with a steeply banded structure, and a smaller type with a diameter of 2-8 nm. Stone and Cronshaw (1973) described as many as four and possibly five morphological types of filaments in the exudate of *Ricinus*. Over 50% of the filaments were 20 nm in diameter with a cross-hatched appearance; a second type measured 14 nm in diameter and consisted of a central filament with lateral projections. The projections were 5 by 14 nm and were spaced at regular intervals of 6.5-10 nm. Interestingly some filaments had structural characteristics of one of these types for part of their length and structural characteristics of the other type for part of their length (Fig. 13.2). It was concluded that this may indicate a structural relationship and possibly an interconvertibility between the two major subfilament types.

13.2.4 Tubular P-Protein

Tubular forms of P-protein have been observed in a large number of species, often in cells at early stages of differentiation where individual tubules are the components of P-protein bodies. Species in which tubular P-proteins have been observed include both dicotyledons and monocotyledons.

Individual tubules occur in a variety of sizes ranging in diameter from 10 to 30 nm. The size of the tubules may be characteristic of a species (Cronshaw 1975 a), but in some cases there is variation within a single plant (Parthasarathy and Mühlethaler 1969). They are all characterized by having an electron-lucent core when sectioned transversely (Fig. 13.3). Observations of P-protein components that appear to be part tubular and part filamentous, and of the apparent replacement of tubular P-protein by filamentous P-protein during cell differentiation, have led to suggestions that the tubules can extend, fray out, or partially

Fig. 13.1. *Acer rubrum*. Section of a mature sieve element containing filaments of P-protein showing characteristic striations. **Fig. 13.2.** *Ricinus communis*. Negatively stained filament from phloem exudate with two structural regions. One region shows the characteristics of a filament type 20 nm in diameter, with a cross-hatched appearance. The other region shows the characteristics of a filament type 14 nm in diameter that has projections spaced 6.5-10 nm apart. **Fig. 13.3.** *Nicotiana tabacum*. Portion of a P-protein body in a diffentiating sieve element. The tubular nature of the P-protein can be clearly seen. Fine projections radiate from the tubules and interconnect some adjacent tubules

unravel to form finer filaments. Detailed studies have been made of P-protein tubules and attempts made to correlate apparent conformational changes to the substructure of the tubules. High magnification images of transections of P-protein tubules and image reinforcement techniques have shown that the wall of the tubule consists of six near-spherical units 6 to 7 nm in diameter (Parthasarathy and Mühlethaler 1969; Cronshaw et al. 1973). Parthasarathy and Mühlethaler (1969) concluded that the transectional view of a P-protein tubule consists of six near-spherical subunits although, as they occasionally observed a slight reinforcement of symmetry at n = 12, they also proposed an alternate structure with 12 oval subunits in the wall. They suggested that P-protein tubules were composed of either a single helical strand of subunits with either 6 or 12 subunits per turn, or of two helical strands of subunits with 12 subunits per turn. They also suggested that other kinds of filaments seen in sieve elements may be formed from elongated and unwound or disaggregated tubules. With additional evidence from observations of P-protein tubules in negatively stained preparations showing two helical subunits, Cronshaw et al. (1973) described a model based on that of Parthasarathy and Mühlethaler consisting of two helical subfilaments composed of near-spherical subunits and with six subunits per turn. A loosening and unwinding of the helical subfilaments and a stretching of the tubule would produce striated structures and represent a conformation change of the P-protein such as that seen in tobacco (Cronshaw and Esau 1967). This conformation change would account for the abundance of striated filaments often seen near sieve-plate pores when surge artifacts have occurred during processing of the tissue (Cronshaw 1975 a). Lawton and Johnson (1976) proposed an alternate model for the P-protein tubules of *Nymphoides peltata* that consists of two filaments wound around a central lumen. Each filament is helical with six 3-nm globular subunits per turn. They suggest that tubules with such a structure could unravel to form the filaments seen in other images and could also explain how the wall of a tubule becomes thinner as the diameter of the tubule decreases. Arsanto (1982), working with *Ecballium, Cicer* and *Polygonum* also described a model for the structure of P-protein tubules that consists of two strands wound around a central lumen, each strand being wound as a varying-pitched helix.

13.2.5 Crystalline P-Proteins

Characteristic of many Fabaceae are conspicuous crystals in phloem sieve elements that have been called irregular bodies (Bouck and Cronshaw 1965), slime bodies (Wark and Chambers 1965), crystalline inclusions (Zee 1968), flagellar

Fig. 13.4. *Phaseolus vulgaris*. Longitudinal section of a sieve element at an early stage of differentiation. A P-protein crystal with a characteristic tail and several granular accumulations of P-protein are developing in the cytoplasm

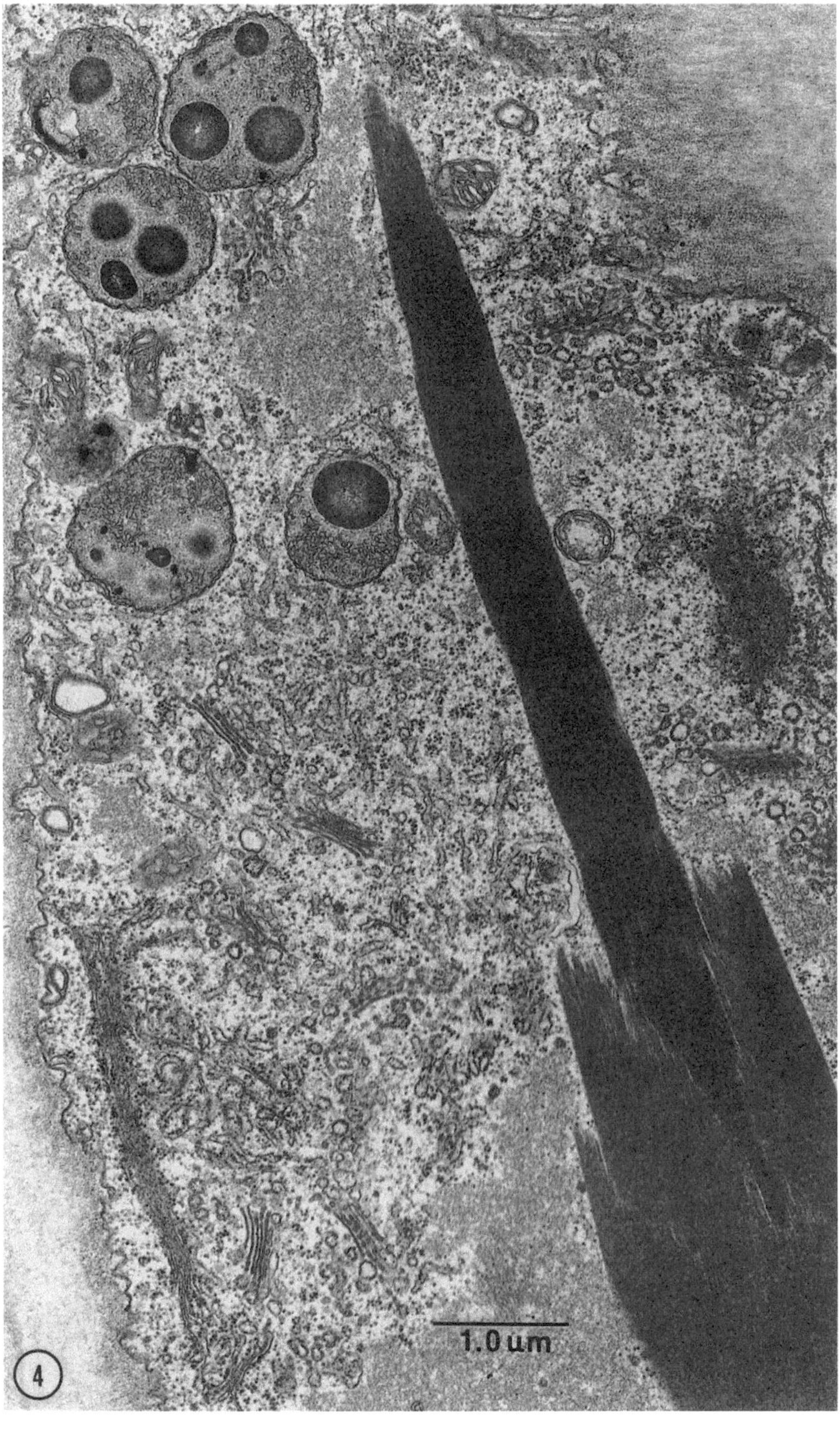
1.0 um
4

inclusions (LaFlèche 1966) and mucoid bodies (Gerola et al. 1969), and are now considered to be forms of P-protein. These inclusions arise in the cytoplasm of differentiating sieve elements in close association with other forms of P-protein such as tubular, fibrillar and granular, and this association had led to the assumption of a developmental relationship between the various components (Palevitz and Newcomb 1971; Wergin et al. 1975). In transection the crystals are often square and in longitudinal section have cross-striations with a center-to-center distance of about 15 nm. In some species characteristic tails are inserted into the main body of the crystal and these tails may have striations with a periodicity that differs from that of the main body of the crystal. Lawton (1978) classified species with these crystals into two groups, depending on whether or not the crystals have tails. In differentiating sieve elements the crystalline inclusions are first seen as small thin crystals that increase in size as the cells mature. There are conflicting opinions as to whether the crystals are formed from other P-protein components such as tubules or fibrils, and also argument as to whether the crystals disperse into other morphological forms during the normal differentiation of the cells. Wergin and Newcomb (1970) and Palevitz and Newcomb (1971) describe the dispersal of crystals into many fine filaments, although in some cases they observed intact crystals in mature sieve elements. Other investigators (Cronshaw 1975 a; Fisher 1975; Lawton 1978) have concluded that the crystals do not disperse and that they are normal components of the cytoplasm of mature cells.

In some dicotyledonous species, the sieve elements contain conspicuous spheroidal inclusion bodies. These were once interpreted as nucleoli extruded from nuclei, but it has now been established that they are a form of P-protein (Deshpande and Evert 1970; Ilker and Currier 1975; Oberhäuser and Kollmann 1977; Esau 1978; Nehls et al. 1978; Behnke and Kiritsis 1983). In *Gossypium hirsutum* the spheroidal bodies consist of a paracrystalline mantle and a compact core. The mantle appears as rods lying parallel to one another or as parts of an irregular mesh (Esau 1978). The spheroidal inclusions of *Xylosma congestum* were shown by Ilker and Currier (1975) to have staining properties similar to those of P-protein. These spheroidal bodies are formed early in the differentiation of the sieve elements and in *Gossypium* are seen in the sieve elements before any other features identifying the cell as a sieve element become evident. Spheroidal bodies do not disperse during the differentiation of sieve elements and are present in mature sieve elements (Esau 1978). The similarity of the spheroidal bodies to the crystalline inclusions of the Fabaceae indicates that they are a form of P-protein.

13.2.6 P-Protein Bodies and Their Formation

P-protein bodies arise as small accumulations in the cytoplasm that enlarge as the cells differentiate. Between species there is considerable variation not only in the morphology of the P-protein components but also in the number of P-protein bodies per cell, the arrangement of the components in the P-protein bodies, and the changes that occur during differentiation. In many plants P-protein first appears in differentiating sieve elements in the form of granules or fine fibrils. Because of morphological associations, a variety of organelles have been implicated in its formation. Helical polyribosomes, not usually prominent in plant cells, have been observed in association with P-protein (Behnke 1974 a; Cronshaw 1985). Behnke (1974 a) surveyed 20 different angiosperms and found that the origin of P-protein was closely linked to the presence of free helical polysomes. Other cytoplasmic components, present at the time of P-protein synthesis, that have most often been implicated in its formation, are the endoplasmic reticulum, dictyosomes, free ribosomes, and spiny and coated vesicles (Bouck and Cronshaw 1965; Northcote and Wooding 1966; Newcomb 1967; Cronshaw and Esau 1967; Behnke 1969; Zee 1969; Cronshaw and Anderson 1971; Cronshaw 1975 a; Deshpande 1974; Hoefert 1979; Arsanto 1982). Cronshaw and Esau (1968 a) described the initial accumulations of P-protein in *Cucurbita* as appearing in the form of small areas of granular material. Observations of similar granular material have been made for *Glycine max*. L. (Wergin and Newcomb 1970), and *Phaseolus vulgaris* (Cronshaw 1975 a). In *Nicotiana* Cronshaw (1975 a) observed finely granular material in association with the endoplasmic reticulum in very young sieve elements although the P-protein of this species is usually first apparent as small accumulations of P-protein tubules.

The development of P-protein bodies in many species is similar to that of *Nicotiana* and *Coleus*, two species that have been studied in detail (Cronshaw and Esau 1967; Steer and Newcomb 1969). In *Nicotiana* and *Coleus*, small accumulations of P-protein tubules gradually increase in size by the addition of more tubules until they reach a size approximating that of the nucleus. Usually there is only one P-protein body per cell and although there is no bounding membrane other organelles are excluded from the accumulations. The bodies are ellipsoidal in shape, and are elongated in the direction of the cell axis. Within them the P-protein tubules are arranged roughly parallel to one another with their long axes in the same direction as the long axis of the P-protein body. A transection of a P-protein body shows most of the tubules in transverse section and between the individual tubules are stained spoke-like projections that appear to connect individual tubules. These connections may be responsible for the structural integrity of the bodies and possibly consist of different proteins from those of the wall of the tubule, in the same way that MAPs are associated with tubulin in microtubules. In some cross sections of P-protein bodies, the tubules are arranged in crystalline aggregates having a hexagonal close-packed arrangement (Steer and Newcomb 1969; Cronshaw and Anderson 1971). In *Cucurbita*, tubules with a similar size and morphology to those of *Nicotiana* may be regularly arranged in crystalline aggregates with quadrangular packing. The bulk of the P-protein in *Cucurbita maxima*, however, is in a second larger type of body that contains P-protein that is mainly filamentous (Fig. 13.5). *Cucurbita* also differs from *Nicotiana* in having numerous P-protein bodies per

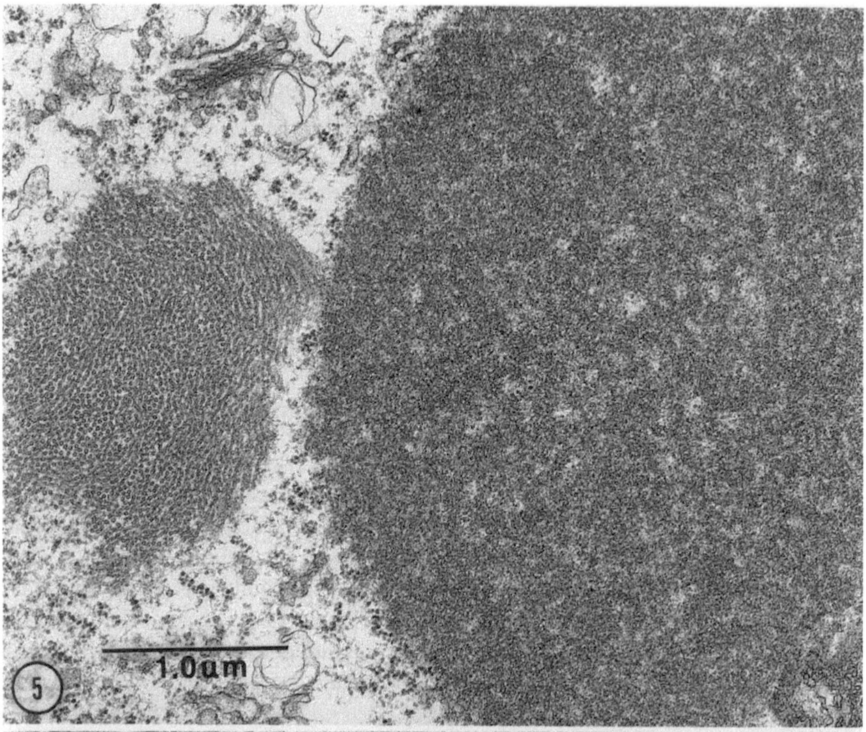
1.0 um
5

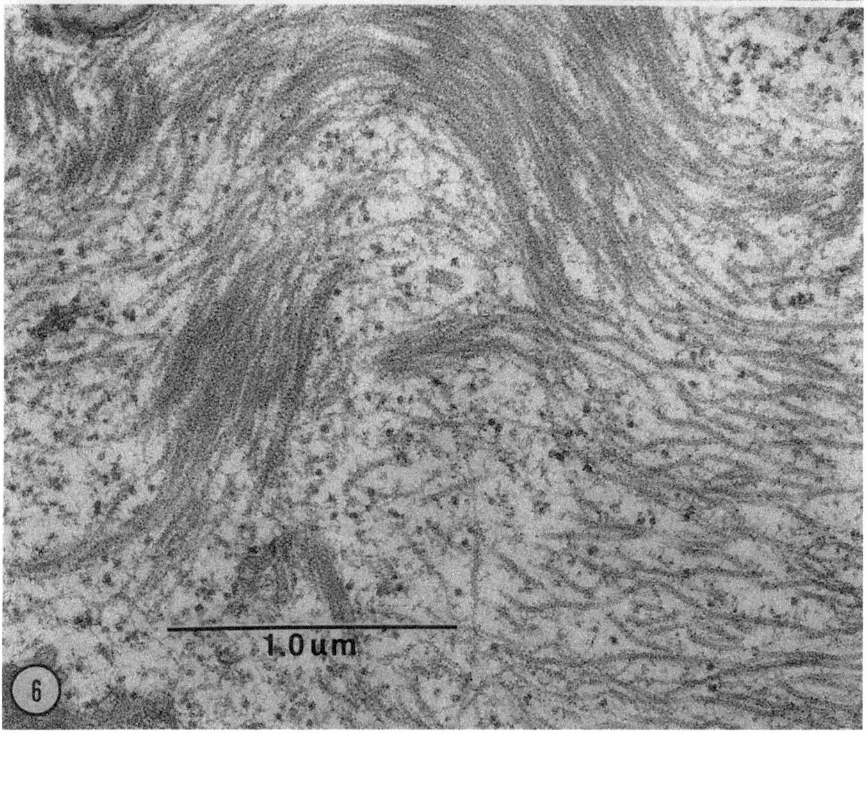
1.0 um
6

cell. In other species, there is a range in the variety of P-protein bodies that reflects the variety of the individual P-protein components. For example, in *Ricinus* the P-protein bodies are loose aggregates and do not have the compact form of those of *Nicotiana* and *Cucurbita* (Cronshaw 1975 a). Sieve elements of this species at early stages of differentiation develop finely filamentous or granular P-protein in the cytoplasm intermixed with ribosomes and vesicles of various sizes. The amount of fibrillar material gradually increases and adjacent bodies may fuse to form larger ones until in some cases a single large filamentous P-protein body is formed that may occupy most of the length of the cell. The most structurally complex P-protein systems that have been investigated in detail are those of the Fabaceae where differentiating cells may contain granular masses and several other structural varieties of P-protein, tubular, filamentous and crystalline, each in separate compact bodies (Cronshaw 1975 a).

13.2.7 Dispersal of P-Protein Bodies

With few exceptions, such as the small P-protein bodies of the extrafascicular sieve elements of *Cucurbita*, the crystalline P-protein bodies of the Fabaceae and the persistent (e.g., spheroidal, globular and other) P-protein bodies of several woody families, the P-protein bodies disperse during the differentiation of the sieve elements. Dispersal starts with the individual components moving apart from one another and becoming dispersed in the surrounding cytoplasm. If there are connections between the components, as in the case of the tubular P-proteins in tobacco, these connections must be broken or solubilized at this stage. In the case of *Cucurbita*, that has a large amount of P-protein in multiple P-protein bodies, the bodies appear to swell during dispersal and eventually fuse as the lumen of the sieve element becomes filled with the products of disaggregation. Disaggregation occurs at those stages of differentiation when degenerative changes are apparent in the nucleus, the endoplasmic reticulum is reorganizing, and the sieve-plate pores are developing. During disaggregation there may be conformational changes of the individual P-protein components, as in *Nicotiana* where the P-protein tubules are transformed into striated filaments. After the disaggregation of P-protein bodies, nucleoside phosphatase acitvity can be demonstrated in association with the P-protein (Gilder and Cronshaw 1973 a, b, 1974; Cronshaw 1980). Nucleoside phophatase activity cannot be demonstrated at early stages of differentiation during the formation of the P-protein bodies. It is not known whether this nucleoside phosphatase

Fig. 13.5. *Cucurbita maxima*. Section through P-protein bodies in a differentiating sieve element. The large body contains filamentous P-protein, the smaller body consists of P-protein tubules. **Fig. 13.6.** *Nicotiana tabacum*. Portion of a sieve element at an intermediate stage of differentiation. Two morphological forms of P-protein, one tubular and another in the form of striated filaments are visible in the dispersing P-protein body

activity is a function of the P-protein subunits themselves or is the property of another protein associated with them. In *Pisum sativum*, however, no nucleoside phosphatase activity is associated with the P-protein at any stage of differentiation of the sieve elements (Bentwood and Cronshaw 1978). Acid phosphatase has also been localized cytochemically in association with P-protein. Catesson (1973), Catesson and Czaninski (1967, 1968), Bentwood and Cronshaw (1976), and Cronshaw (1980), found activity associated only with the dispersed P-protein of mature sieve elements, Esau and Charvat (1975) found P-protein of both immature and mature sieve elements to have acid phosphatase activity.

13.2.8 P-Protein in Mature Sieve Elements

The individual components of P-protein derived from the P-protein bodies become dispersed in the cytoplasm as the sieve elements differentiate. Because of the difficulties of preparing mature sieve elements that have a high hydrostatic pressure for electron microscopic examination, however, the precise distribution of P-protein in these cells remains controversial. Some workers have described a parietal distribution of the P-protein where it is closely applied to the plasma membrane for mature sieve elements. This type of distribution has been described for several dicotyledon species, while other accounts describe the P-protein as having a more or less even network distributed throughout the lumen of the mature sieve element (see discussion in Chap. 6). With either of these arrangements of the P-protein in the mature sieve element, it is most likely that the arrangement is stabilized by attachment to the plasma membrane. It has been concluded from observations of the distribution in mature sieve elements throughout entire plants that P-protein is relatively stationary in the sieve element protoplast and does not move in the assimilate stream (Cronshaw 1975 a). Lectin bridges may serve to anchor the P-protein components to any of the membrane components present in the peripheral cytoplasm of the mature sieve elements (Sabnis and Hart 1979; Kollmann 1980).

13.2.9 P-Proteins in Sieve-Plate Pores

There have been many attempts to determine the nature of the contents of sieve-plate pores and to determine whether or not the plugging of sieve-plate pores by P-protein seen in many electron micrographs is an artifact of preparation procedures. It is now generally agreed that sieve-plate pores in vivo in mature functioning sieve elements are unplugged (Cronshaw 1975 b). Plugging of sieve-plate pores with P-protein, however, occurs as a very rapid response to pressure release. This rapid plugging of sieve-plate pores on injury has been well documented (Esau 1969; Crafts and Crisp 1971; Cronshaw 1975 b) and is thought to be a response that would serve to prevent the rapid loss of nutrients on wounding (see Chap. 6, for further discussion).

13.3 Other Phloem-Specific Proteins

13.3.1 Nuclear Inclusions

Nuclear protein inclusions, amorphous and crystalline, have been described in a variety of plant cells, and their occurrence and structure are sufficiently characteristic to be used for taxonomic purposes. In a few species crystalline nuclear protein inclusions have been found to be characteristic of sieve elements. Hébant (1969) described crystalloids in sieve elements of four species of fern; Evert and Eichhorn (1974) observed nuclear crystalloids in sieve elements of the fern, *Platycerium bifurcatum*. Nuclear crystalloids have also been observed in *Amsinckia, Ehretia, Myosotis* and *Trichodesma* (Esau and Magyarosy 1979 a, b), and in 15 species of *Echium* (Esau and Thorsch 1982). *Echium* crystalloids consist of extremely narrow tubular subunits that develop to be the same length as the nucleus containing them. *Amsinckia* crystalloids have two components, one of these has the same substructure as the *Echium* crystalloid, the other has wider tubules arranged less precisely than the narrower tubules of the first component. In all cases these protein components are released from the nucleus during the differentiation of the sieve elements so that in the mature cells they are free in the cytoplasm where the P-protein is located. In most cases these crystalloids persist after the disintegration of the nucleus, but in the case of *Amsinckia,* upon release from the nucleus, the component consisting of the narrower tubules becomes more or less fractured, while the second component separates into individual tubules. These nuclear crystalloids resemble the P-protein inclusions of the Fabaceae but have not so far been considered P-protein because of their different origin. Tubular inclusions in the nuclei of *Tilia,* however, were described as P-protein by Evert and Deshpande (1970).

13.3.2 Plastid Inclusions

Sieve element plastids have characteristic inclusions and two major types can be recognized: S-type plastids that store starch and P-type plastids that contain elaborate protein inclusions. P-type plastids often also contain starch grains. The proteinaceous inclusions in P-type plastids are present in different types of inclusion bodies, some of which are typically restricted to different plant taxa and have been defined as subtypes of P-type plastids. These inclusion bodies may contain filamentous material (often in ring-shaped bundles), globular crystals, polygonal crystals, or combinations of crystals and filaments. The structural variety of sieve-element plastids has provided useful characteristics for plant classification and phylogenetic interpretations. P-type plastids are found in many dicotyledons and some gymnosperms, and are uniformly distributed in the monocotyledons (Behnke 1972, 1974 b, 1981; Behnke and Schulz 1983). Filaments from P-type plastids can be liberated into the sieve-element protoplast in response to unfavorable conditions during the preparation of material for observation with the electron microscope, and filaments liberated in this way have been erroneously interpreted as P-protein (Campbell 1972). Although some authors consider a breakdown of plastid envelopes and release of their contents to be a normal process during sieve-element ontogeny (Parameswaran 1971; Gamaley 1973), most agree that such a breakdown and release is an artifact of specimen preparation (Behnke 1974 b; Cronshaw 1975 a).

13.3.3 Refractive Spherules and Other Vesicles

Single membrane-bound vesicles called refractive spherules (they are highly refractive when viewed with the light microscope) and containing proteinaceous inclusions have been observed in all groups of seedless vascular plants with the exception of the lycopods.

In *Selaginella,* substantial quantities of crystalline and fibrillar proteinaceous material occur in numerous relatively small vacuoles; in *Isoetes,* vesicles containing proteinaceous material are derived from the endoplasmic reticulum (see Chap. 3).

13.3.4 Proteins Associated with the Endoplasmic Reticulum

Sieve elements at early stages of differentiation contain endoplasmic reticulum that is mostly in the rough form. During differentiation, the rough endoplasmic reticulum gradually loses its ribosomes and in many species an electron-opaque component becomes associated with the external surfaces of the cisternae (Cronshaw 1981). As the endoplasmic reticulum is reorganized into the parietal system and the stacked system, this electron-opaque component often appears as dense fibrillar material that resembles P-protein and is located between adjacent endoplasmic reticulum cisternae and between endoplasmic reticulum cisternae and the plasma membrane. This material is possibly involved in the adhesion of membranes and of membranes to other organelles, such as plastids and mitochondria, which in the mature sieve element occupy a peripheral position (Cronshaw 1985).

13.4 Biochemistry of Phloem Proteins

13.4.1 Isolation and Chemical Properties

Until recently, biochemical studies on phloem-specific proteins were restricted to those species from which phloem exudate containing proteins in quantities adequate for analysis can be collected. Many cucurbit species possess sieve tubes with wide lumens and large sieve pores; such plants bleed relatively copiously from cut stems and incisions into the fruit. Under the electron microscope, negatively stained phloem exudate from the stems of *Cucurbita maxima* showed an abundance of protein filaments of about 4 and 9 nm width, similar to those seen in thin sections (Kollmann et al. 1970). The polymeric proteins could be solubilized in the presence of sulfhydryl reducing agents such as dithiothreitol (DTT). Eschrich et al. (1971) confirmed that exudate droplets from *Cucurbita* truly arose from severed mature sieve tubes and contained filaments of P-protein but no mitochondria or vesicles of endoplasmic reticulum. The P-protein concentration in phloem exudate from most cucurbits is high (10-60 mg/ml^{-1}) and far in excess of any protein contamination from cut cell surfaces. Thus the cucurbits have provided ideal material for initial studies on the properties of phloem-specific proteins.

Biochemical investigations of phloem-specific proteins began with publications from three groups in the same year (Eschrich et al. 1971; Kleinig et al. 1971 a, b; Walker and Thaine 1971; Weber and Kleinig 1971). Weber and Kleinig used sodium dodecyl sulfate polyacrylamide gel electrophoresis (SDS-PAGE) to determine molecular mass values for soluble proteins present in *Cucurbita maxima* exudate. The subunit M_r values obtained ranged from 15 K to 220 K and seemed to double from protein to protein, suggesting some form of dimerization to the authors. In a subsequent publication (Weber et al. 1974), eight polypeptides with M_r values of 14, 17, 28, 44, 59, 70, 115 and 158 K were reported. Major bands were present at positions corresponding to 28 and 115 K. Eschrich et al. (1971) detected at least 12 proteins in *Cucurbita* exudate using nondenaturing PAGE. Working with the related *Cucurbita pepo,* Walker and Thaine (1971) showed that stem phloem exudate gelled on exposure to air and that this gelling could be prevented by sulfhydryl reducing agents such as 2-mercaptoethanol. Removal of the latter permitted gelling, indicating that the response involved the formation of intermolecular disulfide bonds. Of the several polypeptides present in phloem sap, one basic protein was responsible for gelling. The authors concluded that protein gelling was the primary factor in the formation of slime plugs. Subsequently, Walker (1972) presented an amino acid analysis of the gelling protein of *Cucurbita* exudate and reported high levels of basic amino acids and cysteine residues, consistent with the electrophoretic (under nondenaturing conditions) and gelling behavior of the protein.

In follow-up work on *Cucurbita* (Beyenbach et al. 1974; Weber et al. 1974), the two main protein (M_r 30 and 116 K) in phloem exudate were isolated and partially characterized. Both were highly basic, with isoelectric points (IEP) above 9.5. The smaller polypeptide dimerized to a form which appeared to be the naturally occurring structure. The larger protein (M_r 116 K) formed a gel in the absence of SH-protecting agents.

An important advance on the early biochemical work was made when Kleinig et al. (1975) were able to show the reversible assembly in vitro of the purified M_r 116 K protein into filaments, 6-7 nm wide, similar to the filaments occuring in vivo. This took place when the 2-mercaptoethanol was dialyzed away and indicated that both filament assembly and gelling were regulated by conformational changes resulting from the formation of inter- and intra-molecular disulfide bridges. Unfortunately, this important observation has not been repeated or extended. The apparent importance of disulfide linkages in maintaining the structural integrity of proteins in the translocating channels was emphasized by the cytochemical demonstration of sulfhydryl/disulfide-containing proteins in mature sieve cells of conifers by Sauter (1972). More recently, use of the fluorescing thiol indicator, Thiolyte (Calbiochem), has similarly indicated very high local concentrations of thiol groups in both angiosperm and gymnosperm sieve cells (Schulz and Sabnis, unpublished results).

During this period in the early 1970's, there was interest among some workers in the possibility that one or more components of P-protein might be related to contractile proteins (a loosely used term, since none of the proteins themselves contract) such as actin or tubulin. This was generated by debate over the presence of transcellular strands in the sieve element and speculations that peristaltic pumping of assimilates might occur (Thaine 1969). Dubious evidence for actin in sieve elements offered by Ilker and Currier (1974) was dismissed by Palevitz and Hepler (1975). Attempts to detect actin-myosin-like proteins in large-scale extracts of isolated phloem tissue from *Heracleum mantegazzianum* (Sabnis and Hart 1974), failed to demonstrate diagnostic superprecipitation or viscosity changes on addition of ATP and divalent cations. Furthermore, isolation, fractionation and characterization of ATPase activity in phloem and xylem extracts from the same plant indicated no similarities between the enzymes present in the plant tissues and myosin or flagellar dynein ATPase activity (Sabnis and Hart 1974; Ezeala et al. 1974a, b). Both Williamson (1972) and Palevitz and Hepler (1975) successfully obtained arrowhead decoration with HMM-subfragment 1 on microfilaments of *Chara* and *Nitella,* but were unable to show any such interaction with P-protein filaments in phloem exudate.

The problem of P-protein was complicated further when Sabnis and co-workers (Sabnis and Hart 1976, 1979; Sloan et al. 1976) demonstrated in comparative analyses, that the protein subunits in exudate from plants of several different species, genera and families were remarkably different in apparent molecular mass and isoelectric point. The variation extended to the numbers, relative quantities and properties of exudate polypeptides. The degree of difference in the phloem protein patterns appeared to reflect the taxonomic relationships of the plants (Sabnis and Hart 1979). Each species displayed a consistent protein pattern, irrespective of the organ from which exudate was obtained or the age and physiological status of the plant. The authors stressed that *any* postulated function of phloem-specific proteins must adequately account for the apparent differences. Nevertheless, the work so far did indicate certain features that were common to the proteins present in cucurbit phloem exudate: (1) Several proteins were of relatively large subunit size (M_r values of 80–130 K or more); (2) the pronounced basicity (high IEP) of major components meant that in vivo these proteins would be positively charged, even at the rather alkaline pH of the sieve-tube sap; (3) many of the proteins also interacted strongly with thiol reagents and readily formed disulfide bridges. These biochemical properties suggest that the differences in protein subunits might account for the reported differences in the ultrastructure of P-protein from different plants (Sabnis and Hart 1979).

13.4.2 Lectin Properties of Phloem-Specific Proteins

An exciting and potentially important diagnostic property of phloem-specific proteins was introduced by Kauss and Ziegler (1974) when they reported hemagglutinating (lectin) activity in the sieve-tube sap of *Robinia pseudoacacia*. Sabnis and Hart (1978, 1979) isolated and identified the lectin of *Cucurbita* as a major basic protein of M_r 22 K (Fig. 13.7; identical to the M_r 30 K protein of Weber et al. 1974). No hapten activity (inhibiting hemagglutination) could be demonstrated with the serveral monosaccharides tested, which included N-acetylgalactosamine (reported to be active against the *Robinia* hemagglutinin) and N-acetylglucosamine (GlcNAc). Similar high lectin activity was also evident with phloem sap from *Cucumis sativus* and *Cucumis melo*. Then Allen (1979) showed that a lectin from *Cucurbita pepo* exudate strongly interacted with chitin

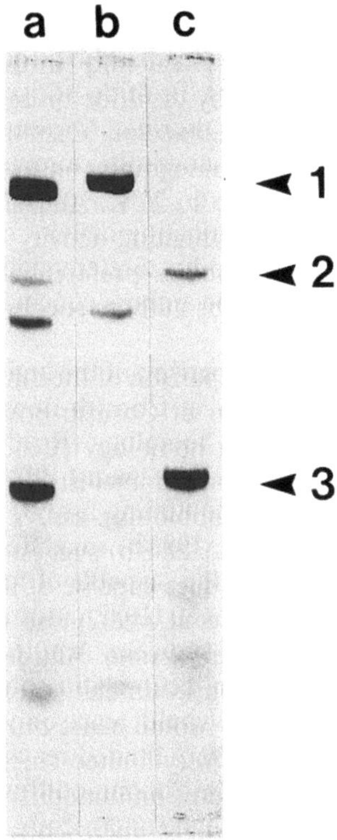

Fig. 13.7 a–c. SDS-PAGE of exudate samples from *Cucurbita maxima*. **a** Whole exudate electrophoresed in the presence of 2-mercaptoethanol; *1* PP1; MW 84000; *2* PP2 (lectin) dimer, MW 48000; *3* PP2 monomer, MW 24000. **b** and **c** Supernatant (**b**) and pellet (**c**) fractions of exudate after freezing, thawing and centrifugation. Very high purification of PP2 is achieved. PP1 and PP2 are the major proteins of *C. maxima* exudate

oligosaccharides, glycopeptides from soybean agglutinin and fetuin. Thus, although GlcNAc itself was only a weak ligand, the lectin interacted with GlcNAc oligomers or internal GlcNAc residues in glycoprotein haptens. Basic lectin have now been identified in other cucurbit genera (Sabnis and Hart 1982; Read and Northcote 1983 a).

The importance of these findings was underlined by Gietl et al. (1979) and Gietl and Zeigler (1980 a, b) who demonstrated both the ubiquity of phloem lectins and their specific affinity for GlcNAc oligomers (the only exception being the *Robinia* lectin). Sixty-three tree species belonging to 21 families were examined. Except for samples from members of the Aceraceae and Oleaceae (which showed pronounced hemolytic activity, possibly due to the presence of saponins, which may have masked hemagglutination), all exudates agglutinated trypsinized rabbit erythrocytes. Results to date indicate that the phloem of each genus appears to be characterized by a single lectin, possibly existing as a dimer in the native state. The monomeric polypeptides can frequently be resolved into closely spaced doublets on SDS-PAGE. However, it is not clear whether the dimer is necessarily a heterodimer. *Pinus* phloem also contains a lectin doublet which is not present in other tissues (Fig. 13.9; Alosi, unpublished data).

Read and Northcote (1983 a) designated the two major exudate polypeptides of *Cucurbita maxima* PP1 and PP2. PP1 has a M_r of 80 K by SDS-PAGE and 96.2 K by sedimentation equilibrium centrifugation, and is identical to the filament-forming protein of Kleinig et al. (1975). PP2, with a reported M_r of 26 K is the lectin. Read and Northcote (1983 a) assert that the dimeric lectin was "incorporated by disulfide bridges into the filament observed in vivo". The lectins of *Cucurbita maxima, Cucumis sativus* and *Citrullus vulgaris* were isolated by affinity chromatography on ovomucoid-Sepharose and shown to vary slightly in M_r from 26.5 to 29 K. Antisera against purified *Cucurbita* lectin immunoinhibited hemagglutinating activity of exudate proteins from all three genera, although the antibodies preferentially recognized the antigen against which they were raised. The authors conclude that the three genera, therefore, were antigenically related.

A comparison of the amino acid composition of PP1 and PP2 using the $S \Delta n$ parameter of Cornish-Bowden (1977, 1979) suggested appreciable amounts of sequence homology (Read and Northcote 1983 a). Nevertheless, none of the antisera raised against PP2 were cross-reactive against PP1 in immunodiffusion or immunoblotting assays. Analysis of the disulfide bridges of PP2 (Read and Northcote 1983 b) suggested that each monomer possessed two accessible cysteine residues capable of forming bonds holding the dimer together.

Smith et al. (1987) also raised rabbit antibodies to purified lectin (PP2) from *Cucurbita maxima*. Anti-lectin IgG was further purified by affinity chromatography on lectin-chitin columns. Immunoblots indicated that the affinity-purified IgG would react only with the lectin monomer and dimer in *Cucurbita* exudate. Very limited reaction was obtained with exudate from other genera in Ouchterlony immunodiffusion tests. Immunocytochemistry, using peroxidase label for light microscopy and colloidal gold-Protein A at the EM level, was

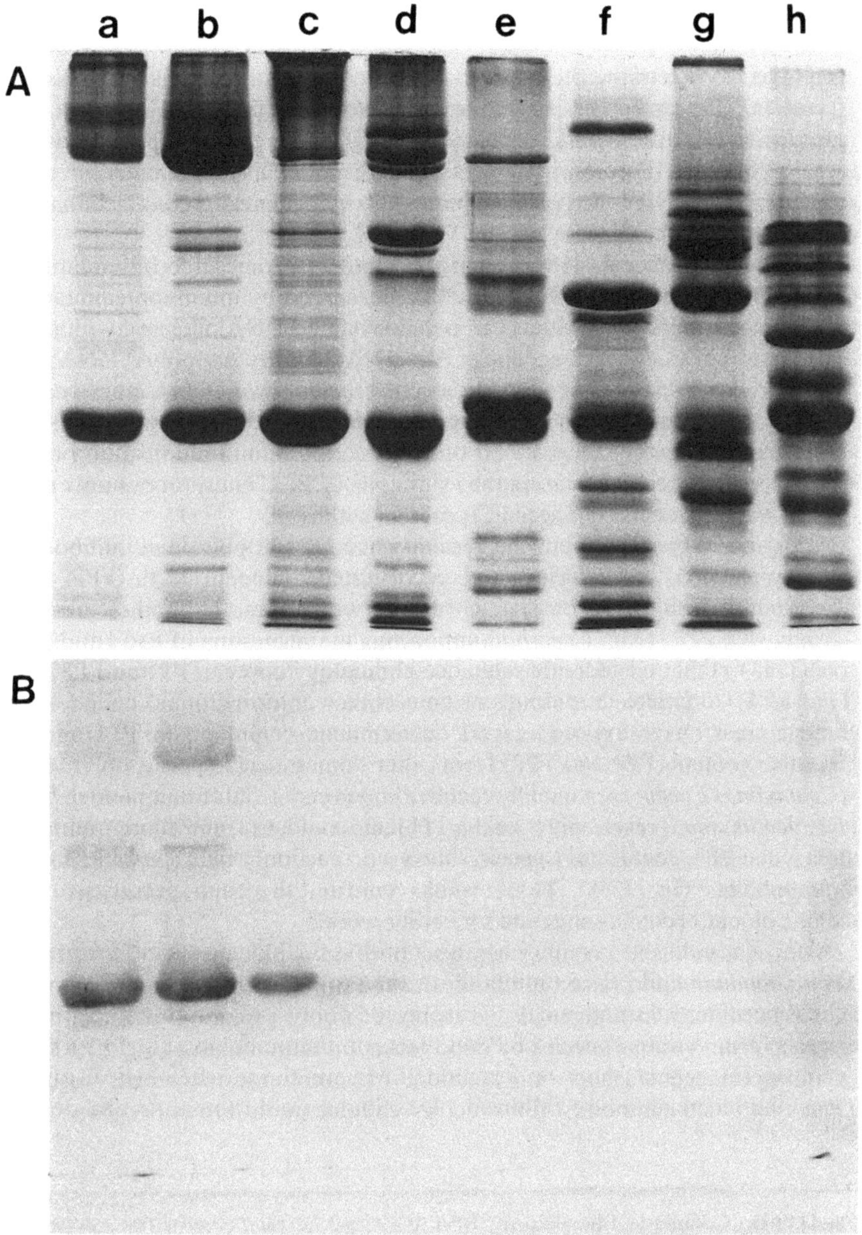

Fig. 13.8 A, B. Comparative SDS-PAGE and immunoblots of phloem exudate from several cucurbit species. The immunoblot (**B**) of gel **A** was incubated with rabbit antibody raised against purified lectin (PP2) of *Cucurbita moschata* (not shown). The antibody reacts with only the lectin monomer and dimer bands of *Cucurbita moschata (a), C. maxima (b)* and *C. pepo (c),* and only very weakly with *Cucurbita foetidissima (d).* There is also some reaction with PP1 of these species. However, no proteins belonging to other cucurbit genera react with the antibody probe. *e Cucumis sativus; f Luffa acutangula; g Lagenaria siceraria; h Benincasa hispida.* The immunoblot was developed with Protein A-gold and silver enhancement. The results of both SDS-PAGE and the immunoblots emphasize the inter-generic differences in phloem-specific proteins

employed to determine the location of the lectin within the cells and tissues of *Cucurbita*. The antibodies bound strongly and specifically to P-protein aggregates in sieve elements and companion cells. Under the EM the antibody was located over the P-protein filaments and the parietally distributed sieve-tube reticulum which lines the plasmalemma. No cytochemical cross-reactions were obtained with other genera.

Sham and Northcote (1987) found that both PP2, measured by quantitative electrophoresis, and the mRNA for PP2, measured by immunoprecipitation of the in-vitro translation products of polyadenylated RNA, increased with age in hypocotyls of *Cucurbita* seedlings. A cDNA library to poly-A RNA from hypocotyls of 3-day-old seedlings was constructed. Two cDNA clones complementary to the mRNA for PP2 have been identified. Preliminary results suggested that regulatory controls on both transcription and modification of transcripts appeared to occur during the synthesis of PP2. Thus, probes now exist to follow developmental changes in P-protein synthesis.

Alosi and Sabnis (unpublished results) have raised polyclonal antibodies to affinity-purified, freeze-dried and reconstituted phloem lectin (PP2) from *Cucurbita moschata*. In contrast to earlier work, these antibodies cross-react strongly with PP1 from *Cucurbita*, supporting the suggestion of Read and Northcote (1983 a) that considerable sequence homology between PP1 and PP2 exists (Fig. 13. 8). Complete unfolding or some other conformational change of the antigen must have exposed several determinants common to PP1 and PP2. Exudate proteins (PP1 and PP2) from other domesticated species of *Cucurbita (C. maxima, C. pepo)* are equally reactive; however, a Californian native, *Cucurbita foetidissima*, reacts only weakly. Phloem exudate from a large number of other cucurbit genera and species shows no reaction whatsoever in sensitive immunoblots (Fig. 13. 8). These results confirm the genus-specificity of the major phloem proteins suggested by earlier work.

Alosi (unpublished results) has also purified a phloem-specific lectin from *Pinus sabiniana* and raised antibodies to the protein. Once again, very marked genus-specificity in antigenicity is displayed; only a single protein in phloem extracts from various species of *Pinus* reacts on immunoblots (Fig. 13. 9). Other gymnosperm genera show no reaction. EM immunocytochemistry using the *Pinus* anti-lectin antibody followed by colloidal gold-Protein A, has demon-

Fig. 13.9 (A). Coomassie Blue-stained SDS-PAGE gel of total protein from gymnosperm xylem *(a)* and phloem tissues *(b–j)*. Lane *a Pinus sabiniana* xylem; *b Pinus sabiniana* phloem; *c Pinus muricata; d Pinus torreyana; e Pinus murrayana; f Pseudotsuga menziessi; g* and *h Abies concolor; i Abies grandis; j Metasequoia glyptostroboides*. (**B**) Corresponding immunoblot to the above gel stained with peroxidase/anti-peroxidase/diaminobenzidine. Polyclonal, monospecific antiserum was raised in rabbits by immunization with purified, denatured 24000–25000 MW phloem polypeptide from *Pinus sabiniana*. All pine phloem samples (lanes *b–e*) bind the antiserum to bands of major phloem-specific polypeptides of similar molecular mass. The antiserum shows no reaction with other gymnosperm phloem extracts, xylem extract from *Pinus*, or phloem exudate from several cucurbit species tested (not shown)

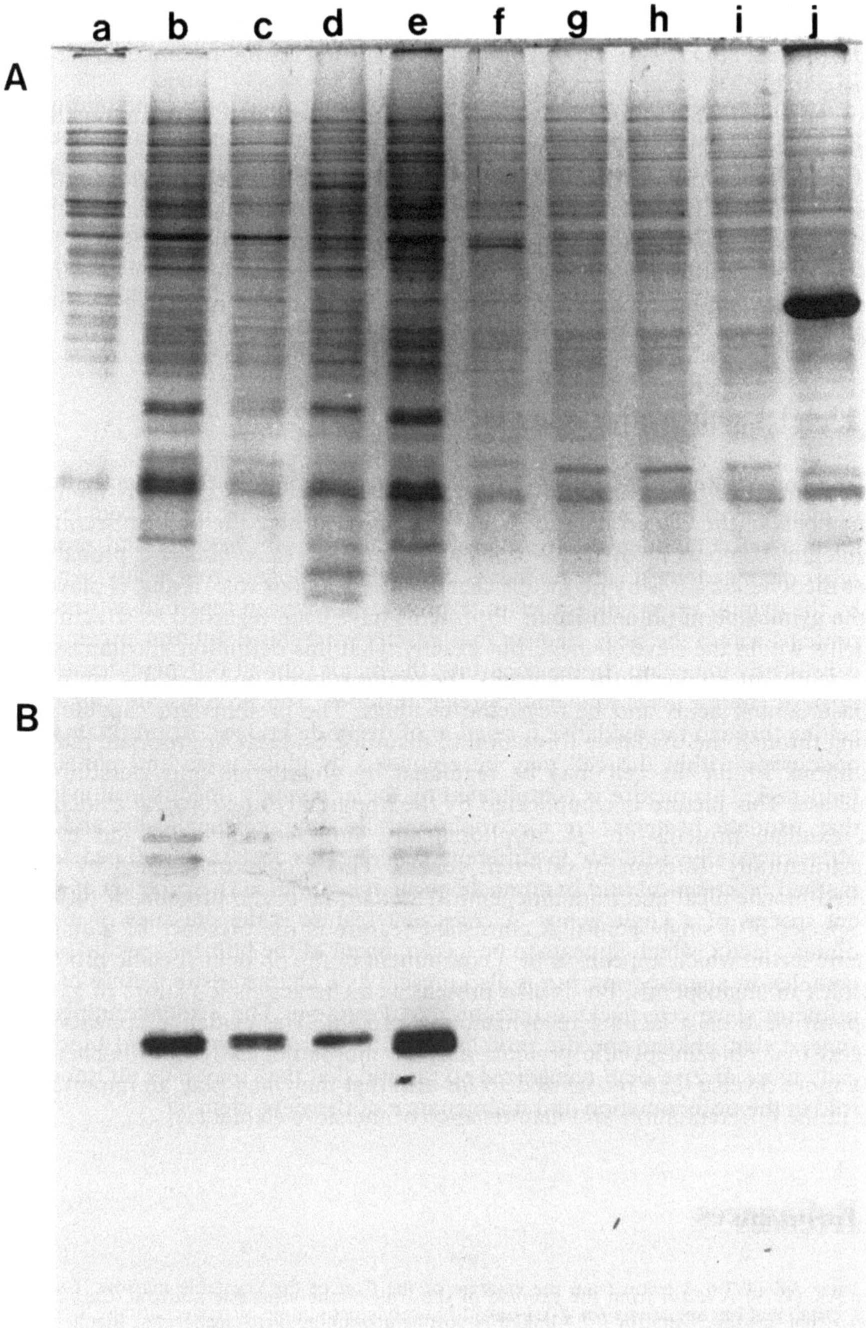
a b c d e f g h i j
A
B

strated precise and specific labeling of the plastid crystals in *Pinus* sieve elements (Schulz et al. 1989). The significance of this interesting finding has yet to be assessed.

An indication of how redox conditions are maintained in vivo, permitting the assembly and stability of P-protein filaments in the functioning sieve element, is provided by the work of Alosi et al. (1988). They have provided evidence for the presence of glutathione and glutathione reductase activity in phloem exudate from *Cucurbita moschata* and suggest that these factors may be responsible for gel/sol phases within the cytoplasm of the sieve element. A breakdown of this regulatory system by injury would account for the complete oxidative gelation of P-protein through the formation of disulfide bonds.

13.4.3 Comparative Aspects

The comparative biochemistry of phloem-specific proteins is presently limited to members of the Cucurbitaceae. While it seems clear that the phloem lectin is an integral component of the P-protein filaments of *Cucurbita* and probably other dicotyledons, it is by no means clear what analogous role, if any, is played by the gymnosperm phloem lectin. P-proteins have been regarded as structural proteins within the sieve element, but exactly what this definition encompasses is presently uncertain. In the cucurbits, the major subunit polypeptides are rich in basic amino acids and half-cysteine residues. The proteins are capable of gelling through the oxidative formation of disulfide bridges. Appropriate redox conditions within the cell may be regulated by glutathione and glutathione reductase. This picture is complicated by the apparently unequivocal evidence that exudate proteins are electrophoretically, immunochemically and even ultrastructurally different in different genera. This fact is underlined by the marked biochemical and immunochemical *similarities* in the proteins of different species of a single genus. A consistent feature is the presence of a single dimeric lectin, which appears to be a component of the phloem-specific protein complex in angiosperms, but is also present as a characteristic feature of gymnosperm sieve cells lacking recognizable P-protein. The evidence appears to suggest that phloem-specific proteins may be morphologically and biochemically more diverse than recognized so far and that they may play an important role in the differentiation and maintenance of the sieve element.

References

Allen AK (1979) A lectin from the exudate of the fruit of the vegetable marrow *(Cucurbita pepo)* that has specificity for β-1,4-linked N-acetylglucosamine oligosaccharides. Biochem J 183: 133-137

Alosi MC, Melroy DL, Park RB (1988) The regulation of phloem exudate from *Cucurbita* fruit by dilution, glutathione and glutathione reductase. Plant Physiol 86: 1089-1094

Arsanto JP (1982) Observations on P-protein in dicotyledons. Substructural and developmental features. Am J Bot 69: 1200-1212

Behnke H-D (1969) Über den Feinbau und die Ausbreitung der Siebröhren-Plasmafilamente und über Bau und Differenzierung der Siebporen bei einigen Monocotylen und bei *Nuphar*. Protoplasma 68: 377-402

Behnke H-D (1972) Sieve-tube plastids in relation to systematics. An attempt towards a classification by ultrastructural analysis. Bot Rev 38: 155-197

Behnke H-D (1974a) Comparative ultrastructural investigations of angiosperm sieve elements: aspects of the origin and early development of P-protein. Z Pflanzenphysiol 74: 22-34

Behnke H-D (1974b) Sieve element plastids of the Gymnospermae. Their ultrastructure in relation to systematics. Plant Syst Evol 123: 1-12

Behnke H-D (1981) Sieve element characters. Nord J Bot 1: 381-400

Behnke H-D, Dörr I (1967) Zur Herkunft und Struktur der Plasmafilamente in Assimilatleitbahnen. Planta 74: 18-44

Behnke H-D, Kiritsis U (1983) Ultrastructure and differentiation of sieve elements in primitive angiosperms. I. Winteraceae. Protoplasma 118: 148-156

Behnke H-D, Schulz A (1983) The development of specific sieve-element plastids in wound phloem of *Coleus blumei* (S-type) and *Pisum sativum* (P-type), regenerated from amyloplast-containing parenchyma cells. Protoplasma 114: 125-132

Bentwood BJ, Cronshaw J (1976) Biochemistry and cytochemical localization of acid phosphatase in the phloem of *Nicotiana tabacum*. Planta 130: 97-104

Bentwood BJ, Cronshaw J (1978) Cytochemical localization of adenosine triphosphatase in the phloem of *Pisum sativum* and its relation to the function of transfer cells. Planta 140: 111-120

Beyenbach J, Weber C, Kleinig H (1974) Sieve tube proteins from *Cucurbita maxima*. Planta 119: 113-124

Bouck GB, Cronshaw J (1965) The fine structure of differentiating sieve tube elements. J Cell Biol 25: 79-96

Buvat R (1963) Sur la présence d'acide ribonucléique dans les corpuscles muqueux des cellules criblées de *Cucurbita pepo* L. CR Acad Sci Paris 257: 733-735

Campbell R (1972) Electron microscopy of the development of needlés of *Pinus nigra* var *maritima*. Ann Bot (London) 36: 711-720

Catesson AM (1973) Observations cytochimiques sur les tubes criblés de quelques angiospermes. J Microsc Paris 16: 95-104

Catesson AM, Czaninski Y (1967) Mise en évidence d'une activité phosphatasique acide dans le reticulum endoplasmique des tissus conducteurs de *Robinia* et de Sycomore. J Microsc Paris 6: 509-514

Catesson AM, Czaninski Y (1968) Localization ultrastructurale de la phosphatase acide et cycle saisonnier dans les tissus conducteurs de quelques arbres. Bull Soc Fr Physiol Veg 14: 165-73

Cornish-Bowden A (1977) Assessment of protein sequence identity from amino-acid composition data. J Theoret Biol 65: 735-742

Cornish-Bowden A (1979) How reliably do amino-acid composition comparisons predict sequence similarities between proteins. J Theor Biol 76: 369-386

Crafts AS, Crisp CE (1971) Phloem transport in plants. Freeman, San Francisco

Cronshaw J (1970) The P-protein components of sieve elements. In: 7me Congr Int Microscopie electronique, Grenoble. Soc Fr Microsc Electron, Paris, pp 429-430

Cronshaw J (1975a) P-proteins. In: Aronoff S, Dainty J, Gorham PR, Srivastava LM, Swanson CA (eds) Phloem transport. Plenum, New York London, pp 79-115

Cronshaw J (1975b) Sieve element walls. In: Aronoff S, Dainty J, Gorham PR, Srivastava LM, Swanson CA (eds) Phloem transport. Plenum, New York London, pp 129-147

Cronshaw J (1980) Histochemical localization of enzymes in the phloem. Ber Dtsch Bot Ges 93: 123-139

Cronshaw J (1981) Phloem structure and function. Annu Rev Plant Physiol 32: 465-484

Cronshaw J (1985) The fine structure of phloem cells. In: Proc 43rd Annu Meet Electron Microscopical Society of America. San Francisco Press, San Francisco, pp 632-635
Cronshaw J, Anderson R (1971) Phloem differentiation in tobacco pith culture. J Ultrastruct Res 34: 244-259
Cronshaw J, Esau K (1967) Tubular and fibrillar components of mature and differentiating sieve elements. J Cell Biol 34: 801-816
Cronshaw J, Esau K (1968a) P-protein in the phloem of *Cucurbita*. I. The development of P-protein bodies. J Cell Biol 38: 25-39
Cronshaw J, Esau K (1968b) P-protein in the phloem of *Cucurbita*. II. The P-protein of mature sieve elements. J Cell Biol 38: 292-303
Cronshaw J, Gilder J, Stone D (1973) Fine structural studies of P-protein in *Cucurbita*, *Cucumis* and *Nicotiana*. J Ultrastruct Res 45: 192-205
Deshpande BP (1974) On the occurrence of spiny vesicles in the phloem of *Salix*. Ann Bot (London) 38: 865-868
Deshpande BP, Evert RF (1970) A reevaluation of extruded nucleoli in sieve elements. J Ultrastruct Res 33: 483-494
Esau K (1969) The phloem. In: Zimmermann W, Ozenda P, Wulff HD (eds) Encyclopedia of plant anatomy, vol 5, pt. 2. Borntraeger, Berlin Stuttgart, 505 pp
Esau K (1978) The protein inclusions in sieve elements of cotton (*Gossypium hirsutum* L.). J Ultrastruct Res 63: 224-235
Esau K, Charvat ID (1975) An ultrastructural study of acid phosphatase localization in cells of *Phaseolus vulgaris* phloem by the use of the azo dye method. Tissue Cell 7: 619-630
Esau K, Cheadle VI (1965) Cytologic studies on phloem. Univ Cal Publ Bot 36: 253-344
Esau K, Cronshaw J (1967) Tubular components in cells of healthy and tobacco mosaic virus infected *Nicotiana*. Virology 33: 26-35
Esau K, Magyarosy AC (1979a) A crystalline inclusion in sieve element nuclei of *Amsinckia*. I. The inclusion in differentiating cells. J Cell Sci 38: 1-10
Esau K, Magyarosy AC (1979b) A crystalline inclusion in sieve element nuclei of *Amsinckia*. II. The inclusion in maturing cells. J Cell Sci 38: 11-22
Esau K, Thorsch J (1982) Nuclear crystalloids in sieve elements of species of *Echium* (Boraginaceae). J Cell Sci 54: 149-160
Eschrich W, Evert RF, Heyser W (1971) Proteins of sieve tube exudate of *Cucurbita maxima*. Planta 100: 208-221
Evert RF, Deshpande BP (1970) Nuclear protein in sieve elements of *Tilia americana*. J Cell Biol 44: 462-466
Evert RF, Eichhorn SE (1974) Sieve-element ultrastructure in *Platycerium bifuracatum* and some other polypodiaceous ferns: the nucleus. Planta 119: 301-318
Ezeala DO, Hart JW, Sabnis DD (1974a) Fractionation of monovalent ion-stimulated nucleoside triphosphatase activity in extracts of petiolar tissue. J Exp Bot 25: 1037-1044
Ezeala DO, Hart JW, Sabnis DD (1974b) Stimulation by monovalent ions of adenosine triphosphatase activity in extracts of petiolar tissue. J Exp Bot 25: 1045-1052
Fischer A (1884) Untersuchungen über das Siebröhren-System der Cucurbitaceen. Bornträger, Berlin
Fisher DB (1975) Structure of functional soybean sieve elements. Plant Physiol 56: 555-569
Freundlich A (1974) No polysaccharide demonstrated in filamentous structures in sieve elements by Thiery's periodic acid-thiocarbohydrazide-silver proteinate method for electron microscopy. Planta 118: 85-87
Gamaley YV (1973) The phloem protein in the conifer sieve cells (according to electron microscopic data). Bot Z 58: 981-985
Gerola FM, Lombardo G, Cataro A (1969) Histological localization of citrus infectious variegation (CVV) in *Phaseolus vulgaris*. Protoplasma 67: 319-326
Gietl C, Ziegler H (1980a) Affinity chromatography of carbohydrate binding proteins in the phloem exudate from several tree species. Biochem Physiol Pflanzen 175: 50-57
Gietl C, Ziegler H (1980b) Distribution of carbohydrate-binding proteins in different tissues of *Robinia pseudoacacia* L. Biochem Physiol Pflanzen 175: 58-66

Gietl C, Kauss H, Ziegler H (1979) Affinity chromatography of a lectin from *Robinia pseudoacacia* L. and demonstration of lectins in sieve tube sap from other tree species. Planta 144: 367-372

Gilder J, Cronshaw J (1973a) The distribution of adenosine triphosphatase activity in differentiating and mature phloem cells of *Nicotiana tabacum* and its relationship to phloem transport. J Ultrastruct Res 44: 388-404

Gilder J, Cronshaw J (1973b) Adenosine triphosphatase in the phloem of *Cucurbita*. Planta 110: 189-204.

Gilder J, Cronshaw J (1974) A biochemical and cytochemical study of adenosine triphosphatase activity in the phloem of *Nicotiana tabacum*. J Cell Biol 60: 221-235

Hartig T (1854) Über die Querscheidewände zwischen den einzelnen Gliedern der Siebröhren in *Cucurbita pepo*. Bot Z 12: 51-54

Hébant C (1969) Observations sur le phloème de quelque Filicinées tropicales. Nat Monspel Ser Bot 20: 135-196

Hoefert LL (1979) Ultrastructure of developing sieve elements in *Thlaspi arvense L*. I. The immature state. Am J Bot 66: 925-932

Ilker R, Currier HB (1974) Heavy meromyosin complexing filaments in the phloem of *Vicia faba* and *Xylosma congestum*. Planta 120: 311-316

Ilker R, Currier HB (1975) Histochemical studies of an inclusion body and P-protein in phloem of *Xylosma congestum*. Protoplasma 85: 127-132

Kauss H, Ziegler H (1974) Carbohydrate-binding proteins from the sieve-tube sap of *Robinia pseudoacacia*. Planta 121: 197-200

Kleinig H, Dörr I, Kollmann R (1971a) Vineblastine-induced precipitation of phloem proteins in vitro. Protoplasma 73: 293-302

Kleinig H, Dörr I, Weber C, Kollmann R (1971b) Filamentous proteins from plant sieve tubes. Nature New Biol 229: 152-153

Kleinig H, Thoenes J, Dörr I, Kollmann R (1975) Filament formation *in vitro* of a sieve tube protein from *Cucurbita maxima* and *Cucurbita pepo*. Planta 127: 163-170

Kollmann R (1980) Fine structural and biochemical characterization of phloem proteins. Can J Bot 58: 802-806

Kollmann R, Dörr I, Kleinig H (1970) Protein filaments - structural components of the phloem exudate. I. Observations with *Cucurbita* and *Nicotiana*. Planta 95: 86-94

LaFlèche D (1966) Ultrastructure et cytochimie des inclusions flagellées des cellules criblées de *Phaseolus vulgaris*. J Microsc Paris 5: 493-510

Lawton DM (1978) P-protein crystals do not disperse in uninjured sieve elements in roots of runner bean *(Phaseolus multiflorus)* fixed with glutaraldehyde. Ann Bot (London) 42: 353-361

Lawton DM, Johnson RPC (1976) A superhelical model for the ultrastructure of "P-protein tubules" in sieve elements of *Nymphoides peltata*. Cytobiologie 14: 1-17

McEuen AR, Hart JW, Sabnis DD (1981) Calcium-binding protein in sieve tube exudate. Planta 151: 531-534

Nägeli CW (1861) Über die Siebröhren von *Cucurbita*. Sitzungsber Königl Bayer Akad Wiss München 1: 212-238

Nehls R, Schaffner G, Kollmann R (1978) Feinstruktur des Protein-Einschlusses in den Siebelementen von *Salix sachalinensis* Fr. Schmidt. Z Pflanzenphysiol 87: 113-127

Newcomb EH (1967) A spiny vesicle in the slime-producing cells of the bean root. J Cell Biol 35: C17-C22

Northcote DH, Wooding RBP (1966) Development of sieve tubes in *Acer pseudoplatanus*. Proc R Soc London Ser B 163: 524-537

Oberhäuser R, Kollmann R (1977) Cytochemische Charakterisierung des sogenannten Freien Nucleolus als Proteinkörper in den Siebelementen von *Passiflora coerulea*. Z Pflanzenphysiol 84: 61-75

Palevitz BA, Hepler PK (1975) Is P-protein actin like? Not yet. Planta 125: 261-271

Palevitz BA, Newcomb EH (1971) The ultrastructure and development of tubular and crystalline P-protein in the sieve elements of certain papilionaceous legumes. Protoplasma 72: 399-426

Parameswaran N (1971) Zur Feinstruktur der Assimilatleitbahnen in der Nadel von *Pinus sylvestris*. Cytobiology 3: 70-88

Parthasarathy MV, Mühlethaler K (1969) Ultrastructure of protein tubules in differentiating sieve elements. Cytobiologie 7: 17-36

Read SM, Northcote DH (1983a) Chemical and immunological similarities between the phloem proteins of three genera of the Cucurbitaceae. Planta 158: 119-127

Read SM, Northcote DH (1983b) Subunit structure and interactions of the phloem proteins of *Cucurbita maxima* (pumpkin). Eur J Biochem 134: 561-569

Rouschal E (1941) Untersuchungen über die Protoplasmatik und Funktion der Siebröhren. Flora 35: 135-200

Sabnis DD, Hart JW (1974) Studies on the possible occurrence of actomyosin-like protein in phloem. Planta 118: 271-281

Sabnis DD, Hart JW (1976) A comparative analysis of phloem exudate proteins from *Cucumis melo, Cucumis sativus* and *Cucurbita maxima* by polyacrylamide gel electrophoresis and isoelectric focussing. Planta 130: 211-218

Sabnis DD, Hart JW (1978) The isolation and some properties of a lectin (haemagglutinin) from *Cucurbita* phloem exudate. Planta 142: 97-101

Sabnis DD, Hart JW (1979) Heterogeneity in phloem protein complements from different species. Consequences to hypotheses concerned with P-protein function. Planta 145: 459-466

Sabnis DD, Hart JW (1982) Microtubule proteins and phloem protein. In: Boulter D, Parthier B (eds) Nucleic acids and protein in plants. Encyclopedia of plant physiology, NS, vol 14a. Springer, Berlin Heidelberg New York, pp 401-437

Salmon J (1946) Différenciation des tubes criblées chez les Angiospermes. Rev Cytol Cytophysiol Veg 9: 55-168

Salmon J (1951) Mise en évidence de ribonucléoproteides dans les cellules criblées au moyen de la méthode de Brachet. C R Acad Sci Paris 233: 495-496

Sauter JJ (1972) Cytochemical demonstration of sulfhydryl disulfide-containing proteins in sieve elements of conifers. Naturwissenschaften 59: 470

Schulz A, Alosi AC, Sabnis DD, Park RB (1989) A phloem-specific, lectin-like protein is located in pine sieve-element plastids by immunocytochemistry. Planta 179: 506-515

Sham MH, Northcote DH (1987) Transcription and translation of phloem protein (PP2) during phloem differentiation in *Cucurbita maxima*. Planta 170: 392-399

Sloan RT, Sabnis DD, Hart JW (1976) The heterogeneity of phloem exudate from different plants: A comparative survey of ten plants using polyacrylamide gel electrophoresis. Planta 132: 97-102

Smith LM, Sabnis DD, Johnson RPC (1987) Immunocytochemical localisation of phloem lectin from *Cucurbita maxima* using peroxidase and colloidal gold labels. Planta 170: 461-470

Steer MW, Newcomb EH (1969) Development and dispersal of P-protein in the phloem of *Coleus blumei* Benth. J Cell Sci 4: 155-169

Stone DL, Cronshaw J (1973) Fine structure of P-protein filaments from *Ricinus communis*. Planta 113: 193-206

Thaine RG (1969) Movement of sugars through plants by cytoplasmic pumping. Nature (London) 222: 873-875

Walker TS (1972) The purification and some properties of a protein causing gelling in phloem sieve tube exudate from *Cucurbita pepo*. Biochim Biohphys Acta 257: 433-444

Walker TS, Thaine R (1971) Proteins and fine structural components in exudate from sieve tubes in *Cucurbita pepo* stems. Ann Bot (London) 35: 773-790

Wark MC, Chambers TC (1965) Fine structure of the phloem of *Pisum sativum*. 1. The sieve element ontogeny. Aust J Bot 13: 171-183

Weber C, Kleinig H (1971) Molecular weights of *Cucurbita* sieve tube proteins. Planta 99: 179-182

Weber C, Franke WW, Kartenbeck J (1974) Structure and biochemistry of phloem-proteins isolated from *Cucurbita maxima*. Exp Cell Res 87: 79-106

Wergin WP, Newcomb EH (1970) Formation and dispersal of crystalline P-protein in sieve elements of soybean (*Glycine max* L.). Protoplasma 71: 365-388
Wergin WP, Palevitz BA, Newcomb EH (1975) Structure and development of P-protein in phloem parenchyma and companion cells of legumes. Tissue Cell 7: 227-242
Wilhelm K (1880) Beiträge zur Kenntnis des Siebröhrenapparates dicotyler Pflanzen. Engelmann, Leipzig
Williamson RE (1972) An investigation of the contractile protein hypothesis of phloem translocation. Planta 106: 149-157
Wooding FBP (1969) P-protein and microtubular systems in *Nicotiana* callus phloem. Planta 85: 284-98
Yapa PAJ, Spanner DC (1972) The effect of protease digestion (in situ) on the slime substance of mature sieve tubes. Planta 107: 89-96
Zee SY (1968) Ontogeny of cambium and phloem in the epicotyl of *Pisum sativum*. Aust J Bot 17: 441-456
Zee SY (1969) Fine structure of the differentiating sieve elements of *Vicia faba*. Aust J Bot 17: 441-56

14 Phloem Evolution: An Appraisal Based on the Fossil Record[1]

EDITH L. TAYLOR[2]

14.1 Introduction

Our knowledge of the evolution of vascular tissue through geologic time has come almost entirely from xylem; no doubt due to the fact that thick-walled tracheids are more easily preserved in depositional environments. As a result, secondary xylem is common in the fossil record from the Devonian to the Recent. In contrast, phloem tissue consists of relatively thin-walled cells that are less likely to be preserved. Conducting elements in the phloem are under high hydrostatic pressure while they are functioning and this physiological constraint probably contributes to their collapse during fossilization and their subsequent rarity in the geologic record. Within the past 10 years, however, there have been a number of contributions that have focused on the anatomy of phloem in fossil plants (Table 14.1) and the data from these studies are beginning to provide a picture of phloem structure and evolution through geologic time.

Although the phloem in a number of major taxonomic groups has been examined, in most instances only one or two genera in each group have been studied in any detail (Table 14.1). Phloem anatomy in fossil angiosperms has not yet been examined in detail, so only vascular cryptogams (lycopods, ferns, sphenophytes) and gymnosperms (seed ferns, cordaites, cycadeoids) will be considered here. The taxa listed in Table 14.1 represent only those that have been studied specifically for phloem anatomy; not included are the many systematic contributions on fossil plants that have involved a brief mention or description of phloem tissue. The majority of the plants listed in this table represent Carboniferous taxa preserved in coal balls. Coal balls are nodular masses of calcium carbonate that contain permineralized plant parts. In this type of preservation, cell lumens and intercellular spaces have been filled with mineral matter, but cell walls, although they are changed chemically, still remain organic. Permineralizations are generally well preserved so that cellular and sometimes subcellular detail can be critically examined. In addition, plants preserved in this way are relatively easy to prepare for light microscopy via the cellulose acetate peel technique (Joy et al. 1956). The details of phloem structure in many of these plants and other fossil examples have been reviewed (Taylor and Taylor 1990) and will not be repeated here. Instead, our current knowledge of

---

[1] Dedicated to Katherine Esau for her inspiring work on phloem structure.
[2] Formerly Edith L. Smoot.

Table 14.1. Phloem structure in selected fossil plants (Modified from Taylor and Taylor, 1990)

Taxon	Phloem composition	Sieve elements		End walls	Sieve area size	Sieve pores
		x. s.	l. s.			
LYCOPODS:						
Lepidodendron (Eggert and Kanemoto 1977)	sieve elements parenchyma	15.7 μm[a]		Horizontal-oblique	4.4 × 15.5 μm[a]	1.5–3.0 μm[b]
Stigmaria appendages (Stewart 1940; 1947)	sieve elements	12.5–21.4 μm[a]	L = 3–8 ẋW	Horizontal	7 × 9 μm[a]	
SPHENOPHYTES:						
Astromyelon (Wilson and Eggert 1974)	sieve elements		330 μm	Horizontal-sl. oblique	Not seen	
Sphenophyllum (Eggert and Gaunt 1973) (Stidd and Ma 1978)	Primary: sieve elements parenchyma (?)			Horizontal-sl. oblique		8.0 μm
	Secondary: sieve elements (fasc.)	20–40 μm	280–370 μm	Horizontal	Not seen	
	sieve elements (interfasc.) parenchyma (?)	91–100 μm	500–1440 μm			
FERNS:						
Anachoropteris (Smoot 1985)	Adaxial: sieve elements	9.8 μm	78–375 μm	Horizontal-sl. oblique	Present	
	Abaxial: small sieve elements:	8–27 μm	195–618 μm	horizontal-sl. oblique	Present	
	and large sieve elements:	48–120 μm	1.23–2.74 + mm	very oblique	5.0–8.5 μm	1.0 μm
Ankyropteris (Smoot 1985)	sieve elements parenchyma	7.5–13.5 μm	167–460 μm	Horizontal-oblique	4.3 × 5.4 μm	0.9–1.8 μm
Botryopteris (Smoot 1979)	sieve elements	21 μm	90–138 μm	Horizontal-sl. oblique	Present, elliptical	Poorly preserved

Etapteris (Smoot 1979)	sieve elements	16 × 31 μm	120–360 μm	Horizontal-sl. oblique	Present, elliptical	Few 1/10ths-1.0 μm
Psaronius (Smoot 1984c)	small sieve elements:	30 μm	238 μm	Horizontal-rounded	3 × 7.5-4.5 × 9 μm	1.0 μm
	large sieve elements:	90–120 μm	595 μm	rounded-sl. oblique	Same as small cells	
Stauropteris (Smoot and Vande Wege 1986)	small sieve elements:	23.5 × 37.5 μm	240–420 μm	Horizontal-oblique	5 × 6-4 × 10 μm	1.0 μm ind. 1.7 μm s. a.
	large sieve elements:	79.7 × 51.2 μm	?			
	parenchyma					
PROGYMNOSPERMS:						
“aneurophytalean”[c] (Wight and Beck 1984)	sieve cells	15–25 μm	0.3 mm	Very oblique	3.8–7.5 × 15–25 μm	0.6 μm
	fibers					
	parenchyma					
	sclereids					
Callixylon[c] (Arnold 1930)	empty cells	L = 3–4 xW	Horizontal			
	(2 types)	L = W	Horizontal			
	tannin cells	L = 2 xW	Oblique			
PTERIDOSPERMS:						
Calamopitys[d] (Galtier and Hébant 1973)	sieve cells (secondary)	57 μm[a]	2.5 mm	Oblique	4–8 × 20 μm	Present
	sieve cells (primary)	20–35 μm				
	parenchyma					
Callistophyton (Russin 1981; Smoot 1984a)	sieve cells	28–99 μm	6.0 mm	Very oblique	11 × 28 μm	25/area
	parenchyma	19–25 μm	104–215 μm			
	secretory canals					
Heterangium (Hall 1952)	sieve cells	44 μm	9 mm	“Blunt”	Present	Numerous
	parenchyma	20 μm	35–135 μm			

Table 14.1. continued

Taxon	Phloem composition	Sieve elements x. s.	Sieve elements l. s.	End walls	Sieve area size	Sieve pores
Medullosa (Smoot 1984b)	sieve cells	66 µm	1.6–4.2 mm	Very oblique	27.3 × 7.8 µm	0.1–0.2 µm 40–50/area
	fibers	35 × 90 µm	4.2 mm			
	parenchyma	34 × 35 µm	206 µm			
CORDAITES:						
cf. *Cordaixylon* (Taylor 1988a)	sieve cells	25 × 36 µm	0.7–1.7 mm	Very tapered	10 × 20 µm	Present
	parenchyma	25 × 36 µm				
CONIFERALES:						
Cupressinocortex[e] (Ramanujam 1970)	sieve cells	25–30 µm			15–24 µm	12–120/ area
	parenchyma	15–20 µm				
	fibers	23 × 40 µm				
	resin canals	0.25–1.5 mm	1.0 cm			
Taxodioxylon bark[e] (Ramanujam and Stewart 1969)	sieve cells	20 × 50 µm			8–18 µm	9–48/area
	fibers	18 × 45 µm	4.0 mm			
	parenchyma		60–150 µm			

[a] Calculated from published photographs
[b] Measurement includes presumed callose deposit
[c] Devonian of New York, U.S.A.
[d] Lower Carboniferous of France.
[e] Cretaceous of Canada.
All other taxa = Carboniferous coal balls from North America.

sl. = slightly
ind. = individual sieve pores
s. a. = sieve pores within sieve areas
L = length; W = width
fasc. = fascicular
interfasc. = interfascicular

the structure of the cryptogam sieve element and the gymnospermous sieve cell as based on fossil evidence will be summarized. The data from fossil phloem will then be compared to extant examples and the implications of these data for understanding phloem phylogeny will be discussed.

14.1.1 Phloem Phylogeny - Background

Although parenchyma and sclerenchyma cells are generally present in phloem tissue, the components that are considered to be the most important in understanding phylogenetic trends are the conducting cells. These have traditionally been defined as consisting of two basic types (Esau 1969): sieve-tube members and sieve cells. With two possible exceptions, sieve-tube members occur only in angiosperms, while sieve cells are thought to be characteristic of gymnosperms and vascular cryptogams. Sieve-tube members are distinguished from sieve cells by the presence of more specialized sieve areas (i. e., with larger sieve pores) on end walls than on side walls (see also Chap. 3).

In certain groups, such as the vascular cryptogams, it is often difficult to determine whether a particular cell represents a sieve cell or a sieve-tube member. Perry and Evert (1975), and Kruatrachue and Evert (1974) discussed this problem in relation to their work on living cryptogams, and Smoot (1985) and Taylor and Taylor (1990) noted a similar problem when dealing with sieve elements in fossil cryptogams. The problem is compounded in fossil material by the vagaries of preservation, thus making it difficult in some cases to determine which aspects of phloem structure represent artifacts of the preservation process (i. e., burial, possible distortion of plant organs, infiltration by minerals, etc.) and which are accurate reflections of the anatomy of the plant. Because of this, it has been suggested (Taylor and Taylor 1990) that the conducting cells in fossil cryptogams be designated by the general term "sieve element." The advantage of this term is that it avoids the suggestion of a phylogenetic relationship between the gymnospermous sieve cell and the sieve element which is found in the ferns and lycopods (cf. Behnke 1986, see Sect. 14.4).

In the past, contributions on the evolution and composition of phloem tissue have been based almost entirely on extant examples (e. g., Cheadle and Whitford 1941, Cheadle 1948, Esau et al. 1953, Zahur 1959; but see Smoot 1984b, 1985). Evolutionary trends have been examined most extensively in the angiosperms, with the primary emphasis on the monocots. Within this group, it has been proposed that evolutionary advancement in sieve-tube members included a change from oblique to more horizontal end walls, concomitant with an increase in specialization of sieve areas on end walls and a decrease in their development on side walls (Cheadle and Whitford 1941; Cheadle 1948). The change toward more specialized sieve areas is defined as a change from compound to simple sieve areas, a tendency that parallels the trends suggested many years ago by Bailey for tracheary elements in the xylem (e. g., Bailey 1953; Carlquist 1975).

Although the majority of contributions on phloem phylogeny have been concerned with angiosperms, Outer (1967) has examined the secondary phloem of a number of gymnosperms from a phylogenetic point of view. His evolutionary grouping of taxa is based on cellular composition, arrangement of cells in the tissue, and type of intercellular connections. The following evolutionary trends have been suggested: (1) increase in the amount of axial parenchyma, (2) decrease in number of albuminous cells in the rays, (3) increase in axial albuminous cells, (4) increase in fibers, (5) increasingly regular arrangement of cell types.

Outer's (1967) work is based entirely on extant gymnosperms and it has been demonstrated that these proposed trends are not supported by the evidence from the fossil record. Recent work on fossil gymnosperms (Smoot 1984a, b; Taylor 1988a, b) indicates that his scheme is probably oversimplified. For example, the arrangement of cell types in repeating, tangential bands appears to be a reflection of particular taxonomic groups within the gymnosperms, rather than a character that has broad phylogenetic significance. The cycadophytes, for example, do not generally possess secondary phloem arranged in regular tangential bands. This is true in fossil cycadeoids (Bennettitales) such as *Cycadeoidea* (Taylor 1988b), as well as extant members of the Cycadales (Esau 1969). However, some seed ferns, such as *Medullosa* (Smoot 1984b), show regularly repeating patterns in the secondary phloem as early as the Carboniferous, and there are many similarities in the patterns between these Paleozoic plants and a number of living gymnosperms (Outer 1967).

Relatively little has been contributed on the phylogeny of cryptogam sieve elements. Esau suggested, and it is generally accepted, that the cryptogam sieve element arose from a parenchyma cell (Esau et al. 1953; Esau 1969). Unfortunately, histology of extant cryptogams has not shed much light on this evolutionary event since most of the living cryptogams examined, although from several different groups, show a general similarity in sieve element structure and development (see Chap. 3). Therefore, it has become necessary to examine phloem tissue in fossil cryptogams in order to evaluate proposed phylogenetic trends.

14.2 Phloem of Vascular Cryptogams

Phloem anatomy has been examined in fossil lycopods (Lepidodendrales), sphenophytes (Sphenophyllales, Equisetales) and ferns (Filicales, Marattiales, Zygopteridales) and some details are enumerated in Table 14.1 and summarized in Table 14.2. It is now generally believed that all of the phloem produced in these groups was primary in origin, although controversy still exists over the extent of cambial activity in *Sphenophyllum* (Eggert and Gaunt 1973; Taylor and Taylor 1990) and *Astromyelon* (Wilson and Eggert 1974; Cichan 1986). Cichan and Taylor (1990) have suggested that the cambium in the Carboniferous vascular cryptogams was predominantly unifacial (producing only second-

Table 14.2. Comparison of fossil cryptogam and gymnosperm phloem

	Vascular cryptogams	*Gymnosperms*
Tissue origin:	Primary	Primary and secondary
Conducting cell:	Sieve elements	Sieve cell
Tissue composition:	Sieve elements, parenchyma	Sieve cells, parenchyma fibers, sclereids, secretory cells
Sieve elements:		
Shape:	Rectangular-elongate	Elongate
Diameter:	10-40 μm (120 μm)[a]	20-99 μm
Length:	< 600 μm (> 2.75 mm)[a]	1.0- > 9.0 mm
End wall:	Horizontal-oblique	Horizontal-very oblique
Sieve areas:	Elliptical-rectangular, small	Elliptical, larger (10 × 30 μm)
Sieve pores:	~1.0 μm diameter	~1.0 μm diameter, many/sieve area
Callose:	Present in some	Present in some
Other:	Sheath usually present	Tangential banding

[a] Numbers in parentheses represent large diameter conducting elements of *Psaronius* and *Anachoropteris*. See text for further information.

ary xylem) and functioned in a fundamentally different way from that in seed plants. Because of the uncertainty surrounding the interpretation of *Astromyelon* and *Sphenophyllum*, data from these taxa are not included in the summary of cryptogam phloem (Table 14.2) or in the discussion below. For a detailed discussion of phloem interpretation in these taxa see Taylor and Taylor (1990) and the sources listed in Table 14.1.

The composition of the primary phloem tissue in the fossils examined to date is relatively simple, consisting either of sieve elements only, or sieve elements with scattered parenchyma cells (Fig. 14.1i) (see Table 14.1 for data and sources). There is no definitive relationship between conducting elements and parenchyma as is the case with seed plants. Additional parenchyma is often present in the form of a sheath separating the xylem from the narrow phloem zone (usually < 10 cells wide).

Typical cryptogam sieve elements are similar to elongate parenchyma cells. They are relatively small in diameter (probable metaphloem as small as 10 μm, rarely larger than 40 μm - presumed protophloem usually smaller), longer than parenchyma cells but considerably shorter than gymnosperm sieve cells (Table 14.2). In a few cases, e.g., *Anachoropteris* (Smoot 1985) and *Psaronius* (Smoot 1984c), larger sieve elements are present in addition to the small cells. These extend from 50-120 μm in diameter, ranging from 600 μm to more than 2.5 mm in length. End walls of cryptogam sieve elements are horizontal or only slightly oblique (Fig. 14.1j) and sieve areas are small and variable in shape (Fig. 14.1h). Pores are present within the sieve areas as well as scattered on vertical walls. Probable callose deposits have been observed in the lycopods and in one fern *(Anachoropteris)*.

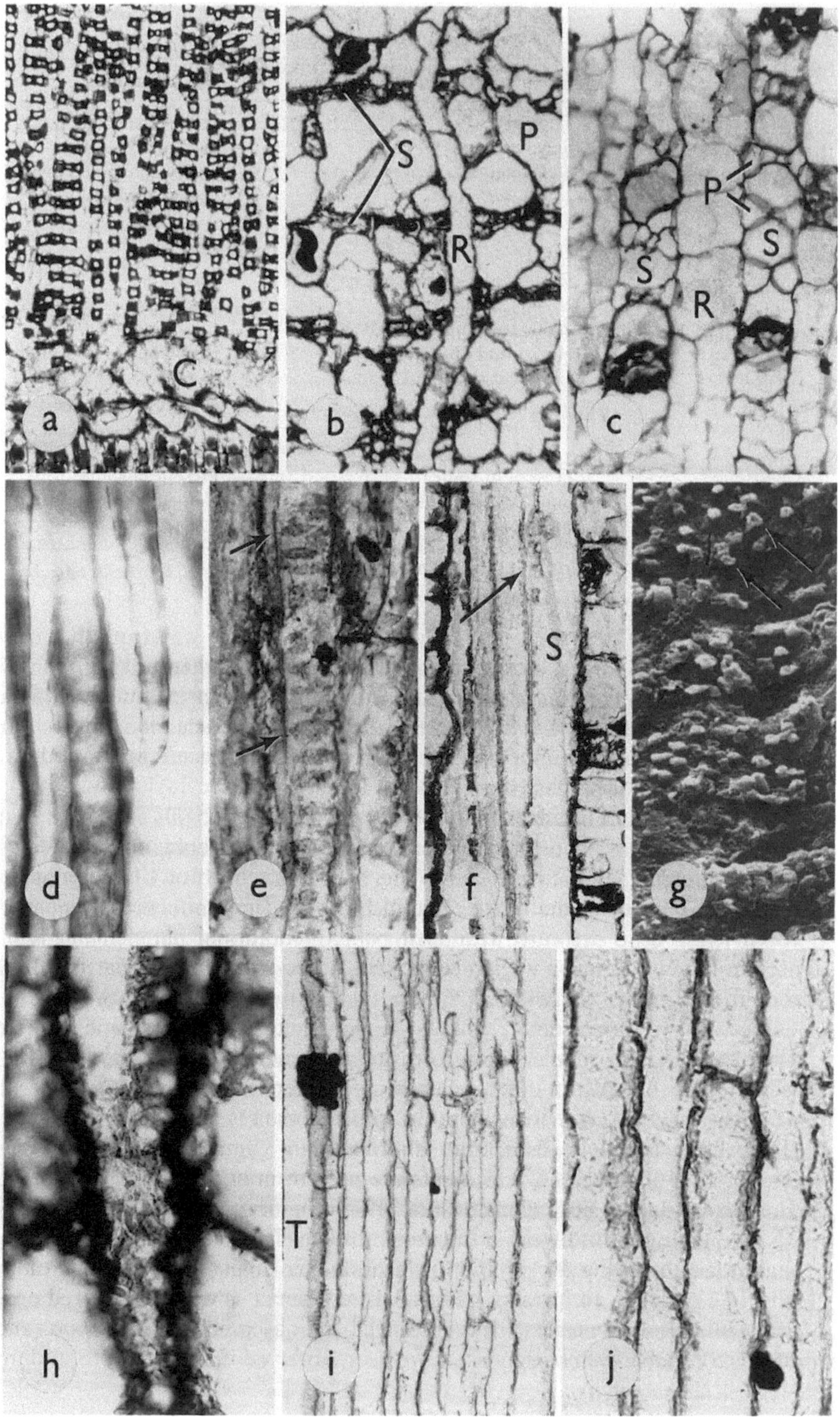

a
C
b
S
P
R
c
P
S
S
R
d
e
f
S
g
h
i
T
j

14.3 Gymnosperm Phloem

Primary phloem is currently very poorly known in fossil gymnosperms. Therefore, the data presented in Tables 14.1 and 14.2 are based only on secondary phloem structure. Phloem tissue has been noted in almost every major group of fossil gymnosperms, but studied in detail in only the cordaites, conifers and Paleozoic seed ferns (Table 14.1). In all of these groups, as well as the cycadeoids (Taylor 1988b), cells of the secondary phloem are arranged in alternating, tangential layers. In one anatomical type, exemplified by the Carboniferous seed fern *Medullosa* and the Cretaceous cycadophyte *Cycadeoidea,* sieve cells alternate with layers of axial parenchyma and fibers (Fig. 14.1a). The cells are arranged with exceptional regularity in *Medullosa,* but are less so in *Cycadeoidea.* In the other anatomical type, fibers are not present and layers of sieve cells and parenchyma alternate (Fig. 14.1b, c). This type of arrangement is found in cordaites and the seed ferns *Callistophyton* and *Heterangium.* Thus, unlike the cryptogams, phloem parenchyma exhibits a constant and regular relationship with the conducting elements in gymnosperms. Since extant Strasburger (albuminous) cells are defined on cytologic features, it is not possible to determine if these cells are present in fossil material, although cells of similar size and shape have been reported in *Cordaixylon* (Taylor 1988a).

Although the composition of the secondary phloem may vary slightly from taxon to taxon, the gymnospermous sieve cell is remarkably constant in its structure. The geologically earliest sieve cell was described from a progymnosperm (Wight and Beck 1984) and is anatomically very similar to those in Carboniferous seed plants, where greater detail is known. These cells are comparable to extant sieve cells in that they are elongate (2.5->9.0 mm) with end

Fig. 14.1a-j. Representative examples of phloem anatomy in fossil gymnosperms **(a-g)** and vascular cryptogams **(h-j)**. **a** Cross-section of xylem *(bottom),* cambium and extensive secondary phloem in *Cycadeoidea paynei. Dark cells* are fibers and sieve cells. *Light areas* represent poorly preserved axial parenchyma. × 50. **b** Cross-section of outer phloem in *Cordaixylon.* Parenchyma cells have enlarged and partially crushed the sieve cells. Some parenchyma cells contain dark contents *(left, center).* × 100. **c** Secondary phloem of *Callistophyton* in transverse section, showing alternation of sieve cells and small axial parenchyma cells; × 150. **d** Longitudinal section of *Cycadeoidea paynei* phloem. Sieve areas appear as *oval clearer areas* on dark sieve cell walls. × 400. **e** Sieve cell of *Medullosa* with elongate oval sieve areas. *Arrows* show two clearly outlined sieve areas. Areas probably appear dark due to the presence of callose in the pores (see **g**). × 224. **f** Tangential section of *Medullosa* phloem showing sieve cells in center. *Arrow* indicates part of tapered end wall. × 70. **g** Scanning electron micrograph of sieve areas in *Medullosa* phloem. *White spots* represent callose deposits associated with individual pores. *Arrows* indicate collars of callose. × 2000. **h** Oblique longitudinal section of sieve element in *Psaronius* showing numerous sieve areas of variable size. × 600. **i** Longitudinal section of phloem tissue in *Stauropteris* showing primary xylem tracheid *(left)* and elongate, narrow sieve elements. × 150. **j** Sieve elements in *Stauropteris* illustrating slightly oblique end wall configurations. × 488. *C* cambium; *P* axial parenchyma; *R* phloem ray; *S* sieve cells; *T* tracheids

walls that gradually taper, so that it is often difficult to distinguish end walls from side walls (Fig. 14.1 f, arrow). Distinct oval sieve areas are present on the radial walls (Fig. 14.1 d, e) that are considerably larger than those seen in the cryptogams (Table 14.2) and sieve pores are more clearly defined, often outlined by callose deposits (Fig. 14.1 g) (Smoot 1984 b).

14.4 Conclusions - Phloem Phylogeny

As more information on fossil phloem and especially conducting cells becomes available, a clearer picture of the significance of these data is beginning to emerge (Taylor 1987). In the cryptogams, overall tissue composition and the basic structure of the sieve element are fairly uniform, not only among fossil taxa, but also when compared to extant examples, despite the fact that taxa from different groups have been examined (e. g., Smoot 1984 c, 1985; see also Chap. 3). The only exceptions to this rule are the large elements that occur in *Psaronius* and *Anachoropteris*. At present, the evolutionary significance of these cells and their role in the evolution of sieve elements is not fully understood. Except for these large cells, cryptogam sieve elements are relatively short and rectangular, with horizontal to slightly oblique end walls. Sieve areas do not exhibit a great deal of specialization, and are of variable size and shape. Overall, however, they are smaller and less regular than those in gymnosperms. Sieve pores are uniformly small, often less than 1.0 μm in diameter. Phloem parenchyma is present but does not exhibit the close relationship with specific conducting cells that is present in the seed plants.

In the gymnosperms the same type of pattern is emerging, although here only secondary phloem has been examined. Regardless of the particular group of gymnosperms the composition and arrangement of the phloem tissues are relatively uniform. There is an intimate and regular relationship between axial parenchyma and conducting elements. Fibers are present either as part of the (presumed) functional (i. e., innermost) phloem or, in many cases, they occur at a later stage of development in the outermost bark (e. g., in *Medullosa* and *Cordaixylon*). In addition, the sequence of histological changes from inner, presumably functional phloem to outer, obviously nonfunctional phloem are comparable in both fossil and living forms (Smoot 1984 a, Taylor 1988 a).

Of perhaps greater significance, especially from the standpoint of the evolution of conducting cells in the gymnosperms, is the remarkable uniformity in sieve cell structure. Sieve cells are very elongate (> 1.0 mm), often with extremely oblique end walls. Sieve areas are discrete and uniform in size and shape, and pores are distinct enough that the number/sieve area can generally be determined (in contrast to the cryptogams). Callose has been observed, both in the form of a collar surrounding the pore (Fig. 14.1 g, arrows) (Smoot 1984 b) and as definitive callose (Russin 1981; Smoot 1984 a).

Thus, within both the gymnosperms and cryptogams there is a similarity in structure, not only in the fossils examined to date, but also when they are compared to extant relatives. It is important to note, however, that not all characters can be utilized when comparing fossil phloem. Cytologic features are often important in comparisons of living phloem; for example, modern cryptogam sieve elements are characterized by the presence of refractive spherules as well as nacreous walls at some stage in development. At the present time, no cytologic or developmental characters have been observed in fossil phloem, although preservation is known to be good enough to preserve sub-cellular detail in some fossils (e.g., Baxter 1964; Niklas et al. 1978).

It is clear from the above data that as early as the Carboniferous the sieve element of the vascular cryptogams was considerably different structurally and perhaps functionally from the sieve cell of gymnosperms. Although it has been proposed (Esau et al. 1953) that the gymnosperm sieve cell was derived from the cryptogam sieve element, by the early Carboniferous these two types had clearly already diverged. In fact it may be argued that these two cell types were evolutionarily separated by the middle Devonian, since the sieve cell described by Wight and Beck (1984) in a progymnosperm already showed characteristics of Carboniferous and later gymnosperm sieve cells. Although Wight and Beck (1984) note similarities between the progymnosperm sieve cells and sieve elements of living pteridophytes, they do not elaborate on these similarities and the above data show that any similarities in tissue or cell structure between these two groups are only superficial. Phloem anatomy in cryptogams prior to the Carboniferous is relatively poorly known, but Kidston and Lang (1917; 1920) and Banks et al. (1975) noted the presence of thin-walled cells with slightly oblique end walls in the position of phloem in the Devonian age plants *Rhynia* and *Trimerophyton* respectively.

In addition to differences in the type of conducting cells, there are other distinctions between gymnosperm and cryptogam phloem. The cellular composition is divergent; however, this may be attributed to the fact that only primary phloem is known from fossil cryptogams, whereas secondary phloem is documented for the gymnosperms. More important, however, is the extent of the functional interrelationship between conducting and parenchyma cells. In cryptogams this relationship is an indistinct one; although parenchyma cells are usually present in the phloem, there is no consistent occurrence of sieve element/parenchyma cell complexes as is present in gymnosperms.

The first group to evolve a distinct relationship between sieve cells and parenchyma was the gymnosperms. Although not developmentally related, Strasburger cells in modern gymnosperms have an important physiological relationship with specific sieve cells. This relationship has been carried one step further in the angiosperm sieve-tube member/companion cell relationship. Perhaps the advent of this relationship in the early gymnosperms or their ancestors signaled not only an evolutionary advancement, but a physiological one as well. Although there is controversy over the exact method of phloem loading (and indeed the method may vary from species to species; e.g., Delrot 1987), there can be little doubt that the role of the companion cell (or its physiological equivalent in the gymnosperms) is an important one. Does the lack of this close relationship in vascular cryptogams indicate a physiological difference in the method of phloem loading? Although this suggestion may be accurate, there are still far too few data on phloem physiology in any groups other than angiosperms to do any more than speculate on structure/function relationships. Ultrastructural data from living plants, however, support a possible physiological divergence in conducting elements between the vascular cryptogams and the seed plants (e. g., Warmbrodt 1980; Behnke 1986) and this may also explain the consistently different phloem structure in cryptogams and gymnosperms.

Whatever the possible physiological causes of the differences between vascular cryptogam and gymnosperm phloem, it is apparent that there are no shared features of phloem anatomy between these two groups by the Late Carboniferous. The fossil evidence thus clearly supports the use of different terminology to describe conducting cells in these two groups. In addition, as far back as fossil evidence is currently available, the data suggest that within large groups of vascular plants, little structural modification has occurred in the phloem over large periods of geologic time, i. e., phloem anatomy in Carboniferous vascular cryptogams is almost identical to that of living cryptogams. The same is true of the gymnosperms. Early in their evolution, these groups each developed a certain "blueprint" for phloem structure (and possibly function) and they have maintained this design relatively unchanged for millions of years. This evolutionary conservatism can perhaps be explained by the enormous physiological constraints on the functioning of phloem, i. e., the need to interact with other tissue systems at both source and sink, to load and unload solutes at these sites and to move materials rapidly through the system. In addition, an essential part of well-adapted phloem function is some type of sealing mechanism in the event of wounding or a failure in part of the system. All of these functions are further complicated by the fact that phloem conducting cells are alive at maturity, thus necessitating the presence of a certain complement of organelles. In contrast, the function and maintenance of xylem tissue, with fewer physiological controls and conducting cells which are dead (and therefore hollow) at maturity, seems relatively simple. In a phylogenetic sense, perhaps it is the precise and complex physiological requirements of the phloem that have necessitated a relatively narrow range of structural and developmental possibilities in this tissue.

Additional work on fossil phloem research is needed in order to elucidate the origin of the gymnosperm sieve cell and the possible point of divergence of the gymnosperm and cryptogam conducting cells. This research will necessitate an examination of older fossils, including some of the earliest known gymnosperms. Recent discoveries of Devonian seed plants and cyptogams with some anatomical preservation (e. g., Gillespie et al. 1981) provide an opportunity to examine early gymnosperm phloem and compare it with cryptogam phloem of a similar age. In addition, the advent of new techniques continues to expand the information that can be extracted from the fossils. In 1969, Katherine Esau wrote, "so far, paleobotany has provided no significant data that could be used in discussions of the evolutionary trends in the (phloem) tissue." It is clear that we have come a long way since Esau wrote those words, but it is equally clear that we still have a long but potentially rewarding way to go in elucidating the origins and subsequent evolution of phloem tissue.

Acknowledgment. Supported in part by the National Science Foundation (BSR-8407835).

References

Arnold CA (1930) Bark structure of *Callixylon.* Bot Gaz: 427-431
Bailey IW (1953) Evolution of the tracheary tissue of land plants. Am J Bot 40: 4-8
Banks HP, Leclercq S, Hueber FM (1975) Anatomy and morphology of *Psilophyton dawsonii,* sp. n. from the late Lower Devonian of Quebec (Gaspé), and Ontario, Canada. Palaeontogr Am 8: 77-127
Baxter RW (1964) Paleozoic starch in fossil seeds from Kansas coal balls. Trans Kansas Acad Sci 67: 418-422
Behnke H-D (1986) Sieve element characters and the systematic position of *Austrobaileya,* Austrobaileyaceae - with comments to the distinction and definition of sieve cells and sieve-tube members. Plant Syst Evol 152: 101-121
Carlquist S (1975) Ecological strategies of xylem evolution. Univ Cal Press, Berkeley
Cheadle VI (1948) Observations on the phloem in the Monocotyledoneae. II. Additional data on the occurrence and phylogenetic specialization in structure of the sieve tubes in the metaphloem. Am J Bot 35: 129-131
Cheadle VI, Whitford NB (1941) Observations on the phloem in the Monocotyledoneae. I. The occurrence and phylogenetic specialization in structure of the sieve tubes in the metaphloem. Am J Bot 28: 623-627
Cichan MA (1986) Vascular cambium and wood development in Carboniferous plants. III. *Arthropitys* (Equisetales; Calamitaceae). Can J Bot 64: 688-695
Cichan MA, Taylor TN (1990) Evolution of cambium in geologic time - a reappraisal. In: Iqbal M (ed) The vascular cambium. Research Studies Press, Taunton (England), (in press)
Delrot S (1987) Phloem loading: apoplastic or symplastic? Plant Physiol Biochem 25: 667-676
Eggert DA, Gaunt DD (1973) Phloem of *Sphenophyllum.* Am J Bot 60: 775-770
Eggert DA, Kanemoto NY (1977) Stem phloem of a Middle Pennsylvanian *Lepidodendron.* Bot Gaz 138: 102-111
Esau K (1969) The phloem. In: Zimmermann W, Ozenda P, Wulff HD (eds) Encyclopedia of plant anatomy, vol 5, pt. 2. Borntraeger, Berlin Stuttgart, 505 pp
Esau K, Cheadle VI, Gifford EM (1953) Comparative structure and possible trends of specialization of the phloem. Am J Bot 40: 9-19
Galtier J, Hébant C (1973) Sur le phloème et le cambium d'une Calamopityacée, ptéridospermale probable du Carbonifère inférieur français. C R Acad Sci Paris Ser D 276: 2257-2259
Gillespie WH, Rothwell GW, Scheckler SE (1981) The earliest seeds. Nature (London) 293: 462-464
Hall JW (1952) The phloem of *Heterangium americanum.* Am Midl Nat 47: 763-768
Joy KW, Willis AJ, Lacey WS (1956) A rapid cellulose peel technique in palaeobotany. Ann Bot (London) 20: 635-637
Kidston R, Lang WH (1917) On Old Red Sandstone plants showing structure, from the Rhynie chert bed, Aberdeenshire, pt 1. *Rhynia Gwynne-Vaughani,* Kidston and Lang. Trans R Soc Edinburgh 51: 761-784
Kidston R, Lang WH (1920) On Old Red Sandstone plants showing structure, from the Rhynie chert bed, Aberdeenshire, pt 2. Additional notes on *Rhynia Gwynne-Vaughani,* Kidston and Lang; with descriptions of *Rhynia major,* n. sp., and *Hornea Lignieri,* n. g., n. sp. Trans R Soc Edinburgh 52: 603-627

Kruatrachue M, Evert RF (1974) Structure and development of sieve elements in the leaf of *Isoetes muricata*. Am J Bot 61: 253-266
Minchin PEH, Thorpe MR (1987) Is phloem transport due to a hydrostatic pressure gradient? Supporting evidence from pressure chamber experiments. Aust J Plant Physiol 14: 397-402
Niklas KJ, Brown RM, Santos R, Vian B (1978) Ultrastructure and cytochemistry of Miocene angiosperm leaf tissues. Proc Natl Acad Sci USA 75: 3263-3267
Outer RW den (1967) Histological investigations of the secondary phloem in gymnosperms. Meded Landbouwh Wageningen 67 (7): 1-119
Perry JW, Evert RF (1975) Structure and development of the sieve elements in *Psilotum nudum*. Am J Bot 62: 1038-1052
Ramanujam CGK (1970) A petrified bark of Cupressaceae from the Cretaceous of Alberta. Can J Bot 48:; 855-858
Ramanujam CGK, Stewart WN (1969) Taxodiaceous bark from the Upper Cretaceous of Alberta. Am J Bot 56: 101-107
Russin WA (1981) Secondary phloem of the Paleozoic pteridosperm *Callistophyton boyssetii* (Renault) Rothwell. Bot Gaz 142: 165-175
Smoot EL (1979) The phloem of *Etapteris leclercqii* and *Botryopteris tridentata*. Am J Bot 66: 511-521
Smoot EL (1984a) Secondary phloem anatomy in *Callistophyton boyssetii* (Renault) Rothwell and histological changes in the outer phloem. Bot Gaz 145: 295-406
Smoot EL (1984b) Phloem anatomy of the Carboniferous pteridosperm *Medullosa* and evolutionary trends in gymnosperm phloem. Bot Gaz 145: 550-564
Smoot EL (1984c) Phloem structure in the Carboniferous fern *Psaronius* (Marattiales). Am J Bot 71: 1104-1113
Smoot EL (1985) Phloem anatomy of the Carboniferous coenopterid ferns *Anachoropteris* and *Ankyropteris*. Am J Bot 72: 191-208
Smoot EL, Taylor TN (1978) Sieve areas in fossil phloem. Science 202: 1081-1083
Smoot EL, Taylor TN (1981) Phloem histology of a Lower Pennsylvanian *Psaronius*. Palaeobotanist 28-29: 81-85
Smoot EL, Vande Wege M (1986) Phloem anatomy in *Stauropteris biseriata* from the Pennsylvanian of North America. Am J Bot 73: 1043-1048
Stewart WN (1940) Phloem histology in stigmarian appendages. Trans Ill State Acad Sci 33: 54-57
Stewart WN (1947) A comparative study of stigmarian appendages and *Isoetes* roots. Am J Bot 34: 315-324
Stidd BM, Ma P (1978) Does *Sphenophyllum* have sieve tubes? Bot Soc Am Misc Ser Publ 156: 59
Taylor EL (1987) Fossil phloem tissue: evidence for evolutionary conservatism. Am J Bot 74: 690-691
Taylor EL (1988a) Secondary phloem anatomy in Cordaitean axes. Am J Bot 75: 1655-1666
Taylor EL (1988b) Secondary phloem anatomy in *Cycadeoidea*. Am J Bot 75 (6, pt 2): 118
Taylor EL, Taylor TN (1990) Phloem anatomy in fossil plants. In: Iqbal M (ed) Growth patterns in vascular plants. Dioscorides Press, Portland (Oregon), (in press)
Warmbrodt RD (1980) Characteristics of structure and differentiation in the sieve element of lower vascular plants. Ber Dtsch Bot Ges 93: 13-28
Wight DC, Beck CB (1984) Sieve cells in phloem of a Middle Devonian progymnosperm. Science 225: 1469-1471
Wilson MA, Eggert DA (1974) Root phloem of fossil tree-sized arthrophytes. Bot Gaz 135: 319-328
Zahur MS (1959) Comparative study of secondary phloem of 423 species of woody dicotyledons belonging to 85 families. Cornell Univ Agric Exp Stn Mem 358: 1-160

Index

Unless otherwise noted cellular structures are referred to those present in sieve elements

MIX
Papier aus verantwortungsvollen Quellen
Paper from responsible sources
FSC® C105338

If you have any concerns about our products,
you can contact us on
ProductSafety@springernature.com

In case Publisher is established outside the EU,
the EU authorized representative is:
Springer Nature Customer Service Center GmbH
Europaplatz 3, 69115 Heidelberg, Germany

Printed by Libri Plureos GmbH
in Hamburg, Germany